英语科技论文语法、词汇与修辞
SCI 论文实例解析和语病润色 248 例

Grammar, Vocabulary and Rhetoric of English Academic Paper: Anatomy of 248 Cases for SCI Papers

梁福军　编著

机 械 工 业 出 版 社

本书内容涵盖：英语基础语法概述；英语科技论文中的词汇分类，各类词汇的用法和常见易混淆词语的用法对比；英语科技论文中标点符号的使用场合；大量实例的分析和点评，英语科技论文常见语病的总结，通过修辞而消除语病或提升语言效果的举例。全书分6章包括：英语语法（词法、动词、句法）、英语科技论文词汇、英语科技论文标点、英语科技论文修辞实例。

本书是已出版教材《SCI论文写作与投稿》的配套书，侧重论文的语言表达，后者侧重论文的内容与结构，语言和内容相结合共同铸就论文的高品质。

本书可作为学术论文、英语论文写作的教材或自学用书，也可作为研究人员和其他科技工作者写作的参考书，还可作为科技写作、编辑工作的培训教材，以及从事英语写作教学、研究的教师、专家和学者的参考资料。

图书在版编目（CIP）数据

英语科技论文语法、词汇与修辞：SCI论文实例解析和语病润色248例/梁福军编著. —北京：机械工业出版社，2020.12（2024.12重印）

ISBN 978-7-111-67354-5

Ⅰ.①英… Ⅱ.①梁… Ⅲ.①科学技术-英语-论文-写作 Ⅳ.①G301

中国版本图书馆CIP数据核字（2021）第017667号

机械工业出版社（北京市百万庄大街22号 邮政编码100037）
策划编辑：韩效杰 责任编辑：韩效杰 佟 凤
责任校对：张 薇 封面设计：鞠 杨
责任印制：郜 敏
北京富资园科技发展有限公司印刷
2024年12月第1版第6次印刷
184mm×260mm · 16.75印张 · 412千字
标准书号：ISBN 978-7-111-67354-5
定价：59.00元

电话服务
客服电话：010-88361066
010-88379833
010-68326294

网络服务
机 工 官 网：www.cmpbook.com
机 工 官 博：weibo.com/cmp1952
金 书 网：www.golden-book.com
机工教育服务网：www.cmpedu.com

序　一

FOREWORD ONE

Science without literacy is just like a ship without sail. To write a quality scientific paper requires not only a logical framework and proper formatting with substantial research information, but also appropriate literacy practice for the storage and transmission of such information. The key to understand a scientific paper is to understand its language, and the failure to recognize the centrality of language and literacy to scientific writing prevents many researchers from publishing their research in highly ranked SCI journals. An uncomfortable fact is that every year, countless manuscripts are rejected because of poor language and grammatical lapses.

As a sequel of Dr. Liang's book *Writing and Submission of SCI Papers*, this book mainly focuses on the language in scientific writing, including the use of proper language, rhetoric, and even punctuation in writing scientific papers. Basic rules of English grammar and syntax are explained in this book to form a language background for first-time writers. Common grammatical mistakes found in rejected manuscripts are discussed with suggested corrections to help the first-time writers to avoid them in their writing practices. Academic vocabulary, useful phrases and sentences that can be readily used in research papers are also completely provided and fully demonstrated through numerous examples.

This book, combined with the book *Writing and Submission of SCI Papers*, provides a well-equipped tool box for non-English researchers to translate their research findings from raw data and results into publishable manuscripts while effectively develop their ability in representing their research and communicating with the audience in an unambiguous manner. This book is strongly recommended for the early-career scholars, including graduate students, who are relatively new to writing and publishing papers in English. It can be used for either teaching or self-learning. For English language scholars and students, inclusive of scientific journal editors, this book is also a most recommended reference.

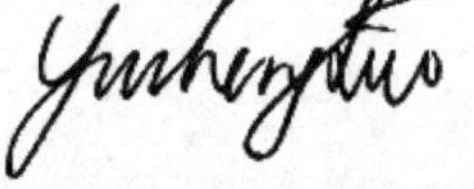

Yucheng Liu, Ph. D., P. E.
Jack Hatcher Chair Professor and Graduate Coordinator
Fellow of American Society of Mechanical Engineers
Department of Mechanical Engineering
Mississippi State University
August 28, 2020

序　二

FOREWORD TWO

SCI论文是科技工作者与国际学术界进行交流、学习、获得评定和发展自己的手段和工具，也是科学技术研究完整体系中不可缺少的一个重要环节。SCI论文写作是硕士、博士研究生和青年科技工作者的基本功。因此，梁博士撰写的SCI论文写作系列教材的出版非常及时，满足了中国科技研发与国际学术交流同时迅猛发展的需要。

记得20年前，我在哈尔滨工业大学开始做“长江学者特聘教授”的时候，就已认识到SCI论文写作是当时博士生和青年教师有待培养和提高的一个主要瓶颈，也可能是一个有效途径。因此，我连续多次在学科和学院里组织了有关SCI论文写作与投稿的讲座，获得了积极的效果，虽然当时我的讲座内容只是个人论文发表经历和经验的总结，远不够系统、规范，但我对广大研究生和青年教师的需求与问题的体会还是相当深刻的。梁博士的著作《英语科技论文语法、词汇与修辞　SCI论文实例解析和语病润色248例》，从英语科技论文写作的角度，对英语语法、词汇及标点进行了系统、规范和创造性的归纳和总结，并以名刊的论文作为实例进行一般或高级修辞，很有代表性和科学性。我相信需要进行SCI论文发表的广大科技工作者一定能认可该书，阅读使用该书一定能够获得事半功倍的效果。希望该书会成为高校、科研院所师生，特别是硕士、博士研究生的重要工具书，并成为许多青年科技工作者的案头必备参考书。

任何学科与学术体系的建立和发展，与其相关学科领域中学术论文的实时发表、高效交流和广泛传播都是分不开的。对每位科技工作者及在校硕士、博士研究生来说，其SCI论文发表、学术著作出版和各自科研学术体系的建立是一个相辅相成、循序渐进、不断完善的发展过程。我坚信梁博士的新著将是一个能帮助我们撰写高水平论文论著、归纳总结科研学术思想的实用教材和工具书。

英国布鲁内尔大学制造系统首席教授
哈尔滨工业大学特聘教授
《国际先进制造技术》（IJAMT）欧洲编辑
程凯（Kai Cheng）

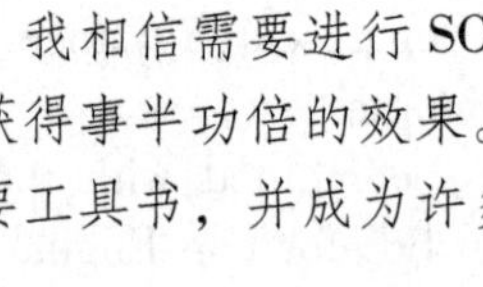

2020-9-30

序　　三

FOREWORD THREE

近年来，我国在科技发展方面投入了大量资金，极大地促进了我国科学研究的快速发展和技术的持续进步，许多领域已由与国际水平跟跑、并跑状态进入并跑、领跑状态，大量优秀科研成果不断涌现。目前，我国科技工作者每年发表的SCI论文数量已经超过美国，成为世界上发表SCI论文数量排名第一的国家，值得国人骄傲和自豪。

然而，过去我国高校通常较少专门设置科研论文写作课程，对科研论文写作的教育投入不足或重视不够，学生在校期间没有受到系统的、专门的论文写作训练，不少学生不论是在校期间撰写科研论文、学位论文，还是毕业后走上工作岗位撰写科研专著、学术论文，通常难以写出合规、得体的论文，论文因写作质量问题常被科技期刊拖延出版或做退稿处理。

同时，我国科技期刊的专业编辑也需全面掌握科技论文的写作规范和技能，以很好地指导作者撰写、修改论文，从而使科技期刊刊登的论文高质量地出版，易于获得更为广泛的交流和更加有效的传播。

我作为长期从事科技期刊出版和传播的专业编辑，深深感受到让作者提高论文写作水平和让科技期刊编辑全面掌握科技论文写作技能的重要性。

梁福军博士编著的《英语科技论文语法、词汇与修辞》一书，主要针对英语科技论文写作，总结了英语基础语法、词汇用法、标点符号用法，并列举了大量国际名刊的实例加以分析、说明和修辞，是一本非常实用的指导英语科技论文写作的工具书。这部书与已出版的《SCI论文写作与投稿》是配套教材，两部书虽各有侧重，却组成一个整体，为读者撰写论文提供指导。

科技导报社副社长、副主编　史永超

2020-8-28

前　言

PREFACE

2018 年 8 月，受机械工业出版社高等教育分社的邀请，笔者开始撰写针对硕士、博士研究生的教材《SCI 论文写作》。SCI 论文的主体是英文科技（学术）论文，笔者曾在 2014 年出版过专著《英文科技论文规范写作与编辑》，有一定的基础，因此欣然接受了这一光荣的任务。然而，着手撰写时，才知道事情并没有想象的那么容易。

SCI 论文与英文科技论文既密切相关又有所区别。如果对其区别搞不清楚，就不知侧重写什么、怎么写，最终难以进行下去。笔者深知，不管干什么，首先得把工作对象搞清楚。写这个教材，不应只是对现有资料的堆积，而要有所突破和创新，难度也不亚于做学问，撰写过程实际上就是做学问。首先要以 SCI 论文为研究对象，搞清它是什么、有什么、怎么样，再研究其各个组成部分的内容、结构及实现方法、步骤。于是笔者进行相关研究，并遴选有代表性的名刊文章作为范例，收获了新认识。

SCI 论文有两层基本含义：内容优（选题恰当，成果优秀，结构合理），语言好（语法正确，逻辑缜密，修辞立诚）。对于一篇科技论文，内容优和语言好都是其成为 SCI 论文的必要条件。因此《SCI 论文写作》的撰写应对 SCI 论文的内容和语言进行区分，二者虽有关联，但区别也很明显。如果将这两部分用一本书来呈现，那么书的篇幅将非常可观，不便作教材。笔者认为有必要将论文的内容和语言分开撰写，用两本书来呈现更适合，并与出版社达成了共识。

于是，先撰写了《SCI 论文写作与投稿》（2019 年 11 月出版），其核心针对论文的内容，而内容与承载它的结构密切相关，因此全书的框架就是由论文的整体结构及其各个部分的内容组成。书名中加了“投稿”，不仅考虑到现实中作者对投稿非常关注，更在于写作与投稿也密切相关。写作前期的准备工作决定了写作进度、效果，虽然发生在写作之前，但若没有准备，写作恐怕就难以进行，写成、写好更是遥不可及。写作犹如做饭，写作素材的准备及文章的内容安排、结构设计是写作前的事，做饭所需的食材及对食材的使用计划就好比写作所需的素材和对素材的安排。试想，若没有食材，也不知要做什么饭，那么这个饭还能做吗？因此讲写作，不应缺少与投稿相关的内容。

随后，撰写配套书《英语科技论文语法、词汇与修辞　SCI 论文实例解析和语病润色 248 例》，其核心针对论文的语言，而语言又与语法、词汇、修辞等密切相关，这就是本书的主题名所反映的。英语科技论文的语法、词汇应是通用的，不存在专门针对某类论文的语法、词汇，因此本书的主题名不再冠以限定语“SCI 论文”，而是代之以“英语科技论文”。此书的书名似乎有些长，但语义明显，至少蕴含两层意思：

第一，对于内容过关的论文，如果语言不过关，也会遭遇退稿的命运。如果说前一本书能引导论文作者撰写出内容过硬的论文，那么后一本书的使命就是帮助作者完善语言和修辞。这两本书如同姊妹，共同成就高水准的论文。

第二，语言过关也有档次，既可以是一般过关（对应普通修辞、消极修辞），也可以是优秀（对应高级修辞、积极修辞），一般过关是论文发表的基础条件，而经过高级润色的优秀语言使论文蹙金结绣。内容过关的论文只有在语言上精益求精才能是一篇优秀的论文。因此，本书书名的后面部分（实例解析和语病润色 248 例）与语言相承接，即启下，是第二本书。

这两层意思解释了本书名的由来！一篇论文之所以高端，缘于其至上的内容结构和优美的语言表述，只有二者完美结合，才能成就一篇 SCI 论文。

笔者再次想起别人的一句话，“对 SCI 论文写作，中国作者大多面临英语能力欠佳的困境，尤其对于年龄大和专业性强的作者来说，内容也许根本不是问题，而英语往往成为影响其成果发表的重要因素。”这种说法既对，又不对，也许说对了一半。说其对，是从强调语言的角度来说的，但忽略了内容，或许对那些母语写作能力较强而英语写作能力相对较弱的作者来说是对的；说其不对，是从强调内容的角度来说的，对内容空洞、结构欠缺的论文来说，语言再好，也不能达标。

书中主要部分辅以大量实例分析和点评，其中较多实例来自国际名刊的文章，有很强的针对性和实用性，对提升论文写作水准具有实践意义。

笔者水平和能力有限，疏漏之处在所难免，诚请广大读者批评指正！

梁博士讲堂公众号（drliangwechat）是笔者分享写作心得的园地，敬请关注。

梁福军

2020-11-18

目　录

CONTENTS

第1章 英语语法：词法

语法是研究词形变化和句子结构的科学。研究词形变化的语法称为词法，研究句子结构的语法称为句法。这一章讲述词法，包括八个词类，分别是名词、冠词、代词、数词、形容词、副词、介词、连词（连接词）。

1.1 名词

1.1.1 专有名词和普通名词

按词汇的意义，名词（noun）分为专有名词（proper noun）和普通名词（common noun）两类。

专有名词是个人、地方、团体、机构、组织、国家等的专用名称，其中实词的首字母大写，如Steve Jobs、Beijing、Apple Inc.、New York University、China。专有名词如果是含有普通名词的短语，则必须使用定冠词the，如the Great Wall、the People's Republic of China。姓氏名称若用复数形式，则表示该姓氏一家人，如the Greens、the Wangs、the Liangs。

普通名词是除专有名词以外的其他名词，是很多事物的共有名称，除位于句首外，通常以小写字母开头，如pupil、family、man、foot、science、research、friend、meeting。

1.1.2 可数名词和不可数名词

按词汇的形式，普通名词分为可数名词（countable noun）和不可数名词（uncountable noun）两类。

可数名词可用简单的数词进行计数，如book、box、child、computer、orange（橙子）、train、watch（表）；不可数名词不可用简单的数词进行计数，如electricity、equipment、information、math、news、orange（橙汁）、oil、population，psychology，water（水），这类词的前面不可用不定冠词或数词来修饰，词尾不能加s。

使用时容易出错的不可数名词常见的有：absence、access、advice、anger、atmosphere、behavior、cancer、capacity、childhood、comfort、confidence、courage、duty、education、energy、environment、equipment、fun、fear、freedom、furniture、growth、hair、homework、information、knowledge、luggage、money、machinery、news、progress、rain、snow、status、time、traffic、trade、weather、welfare。

1.1.3 兼做可数和不可数名词

有些名词既可做可数名词，也可做不可数名词，但表义不同。判断一个名词是否可数的依据是其表义，但一个名词表达同一意思时不可能既是可数名词又是不可数名词。

1）一些表动物的词一般可数，但指肉时不可数，如chicken、duck、fish、lamb。

以fish为例，在“I'll buy *a fish* for dinner tonight.”中，*a fish*指*the whole body of fish*，*fish*是可数名词，而在“Would you like *some fish* for dinner?”中，*some fish*指*the part of fish*或*the dish of fish*，*fish*是不可数名词。

2）一些表饮品的词，表用器具盛装并计量时一般可数，但指液体本身时不可数，如beer、coffee、lemonade、tea。

以beer为例，在“I'd love *a beer*, please.”中，*a beer*指*a cup of beer*，*beer*是可数名词，而在“Do you drink *beer*, please?”

中，*beer* 指 *the liquid of beer*，是不可数名词。

3）物质名词不可数，如 glass（玻璃）、iron（铁）、rock（岩石）、rubber（橡胶）、sculpture（雕刻、雕塑）、stone（石材）；用作可数名词时词义有变化，如 a glass（一个玻璃制品，如玻璃杯）、a pair of glasses（一副眼镜）、an iron（一个熨斗）、a rock（一粒石子）、a rubber（一块橡皮）、a sculpture（一件雕刻品/雕塑作品）、a stone（一块石头）。

以 stone 为例，在"The man cut a block of *stone* out of the hillside."中，*stone* 指 *the material*，是不可数名词，而"A boy threw *a stone* at our car."中，*a stone* 指 *one item*，*stone* 是可数名词。

4）抽象名词不可数，如 drawing（绘画）、experience（经验）、paper（纸）、painting（绘画）、success（成功）、（in）surprise（一种情感）；但指具体意义时是可数的，如 a drawing（一件绘画作品）、a painting（一件绘画作品）、a paper（一套试卷/一个文件/一张报纸）、a success（一件成功的事）、an experience（一种体验/经历）、a surprise（一次惊奇/一件情感事情）。

以 drawing 为例，在"My daughter is very good at *drawing*."中，*drawing* 等同于 *the activity*，是不可数名词，而在"This is *an amazing drawing* by Leonardo."中，*an amazing drawing* 指 *a picture*，*drawing* 是可数名词。

1.1.4　名词复数构成规则

可数名词有单数和复数两种形式。名词由单数变复数的基本方法如下。

1）多数名词，词尾加 s，如 book→books，boy→boys，computer→computers，horse→horses，map→maps，table→tables，train→trains。

2）以 s、o、x、ch、sh 结尾的名词，词尾加 es，如 class→classes，glass→glasses；hero→heroes，tomato→tomatoes，potato→potatoes；box→boxes；bench→benches；brush→brushes 等。但少数以 o 结尾的词，变复数时只加 s，如 photo→photos，piano→pianos，radio→radios，zoo→zoos。（有生命的以 o 结尾的名词变复数时，词尾加 es；无生命的以 o 结尾的名词变复数时，词尾加 s。）

3）以辅音字母加 y 结尾的名词，变 y 为 i，再加 es，如 baby→babies，city→cities，company→companies，family→families，party→parties，secretary→secretaries 等。但以元音字母加 y 结尾的名词，结尾加 s，如 tray→trays，journey→journeys，monkey→monkeys。

4）以 f 或 fe 结尾的名词，多数变 f 为 v，再加 es，如 knife→knives，life→lives，shelf→shelves，wife→wives，wolf→wolves 等。但有些词只加 s，如 chief→chiefs，proof→proofs，roof→roofs。

5）有些名词的复数形式是不规则的，没有规律，需要记住，如 child→children，deer→deer，fish→fish，foot→feet，goose→geese，man→men，mouse→mice，ox→oxen，sheep→sheep，tooth→teeth，woman→women。

6）某些外来词的复数，或许有规律，记住就好，如 bacterium→bacteria，curriculum→curricula，criterion→criteria，datum→data，medium→media，phenomenon→phenomena（um，on→a）；analysis→analyses，basis→bases，crisis→crises，diagnosis→diagnoses（is→es）。

7）复合名词可由两个名词，或一个名词加虚词，或两个非主体词构成，也可由短语动词转化而来，还有别的构词方式。其复数构成具体有以下几种情况：

① 以可数名词结尾的复合名词有复数

形式，如 bedrooms、bookcases、firemen。

② 以不可数名词结尾的复合名词没有复数形式，如 homework、moonlight。

③ 以 man 或 woman 为前缀的复合名词，前后两个名词都变成复数，如 man servant → men servants，woman student → women students。

④ 两个名词由 of 或 in 连接，或一个名词后接 to be，第一个名词用复数，如 bird of prey→birds of prey，editor in chief→editors in chief，brother in law→brothers in law，bride-to-be→brides-to-be。

⑤ 可数名词 + 副词的形式，名词后加 s，如 looker-on→lookers-on（旁观者），passer-by→passers-by（过路人），runner-up → runners-up（亚军）。

⑥ 两个非主体词的形式，即没有主体的合成词，在词尾直接加 s，如 go-between→go-betweens（中间人），grown-up→grown-ups（成年人），shoe-maker→shoe-makers（鞋匠）。

⑦ 短语动词转化的复合名词后加 s，如 show-offs。

⑧ 在数词、年份、字母、缩写词后加 s，如 8s、2019s、Bs、CPAs；有时则加 "'s"，这样做是为了避免单词变为复数后，可能由读音造成的误解，如 A's、I's、M's、U's、i's。

不可数名词一般没有复数形式，说明其数量时，需要用有关的计量名词，如 a bag of rice→two bags of rice，a piece of paper→three pieces of paper，a bottle of milk→five bottles of milk。

1.1.5　名词所有格

名词所有格表所属关系，相当于物主代词，在句中做定语、宾语或主语，其构成法如下。

1）表示有生命事物的名词，常在词尾加 "'s"，如 Childern's Day、my daughter's book。

2）以 s 或 es 结尾的复数名词，只在词尾加 "'"，如 Teachers' Day、the heroes' name。

3）有些表示时间、距离及世界、国家、城镇等无生命事物的名词，可在词尾加 "'s"，如 today's newspaper、ten kilometers' run、China's population。

4）无论表示有生命还是无生命事物的名词，一般可用介词 of 短语来表示所有关系，如 a fine daughter of the Party。

5）"'s" 还可表示某人的家或某个店铺，如 my aunt's、the doctor's。

6）两人共有某物，可用 A and B's 的形式，如 Lucy and Lily's bedroom。

7）of + 名词所有格 / 名词性物主代词，称双重所有格，如 a friend of my mother's、a friend of mine。

1.1.6　集体名词

集体名词（collective noun）指由个体组成的一组事物。做主语时，如果把这一名词看作一个整体，谓语用单数动词；如果着眼点在于整体中的每个个体，则谓语用复数动词。

这类词常见的有：army、audience、bacteria、committee、community、company、crew、data、enemy、family、flock、gang、government、group、herd、media、navy、press、public、staff、team。

1.2　冠词

冠词（article）是一种虚词，在句中不重读，不能独立使用，只能放在名词前，说明该名词所指的事物。这是词性中最小的一类，只有定冠词 the 和不定冠词 a/an。定冠词用来对其后名词加以特指限定，指明该名词是某一特定的事物；不定冠词用来对其后名词加以宽泛的限定，指明该名词是某一类特定事物中的一个，但具体是哪一个并不重要。

1.2.1 不定冠词的用法

1）不定冠词用在读音以元音开头的字母前面。在26个英语字母中，a、e、i、o、f、h、l、m、n、r、s、x等12个字母的读音是以元音开头的，这些字母单独使用时，其前面用an，其余字母则是以辅音开头的，前面用a。例如：

- There is *an* "*a*", *an* "*n*", and *a* "*d*" in the word "and".
- Please pay attention to your spelling. You have dropped *an* "*m*" here.

这12个字母里不包括字母u，因为u虽是元音字母，但它的第一个音素不发元音。因此以上所谓以元音开头，是指an后面紧跟的字母的第一个音是元音，而与该字母本身是否为元音字母无关。例如：uncle中的u读[ʌ]，是元音音素，因此说an uncle；而university中的u读[juː]，[j]不是元音音素，因此说a university。再如：a uniform、a university、a useful book、a European、a one-way street。

2）不定冠词用在读音以元音开头的特殊单词的前面，这类单词的首字母是辅音字母，但该词发音以元音开始。例如：an hour、an honest boy。

3）不定冠词用在序数词前，表示再一、又一之意。例如：

- I have *a third* computer.

4）不定冠词用在可数名词的单数形式前表示一类事物或其中的任意一个。例如：

- *An ear* is *an organ* for listening.
- Be sure to bring me *a dictionary*.

5）不定冠词有时用在专有名词前表示一个、一种、一类或相关意思（专有名词通常是不可数的，除特殊情况外，其前面一般不用不定冠词）。

①用于指人的专有名词前，表示"名叫……的人；……似的人；……的作品"。例如：

- There wasn't *a single Jones* in the village.
- Mr. Liang is *a Lei Feng* of today.
- There's *a Rembrandt* in her collection.

②用于指地方的专有名词前，表示"像……一样的地方"。例如：

- To read Dickens you would never know there would be *a British Isles* that is not fogbound.（你如读狄更斯的书，就不会不知道一个浓雾笼罩下的英伦三岛。）

③用于表示品牌或商标的专有名词前，表示相应的产品。例如：

- I am going to buy *an Apple* or *a Huawei*.

④用于表示星期或月份的专有名词前，表示"某一个"（非特指用法）。例如：

- Christmas Day falls on *a Monday*.
- You won't catch me working on *a Sunday*!
- I came here in *a very wet July* in Beijing.

1.2.2 定冠词的用法

1）定冠词用在加有介词短语的名词前面。并非所有被定语修饰的名词都要加定冠词，如果该定语不能或不用于明确该名词的所指对象，则不加定冠词。例如：

- Show her *the photo of my family*.（特指某张照片）
- Show her *a photo of my family*.（照片可能有很多）
- His attention was attracted by *a picture on the wall*.

2）定冠词用在重新提到的事物前面。例如：

• I met a lovely girl at the gate. Look, this is *the girl who is coming*.

3）定冠词用在谈话双方都知道的事物面前。例如：

• Please fill in *the form* and sign it.

4）定冠词用在单数可数名词前，表示某一类事物。例如：

• *The tiger* is a wild animal.

5）定冠词用在世上独一无二的事物或方位名词前面。例如：

• *The sun* rises in *the east* every day.

• What does *the universe* look like to you?

6）定冠词用在序数词、形容词最高级及 only 所修饰的词前面。例如：

• It's *the first country* we will visit in Asia.

• Autumn is *the best season* in Beijing.

• He is *the only student* who didn't pass the exam.

7）定冠词用在江河、湖海、山脉、岛屿等名称前面。例如：

• Paris is the capital and largest city of France, situated on *the river Seine*.

• This is the peak of *the Himalaya Mountains*.

8）定冠词用在由普通名词构成的专有名词前面。例如：

• *The Clock Tower* looks spectacular at night when the four clock faces are illuminated.

• *The Great Wall* is symbolic of such defensive concept.

• Some time ago, an interesting discovery was made by archaeologists on *the Aegean island*.

9）定冠词用在姓氏的复数形式前面。例如：

• *The Greens* came from the England.

• There are four members of *the Liangs* family.

10）定冠词用在乐器名词前面。例如：

• My daughter likes playing *the piano* and *the cello*.

11）定冠词与某些形容词连用表示一类事物，是复数概念。例如：

• *The poor* and *the rich*, *the old* and *the young* are all equal.

1.2.3 不用冠词的情况

1）专有名词、物质名词、抽象名词前不用冠词，如 China、Beijing、December、Tom。

2）表示一类事物的复数名词前不用冠词。例如：

• They are all *editors* of *Nature*.

3）由物主代词、指示代词、不定代词或名词所有格修饰的名词前不用冠词。例如：

• *His company* is over there.

• *That company* is over there.

• *Some companies* have been moved to that city.

• *Mr Wang's company* is over there.

4）表示特别含义的名词前面不用冠词。例如：

• I usually have my *supper* at home with my daughter.

5）称呼、家庭成员的名称或只有一个人担任的职务名词前不用冠词。例如：

• *Sir*, please show me another picture and explain it.

• *Father* is usually at home, but *Mother* isn't.

• We all recommended him for *Chief Editor* of the new journal.

6）与 by 连用的交通工具名称前面不用冠词，如 by car、by taxi、by high-speed train。

7）由两个相同或相对的名词构成的平行结构前不用冠词，如 year by year、day and night。

8）有些词组中的名词有无冠词的含义是不同的，如 school、market 等表示处所的名词，当不指具体地点时不用冠词，但当指明地点或为了某种目的去这些处所时就要加上冠词。

1.2.4　冠词短语

冠词短语是含有冠词的短语，使用较为固定，包括含 a 和 the 的短语。例如：have a word with sb.，have a drink，have a good time，have a swim，have a bath/shower，have a look（at），have a fever，have a talk，have/take a walk，have a nice trip，have a rest，give a lesson，in a/one word，many a time；by the way，in/at the front of，in the daytime，at/in the beginning，in the middle night，in the end，in the morning/afternoon/evening，on the right/left，the same as。

1.2.5　冠词的省略

名词前本来应该有冠词，但有时在某种语境下可以省略冠词，进而简化语言。

1）为避免重复而省略冠词。例如：

- *The lightning* flashed and *thunder* crashed.（thunder 前省去 the）
- The noun is the name of *a person* or *thing*.（thing 前省去 a）

2）句首的定冠词可省去。例如：

- （*The*）*Project* is dismissed.
- （*The*）*Fact* is that they do not understand the science.

3）在 the next day 和 the next week 之类的短语中，常省去定冠词。例如：

- （*The*）*Next day* we will attend the meeting together early.
- The road show is coming here（*the*）*next week*.

4）信函地址常省去定冠词或不定冠词。例如：

- English Dpt.
- Foreign Studies University
- Xicheng District，Beijing，China

1.2.6　冠词的位置

（1）不定冠词通常位于名词或名词修饰语前面。

1）不定冠词位于形容词 such、what、many、half 之后，如 *such an* animal、*many a* man。

2）名词前的形容词被副词 as、so、too、how、however、enough 修饰时，不定冠词放在形容词之后。例如：

- It is *as pleasant a day* as I have ever spent.
- It is great to finish writing this book in *so short a time*.
- It is *too long a distance* between the two cities.

3）当 quite、rather 与单数名词连用时，冠词放在其后面；当 rather、quite 后仍有形容词时，不定冠词放其前面、后面均可。例如：

- I have *quite a lot* of friends.
- It's *rather a cold day/a rather cold day*.
- She is *quite a good teacher/a quite good teacher*.

4）在 as、though 引导的让步状语从句中，当表语为形容词修饰的名词时，不定冠词放形容词后。例如：

- *Brave a man* though he is，he trembles at the sight of snakes.

（2）名词前有 all、both、double、half、twice、three times 之类的修饰词时，定冠词通常位于这类词之后，名词之前。例如：

- *All the students* in the class went out.
- *Both the boys* have been to London.

1.2.7　冠词的特殊用法

1）两个形容词用来修饰不同名词并连用表示不同事物时，两形容词前都用冠词。例如：

- He raises *a black* and *a white cat*.
- *The black* and *the white cats* are hers.
- I want *a cup of tea* with milk，or *a cup of lemon tea* without milk.

2）两个形容词连用来修饰同一名词表同一事物时，后一个形容词前不用冠词。例如：

- He raises *a black and lively cat*.
- *The black and lively cat* is hers.

1.2.8　冠词的活用

1）表示世界上独一无二的事物的名词前一般加定冠词，但如果名词前有修饰语，也可能用不定冠词，如 the world、a peaceful world，the moon、a bright moon 等。独一无二的事物的名词前有修饰语时，该词就被泛指化，moon 尽管独一无二，但当用 bright 来修饰时，不同状态的 moon 就会构成多样化的 moon，这样它就泛指了。

2）表示一日三餐的名词前一般不用冠词，但前面有定语时，可能用不定冠词。例如：

- We often have *supper* at 7pm.
- We had *a wonderful supper* yesterday.

3）表示乐器的名词前一般用定冠词，但前面有定语时，可能用不定冠词。例如：

- My daughter starts her day by playing *the violin*.
- My daughter is playing *a borrowed violin*.

4）介词与表示交通工具的名词连用表示笼统的方式，前面一般不用冠词，但名词前面有定语时，需要加冠词。例如：

- I went to the station *by car*.
- I went to the station *by a black car*.

5）表示语言的名词前一般不用冠词，但后面若出现 language，则前面加定冠词。例如：

- I like *English* very much.
- I like *the English language* very much.

6）当 turn 用作系动词时，后面做表语的单数名词前不用冠词。例如：

- I *turned writer* many years later.
- I *became a writer* many years later.

1.3　代词

代词（pronoun）是起替代作用的词，用来替代名词、名词短语、分词或句子，替代名词时又叫代名词。多数代词具有名词和形容词的功能。代词按意义、特征及在句中作用分为以下十种。

1.3.1　人称代词

人称代词（personal pronoun）表示自身或人称，有人称、性、数与格之分，见表 1-1。

表 1-1　人称代词分类

类别	单数		复数	
	主格	宾格	主格	宾格
第一人称	I	me	we	us
第二人称	you			
第三人称	he	him	they	them
	she	her		
	it			
	one		ones	

人称代词主格做主语，宾格做表语、宾语（分动词宾语和介词宾语两类，分别简称动宾和介宾）。在现代英语中，系动词 be 后面常用人称代词的宾格，而不用主格。例如：

- *I* am an editor，and work in the editorial office of *CJME*.（第一人称代词 I 做

主语）

- *You* are a good researcher and scientist.（第二人称代词 You 做主语）
- *It*' s a good book, *I*' m reading it interestingly.（第三、第一人称代词 It、I 均做主语）
- *It*' s *me*. Please open the door quickly.（第三人称代词 It、me 分别做主语、表语）
- Don' t tell *him* about *it*.（第三人称代词 him、it 分别作动宾和介宾）
- *She* is always ready to help *us*.（第三人称代词 she、us 分别做主语和动宾）
- Our leader is very strict with *us*.（第三人称代词 us 做介宾）

正确使用人称代词有以下几种情况需要注意。

1）第一人称单数代词 I 不论在什么地方均要大写。例如：

- *I* write English writing book every day.
- *She* and *I* often go to the Beihai Park to communicate with each other.

2）We（we）通常可替代 I，表示同读者、听众或观众之间的一种亲密关系。例如：

- *We* shall do *our* best to help the poor.

3）She（she）通常可替代国家、城市和宠物等，表示一种亲密或爱护的感情。例如：

- I live in China. *She* is a great country. I like *her* very much.

4）It 可指人、身份、天气、环境、时间等，做形式主语、宾语，或构成强调句型。例如：

- *It*' s me. Open the door, please.
- *It*' s very important for me to write so many scientific writing books.
- *It*' s on the end of 2019 that I finished the writing of *SCI Paper Writing*（*English Edition*）.

5）They（we）有时用来替代一般人或普通大众。例如：

- *They* say you are good at writing.
- *We* all think that the new emerged influenza viruses may cause severe human infection and deaths.

1.3.2　物主代词

物主代词（possessive pronoun）表示所有关系，也称代词属格，分为形容词性和名词性两类。形容词性物主代词也属限定词，包括 my、your、his、her、its、our、their；名词性物主代词在句中可做主语、动宾、介宾和主语补语（简称为主补），包括 mine、yours、his、hers、its、ours、theirs。物主代词分类见表 1-2。

表 1-2　物主代词分类

类别	我的	你的	他的	她的	它的	我们的	你们的	他们的
形容词性物主代词	my	your	his	her	its	our	your	their
名词性物主代词	mine	yours	his	hers	its	ours	yours	theirs

形容词性物主代词可以做定语。例如：

- *My* daughter is a clever little girl.
- I love *my* countryside.
- Is this *your* notebook?

名词性物主代词可以做主语、动宾、表语（主补）及介宾（与 of 连接的定语）等成分。例如：

- That isn' t my bag. *Mine* is left at home.（主语）
- Ann showed him her tickets, then Ian

showed her *his*. （动宾）

- That car is *hers*, not *yours*. （表语）
- Yesterday I met a friend of *mine* in the Beihai Park. （介宾）

注意，形容词性物主代词 its 与 it is 或 it has 的缩略形式 it's 是不同的，不要混淆。

1.3.3 指示代词

指示代词（demonstrative pronoun）表示这个、那个、这些、那些，以及它、如此、同一，包括 this、that、these、those、it、such、same，在句中做主语、宾语、表语、定语。这类代词有单数（it、this/that）和复数（these/those）两种形式，与定冠词和人称代词一样，都有指定的含义。例如：

- *This* is a grey car. （主语）
- *Those* are my classmates. （主语）
- "Which do you like?" "I like *that*."（宾语）
- It's not *that*. （问题不在那儿）（表语）
- I should say I know *these* things. （定语）
- *Those* persons are my classmates. （定语）

注意：this 和 these 表示在时间或空间上较近的人或物，而 that 和 those 表示在时间或空间上较远的人或物。例如：

- *This* is a new bought book.
- *These* are my good friends.
- *That* is not a new city.
- I was busy *those* days in June and July.

此外，that 和 those 还可指上文中提及的事物，this 和 these 指下文中将要讲到的事物，它们起着一种承上或启下的作用。例如：

- I often write late into the night, *that*'s why I can write so many books successfully.
- *This* is what I will do. I will apologize first, and then ask for help.

1.3.4 反身代词

反身代词（reflexive pronoun）表示反射和强调，如我自己、你自己、他（它）自己、我们自己、你们自己、他（它）们自己等。这类代词的第一、第二人称由形容词性物主代词加 self（复数加 selves）构成，第三人称由人称代词宾格形式加 self（复数加 selves）构成。反身代词的分类见表 1-3。

表 1-3　反身代词的分类

类别	第一人称	第二人称	第三人称
单数	myself	yourself	himself/herself/itself
复数	ourselves	yourselves	themselves

反身代词放在名词、代词的后面或句末，但省去时并不影响句子的语法正确性。例如：

- The boys (*themselves*) can't explain how this accident happened.
- My wife (*herself*) told me all the news.
- The book was written by professor Liang (*himself*).

反身代词可做宾语（直接宾语、间接宾语和介词宾语）、表语、同位语（主语同位语和宾语同位语）。用做同位语时表示强调"本人，自己"。例如：

- You mustn't blame *yourself* for these mistakes. （直接宾语）
- I am teaching *myself* computer. （间接

宾语）

- Take good care of *yourself*.（介词宾语）
- The child *himself* drew this picture.（主语同位语）
- You should ask the children *themselves*.（宾语同位语）

反身代词常用在介词 by 后面，表示主语在无任何帮助的情况下做了某事，也可表示独自一个之意。这种结构也可用 on my own 和 on your own 等来替代。例如：

- Can you do it by *yourself*?
- I have stayed in Beijing for about 26 years by *myself*.
- I can bear it *on my own*, because it's not so serious.

1.3.5 相互代词

相互代词（reciprocal pronoun）表示相互关系，指出句中动词所表述的动作或感觉在涉及的各个对象之间是相互存在的。相互代词与其所指代的名词或代词是一种互指关系，因此它们是复数，或指二者及以上。这类代词只有 each other 和 one another 两个，正式文体中多用 each other 指两者，one another 指两者以上，现代英语中两组词交替使用的情况也不少。相互代词应当看作复合代词，即使在分开使用时，分开的两部分也是相互关联的。

相互代词在句子中做宾语、定语。例如：

- We often help *each other* in writing academic papers.（动宾）
- They see *one another* every summer vacation.（动宾）
- Dogs bark, cocks crow, frogs croak to *each other*.（犬吠、鸡鸣、蛙儿对唱）（介宾）
- The students borrowed *each other*'s notes.（定语）（做定语时须用所有格）

1.3.6 疑问代词

疑问代词（interrogative pronoun）用来直接表示疑问而构成特殊疑问句（直接问句），或用来间接表示疑问而引导从句（间接问句），包括 who、whom、whose、what、which。

疑问代词在特殊疑问句中一般位于句子前面，可做主语、宾语、表语和定语。例如：

- *Who* was here just now?（主语）
- *Which* came first, the chicken or the egg?（主语）
- *Whom* are you looking for?（动宾）
- For *whom* are they supposed to do for the project?（介宾）
- *What* is this?（表语）
- *Whose* notebook computer is this?（定语）
- *Which* one do you like, this one or that one?（定语）

疑问代词还可做连接代词，引导主语从句、宾语从句、表语从句。例如：

- *What* I should do is still unknown.（主语从句）
- He wondered *what* I would do now.（宾语从句）
- I know *whom* she is looking for.（宾语从句）
- Tell me *who* she is.（宾语从句）
- She is *whom* I am looking for.（表语从句）

1.3.7 关系代词

关系代词（relative pronoun）是用来引导定语从句的关联词，主要有 who、whose、whom、that、which、as 等。这类词在定语从句中同时起两个作用：像其他代词一样，做定语从句的成分，如主语、表语、宾语和定语；像连词一样，代表主句中为定语从句所修饰的那个名词或代词（先行词），把定语从句与主句连接起来。例如：

- The girl *who* was lost ten years ago appeared today.（主语）
- He never hesitates to make such criticisms *as* are considered helpful to others.（他从不犹豫去进行有助他人的批评）（主语）
- She is the girl *whom* I have been looking for.（宾语）
- She is the only one *whose* advice they might listen to.（定语）

关系代词 which 的先行词可以是一个句子，that 在从句中做宾语或表语时可省略。例如：

- He said he saw me in the park yesterday, *which* was a big lie.（先行词是句子）
- I've forgotten much of the German (that) I once learnt.（做宾语的关系代词 that 省略）
- I'm changed greatly. I'm not the man (that) I was.（做表语的关系代词 that 省略）

1.3.8 连接代词

连接代词（conjunctive pronoun）是引起从句的疑问代词，包括 who、whom、whose、what、which、whoever、whomever、whichever、whatever，共九个，除了 whose 后不能加后缀 ever 之外，其余都可以加。连接代词可引起主语从句、宾语从句、表语从句、状语从句和不定式，在句中做主语、宾语、表语、定语等（who、whom、whoever 等不能用于名词前做定语）。例如：

- It hasn't been announced *who* won the prizes.（主语从句的主语）
- It is clear enough *what* she meant.（主语从句的宾语）
- I'll take *whoever* wants to go.（宾语从句的主语）
- I don't care *what* they think.（宾语从句的宾语）
- The question is *who* (*m*) we should trust.（表语从句的宾语）
- What I want to know is *which* road we should take.（表语从句的定语）
- Wherever you go, *whatever* you do, I will be right here waiting for you.（状语从句的状语和宾语）
- I can't decide *which* to choose.（不定式的宾语）
- They exchanged views on the question of *whom* to elect.（不定式的宾语）

1.3.9 不定代词

不定代词是没有明确指定替代任何特定名词或形容词的代词。常见的有：all、any、another、anybody、anyone、anything、both、each、either、every、everybody、everyone、everything、few、little、many、much、neither、no、none、no one、nobody、nothing、one、other、some、somebody、someone、something 等。其中一些是由 any、every、no、some 和 body、one、thing 构成的复合词。

不定代词大部分可替代名词或形容词，可做主语、宾语、表语和定语，但 none 及由 some、any、no 等构成的复合不定代词只能做主语、宾语和表语，every 和 no 只能做定语。例如：

- *Everybody* should be here in time tomorrow.（主语）
- I know *nothing* about her.（宾语）
- That's *all* I know.（表语）
- I read book *every* day.（定语）

any 多用在否定或疑问句中，在句中做主语、宾语、定语。any 做定语时，所修饰的名词没有单复数限制，一般多用复数，用在肯定句中时，表“任何”。例如：

- Do you have *any* books?
- You can come *any* time.

some 多用在肯定句中，表示邀请，或用在对方可能给予肯定回答的疑问句中等。例如：

- There are many beautiful flowers in the garden, *some* are red, *some* are white.
- I am going to get *some* water.
- Will you have *some* tea, please?

no 表示否定，语气比 not any 强，在句中做定语。例如：

- She knows *no* China.
- I have *no* house in Beijing.

none 指人或物，后面通常接 of 短语，做主语时，若指不可数名词，则谓语只能用单数，若指复数名词，谓语可用单数（正式文体），也可用复数（非正式文体）。例如：

- *None* of the milk can be used.
- *None* of the games is/are worth seeing.

many 可替代可数名词，做主语、宾语、定语。例如：

- *Many* of the students like math very much.
- I have *many* books to publish.

much 可替代不可数名词，做主语、宾语、定语。例如：

- *Much* of the time was spent on traveling.
- There is not *much* ink in the bottle.

a few、a little、few、little 表示少量、不多、几个，只是主观上的一种相对说法，无具体数量标准，做主语、宾语、定语，其中 a few 和 few 替代可数名词，a little 和 little 替代不可数名词。例如：

- *A few* friends came to see me the day before yesterday.
- *Few* of the houses are cheap now.
- They have *a little* money to buy the car.
- There is *little* ink in the bottle.

one、ones 可替代一个上文出现过的可数名词，前面可用限定词修饰。one 也可做不定人称代词，有属格形式 one's 和反身代词 oneself，常见于正式文体。例如：

- Both man and woman have an equal right to *a career* if they want *one*.
- They created *a single strong organization* instead of *two weak ones*.
- One wanted only the best for *one's* children.
- *One* can't enjoy *oneself* if one is too tired.（一个人如果玩得太累就不会玩得尽兴。）

不定代词做主语时，用单数谓语动词。用人称代词替代不定代词时，常用复数代词 they、them、their、themselves。不定代词用于指人时可用“'s”，但指物时不可以。例如：

- *Everyone* recognizes the importance of a stronger body.
- Somebody lost *their* wallets.
- Everything has been arranged to *everybody's* satisfaction.

1.3.10 替代词

替代词用来避免重复而替代前文所提及的可数名词。这类词包括 one 和 ones，前者用来替代可数名词的单数，后者用来替代可数名词的复数。注意，中心名词只有在可以用 one 或 a、an 修饰表示“一”的情况下，才可以使用替代词来替代。例如：

- There are many cars in the park, which *one* is yours?（one 替代 car）
- There are many computers in the office, and those black *ones* are Lenovo.（ones 替代 computers）

1.3.11 代词与先行词一致

1）当先行词为 anyone、anybody、anything、everyone、everybody、no one、nobody、someone、somebody 等复合词时，代词及相应的限定语通常按照语法一致原则用单数形式。例如：

- *Anything* on the table can be thrown away，can't *it*?
- *Everyone* talked at the top of *his* voice.

2）当先行词为某种并列结构时，一般根据该并列结构的单、复数意义来确定代词及相应限定词的单、复数形式。例如：

- *My friend and colleague has* agreed to lend me *her* car.
- *My friend and my colleague have* agreed to lend me *their* cars.

3）当先行词为复数名词或代词做句子主语，并带有 each 做同位语时，如果 each 出现在动词之前，随后的人称代词或相应的限定词用复数；如果 each 出现在动词之后，随后的人称代词或相应的限定词用单数。例如：

- *They* each had *their* problems.
- *They* had each *his* own problems.

4）当先行词为单数中性名词，即表示无生命物的名词时，后面的代词及相应的限定词通常用中性代词（it、its、itself）。例如：

- My notebook *computer* has lost *its* original color.
- This *machine* works by *itself*.

5）先行词的人称及单、复数决定着其后代词的人称形式。例如：

- *The boy* next door stole a toy from my daughter. *His* mother told *him* to return it，but *he* said it was *his*.
- *I and my friend Mr. Zhang* will move to Hainan. *We* know that the climate there is healthful.

1.4 数词

数词（numeral）是表示数目多少或顺序先后的词。表示数目多少的数词叫基数词，如 one、five、ten、fifty、sixty-five、one hundred、three hundred and sixty-five 等；表示顺序先后的数词叫序数词，如 first、fifth、tenth、fiftieth、sixty-fifth、one hundredth、three hundred and sixty-fifth 等。数词的用法相当于名词和形容词，在句中做主语、宾语、定语、表语和同位语等。

1.4.1 基数词的用法

1）基数词在句中主要做主语、宾语、定语、表语、同位语。例如：

- *Ten* of them are Party members.（主语）
- "How many would you like?" "*Five*, please."（宾语）
- The *six* boys are from Jinan.（定语）
- Six plus eight is *fourteen*.（表语）
- We *three* will go with you.（同位语）

2）当 hundred、thousand、million、billion 等表示一个具体数字时不用复数，但表示一个不确定数字时，需要用复数的形式（即结尾加 s）。例如：

- Our country has a population of about *1.4 billion* people.
- There are *two thousand* boy students in that school.
- After the war，*thousands* of people became homeless.
- Maize is the most important food crop for *millions* of people in the world.
- They arrived in *twos and threes*.

3）表示"……十"的数词的复数形式可用来表示人的岁数或年代。例如：

- He is in his early *twenties*.

- She died still in her *forties*.
- This war took place in *1940s*.

4）表示时刻用基数词。例如：

- I usually get up at *six*, have breakfast at *seven*, and begin work at half past *eight*.

表示几点过几分，用介词 past，分数须在半小时以内（含半小时），如 ten past ten、a quarter past nine、half past twelve；表示几点差几分，用介词 to，分数须在半小时以内，如 twenty to nine、five to eight、a quarter to ten；表示几点几分，可直接用基数词，如 seven fifteen、eleven thirty、nine twenty。

1.4.2 序数词的用法

1）序数词主要做定语，前面加定冠词。例如：

- *The first* truck is carrying a small car.
- My mother lives on *the twelfth* floor.

2）序数词前面有时可加不定冠词来表示再一、又一的意思。例如：

- We'll have to do it *a second* time. Shall I ask him *a third* time?
- When she sat down, *a fourth* man rose to speak.

3）有几个序数词与其相应的基数词拼写不对应，容易写错。例如：one-first、two-second、three-third、five-fifth、eight-eighth、nine-ninth、twelve-twelfth。

4）表示年、月、日时，年用基数词表示，日用序数词表示。例如：

- October (*the*) *first*, nineteen forty-nine.
- September (*the*) *tenth*, two thousand and twenty (twenty twenty).

5）表示分数时，分子用基数词表示，分母用序数词表示。分子大于 1 时分母加 s。例如：one second, three fourths, two fifths。

6）序数词有时用缩写形式。例如：first→1st, second→2nd, third→3rd, fourth→4th, twenty-second→22nd。

1.5 形容词

形容词（adjective）用来修饰、描述名词或不定代词，主要做定语、表语和补足语等，表示事物的性质、状态和特征。

1.5.1 形容词的分类

形容词分为性质形容词和叙述形容词两大类。

性质形容词用来直接说明事物的性质或特征，有比较级的变化，可用程度副词修饰，可做定语、表语和宾补。大部分形容词属于这一类。例如：academic、beautiful、common、collective、good、hot、interesting、new、rich、wonderful。

叙述形容词只能做表语，又称表语形容词，没有比较级的变化，不可用程度副词修饰。大多数以 a 开头的形容词属于这一类。例如：afraid、alike、alive、alone、asleep、awake、faint、ill、unable、unwell、well、worth。

1.5.2 形容词的用法

1）做定语，放在名词前（形容词 + 名词），但修饰以 thing 结尾的不定代词（如 anything、nothing、something）时，要放在这些词的后面（不定代词 + 形容词）。例如：

- My wife often tells my daughter an *interesting* story before going to bed.
- There is nothing *new* in this book except for several new ideas.

2）做表语，放在系动词的后面（be + 形容词）。例如：

- Lucy was *unable* to find out what had happened.
- The book about writing of academic

papers written by me is *best-selling*.

3）做宾补，放在宾语后，常与 make、leave、keep 等动词连用。例如：

• We are trying to do our best to make our life *rich*.

• The story in yesterday's dream made me *excited*.

4）做主语补足语，放在谓语后，补充说明主语的特征、性质。例如：

• He died *young* because of the disease.

• Many people were buried *alive* in the earthquake.

5）放在定冠词后，变成名词性的词（the + 形容词），表示类别和整体，泛指一类人或事物（如 the dead、the living、the rich、the poor、the blind、the hungry），有关国家和民族的形容词加上定冠词时指这个民族的整体（如 the British、the English、the French、the Chinese），做主语时谓语用复数。例如：

• In modern China *the rich* are becoming more and more, and *the poor* smaller and smaller.

• *The Chinese* have always been hardworking and brave.

6）alone、asleep、alike、alive、awake、ashamed、afraid 等以字母 a 开头的形容词通常不能做前置定语，但可做表语或后置定语。例如：

• We are *alike* in the way we work.

• They are the most miserable children *alive*.

7）elder、little、only、wooden、woolen 等只能做定语，不能做表语，这类形容词称为定语形容词。例如：

• Her *elder* sister has been in London since 2008.

• Whenever I saw *woolen* products advertised, I stopped to see what was on offer.

1.5.3　形容词的形式

（1）以 ly 结尾

形容词加 ly 后通常构成副词，但有些加 ly 后仍是形容词或形容词兼副词。例如：comradely、fatherly、lively、lovely、lonely、silly、sisterly、ugly（形容词）；brotherly、daily、deadly、early、friendly、likely、monthly、weekly（形容词兼副词）。

（2）以 ing/ed 结尾

1）以 ing 结尾的形容词常用来描绘人或事物对人的情绪产生了作用。例如：

• I want to live in a *charming* house just outside Beijing city.

• Many *surprising* tales have been told my daughter by her mother.

2）有些以 ing 结尾的形容词用来修饰名词，表示持续一段时间的过程或状态。例如：

• *Increasing* prices are making people get poorer.

• There is rapidly *rising* population in that countryside.

3）以 ed 结尾的形容词用来描绘人的情绪，表示主语作为动作的承受者处于某种情绪之下，具有被动意义。例如：

• She appeared *alarmed* and *upset* by the chaos around her.

• I shall be *delighted* to meet Mrs. Wang at the university.

4）有些以 ed 结尾的分词（ed 分词）本身不具形容词的特性，不能在名词前做修饰语，但当其被副词修饰时可构成“副词 + ed 分词”的复合形容词，这类形容词可放在名词前。例如：well-behaved、badly-behaved、far-fetched、well-defined、well-dressed、newly-invented。

5）有小部分以 ed 结尾的形容词不能做前置修饰语，只能用在系动词 be、become、

feel 之后，后面跟介词短语、不定式短语或 that 从句做补语。例如：

- She was *thrilled* by the publication of my monograph about academic paper writing.
- The little boy was *scared* to see such a horrible scene in the dark.
- I felt *scared* that I dreamed I could fly freely as a bird.

1.5.4　形容词的顺序

多个形容词修饰一个名词的先后顺序通常是，先放表形状、大小的形容词，后放表颜色、国家、材料的形容词，如 a *big brown* cow、a *little red* bag、a *little Chinese* girl、a *nice long new black British plastic* pen 等。

有一个杜撰的词 OPSHACOM 可帮助确定形容词的顺序。其中，字母组合或字母表示某类形容词：OP 是 opinion，感觉、观点类，如 beautiful、horrible、lovely；SH 是 shape，大小、形状等类，如 long、round、narrow；A 是 age，年龄、时代类，如 old、new、young；C 是 colour，颜色类，如 red、black、orange；O 是 origin，国籍、地区类，如 Chinese、Canadian、German；M 是 material，材质、材料类，如 plastic、metal、aluminium。

名词前的修饰语的具体顺序通常如下：

①冠词、指示代词（this、those 等）→②所有格（my、Tom's 等）→③序数词（first、second 等）→④基数词（one、two 等）→⑤特征、特性（常含主观看法，old、good 等）→⑥大小、长短、高低→⑦年龄、温度、新旧→⑧形态、形状→⑨颜色→⑩国籍、地区、出处、来源→⑪物质、材料→⑫用途、类别、目的→被修饰的名词。

1.5.5　形容词的级的变化

大多数性质形容词有原级、比较级和最高级三种形态，表示事物的等级差别。原级是形容词的原形；比较级和最高级是在原级的基础上变化产生的，有规则变化、不规则变化两种。

（1）规则变化

单音节和少数双音节形容词，结尾加 er、est 来构成比较级和最高级。

1）一般单音节词，加 er、est。例如：cold、colder、coldest，high、higher、highest，rich、richer、richest，small、smaller、smallest。

2）以不发音的 e 结尾的单音节词和少数以 le 结尾的双音节词，只加 r、st。例如：large、larger、largest，nice、nicer、nicest，able、abler、ablest。

3）以一个辅音字母结尾的闭音节单音节词，双写结尾辅音字母，再加 er、est。例如：big、bigger、biggest，hot、hotter、hottest。

4）以辅音字母 + y 结尾的双音节词，改 y 为 i，再加 er、est。例如：busy、busier、busiest，easy、easier、easiest，happy、happier、happiest。

5）少数以 er、ow 结尾的双音节词，加 er、est。例如：clever、cleverer、cleverest、narrow、narrower、narrowest。

6）部分双音节词和多音节词，在前面加 more、most。例如：careful、more careful、most careful，beautiful、more beautiful、most beautiful，important、more important、most important。

7）分词形容词，一般在前面加 more、most。例如：tiring、more tiring、most tiring。

（2）不规则变化

不规则变化没有规律，需要记住。例如：good（well）、better、best，bad（ill）、worse、worst，old、elder（规则变化是 older）、eldest（规则变化是 oldest），much/many、more、most，little、less、least，far、farther/further、farthest/furthest。

1.5.6 形容词的级的用法

（1）原级用法

1）肯定句中常用“A… + as + 原级 + as + B”表示“A 与 B 在某一方面相同”。例如：

- Tom's handwriting is *as beautiful as* yours.

2）否定句中常用“A…not + as/so + 原级 + as + B”表示“A 在某一方面不如 B”。例如：

- I am not *as/so busy as* I used to be.

3）用“A… + 倍数词/half + as + 原级 + as + B”表示“A 是 B 的……倍/一半”。例如：

- This university is *twice/half as large as* that one.

4）第一个 as 后的形容词做定语修饰名词，将该名词及修饰语全放在第一个 as 后。例如：

- I don't have *as much money* as you do.

（2）比较级用法

1）用“A… + 比较级 + than + B”表示“A 比 B 更……”。例如：

- Mrs. Xiao Li is *younger than* Mrs. Xiao Wang.

2）用“倍数词 + 比较级 + than”表示“比……多/少几倍”。例如：

- This machine is *four times heavier than* that one.
- This country is *three times smaller than* that one.

3）用“比较级 + and + 比较级”表示“越来越……”，多音节词和部分双音节词用“more and more + 原级”表示此义。例如：

- The car is running *faster and faster*.
- Reading and learning become *more and more important* today.

4）用“the + 比较级，the + 比较级”表示“越……就越……”。例如：

- *The more careful* you are, *the fewer mistakes* you'll make.

5）用“the + 比较级 + of…”表示“两者中更……的那个”。例如：

- Building A is the *taller of* the two.

6）用“could't + 比较级…”表示“不可能更……”。例如：

- She *couldn't* get a *better* result.

7）用“完成时 + 比较级…”表示最高级。例如：

- I *have never been to a better place* than this.

8）用“who/which is + 比较级，A or B?”表示“A 和 B，哪一个更……”。例如：

- *Who* is *more excellent*, *Liu or Wang*?
- *Which* is *faster*, *a high speed rail or a plane*?

（3）最高级用法

1）用“the + 最高级 + in/of 短语或从句”表示三者或三者以上程度最高。例如：

- The central building is *the tallest in/of* those buildings of the university.

2）用“who/which is + the + 最高级，A, B or C?”表示在三者或三者以上事物中选择“哪一个最……”。例如：

- *Which* is *the most important*, house, money or love?

3）用“one of the + 最高级”表示“最……的……之一”。例如：

- Professor Liang is *one of the most popular* researchers in scientific area.

4）形容词最高级前可以加序数词，表示“第几最……”。例如：

- The Yangtze River is *the first longest*

river in China.

1.6　副词

副词（adverb）用来修饰动词、形容词、副词、介词（短语）或全句，在句中做状语、表语、定语和宾补，说明时间、地点、程度、方式等。它同形容词一样，在句中主要做修饰成分，但不用来修饰名词或不定代词，而且在句中的位置比较灵活，主要取决于在句中的作用。

1.6.1　副词的分类

副词大体上分为以下八类。

（1）时间和频度副词

时间和频度副词表示时间或动作发生的频度。常见的有：ago、already、always、before、early、ever、finally、frequently、generally、hardly、immediately、last、lately、now、next、never、often、soon、seldom、shortly、sometimes、too、then、today、usually、yet、yesterday。

（2）地点副词

地点副词表示地点或位置关系。常见的有：across、above、around、away、anywhere、along、back、below、down、downstairs、everywhere、forward、here、home、in、inside、near、off、on、out、outside、past、round、there、up、upstairs。

（3）方式副词

方式副词表示行为动作发生的方式，常可回答 how 引导的问句，大多以 ly 结尾。常见的有：anxiously、calmly、carefully、fast、loudly、luckily、normally、properly、politely、proudly、quickly、softly、suddenly、warmly、well。

（4）程度副词

程度副词对一个形容词或副词在程度上加以限定或修饰，常位于被修饰词之前。常见的有：almost、enough、extremely、entirely、little、much、perfectly、quite、rather、so、still、slightly、too、very。

（5）疑问副词

疑问副词用来引导特殊疑问句，表示时间、地点、方式、原因等。常见的有：how、how long、how often、when、where、why。

（6）焦点副词

焦点副词强调使之成为人们注意的焦点。常见的有：also、alone、even、exactly、especially、just、merely、mainly、only、simply、too。

（7）关系副词

关系副词引导定语从句，修饰句中某一名词或代词，被修饰词称作先行词，关系副词在先行词后。常见的有：how、when、where、why（when = on which，where = in which，why = for which）。

（8）连接副词

连接副词连接句子，相当于并列连词，或引导名词从句、不定式。常见的有：besides、however、otherwise、moreover、meanwhile、still、thus、therefore、wherever（连接句子或从句）；how、when、where、why、whether（引导从句或不定式）。

1.6.2　副词的用法

（1）修饰动词

副词修饰动词时，通常位于动词后面，也可放在动词前面，动词带有宾语时，副词应放在宾语后面。副词也可放在系动词（be）、助动词、情态动词之后（但在省略或强调句中必须放在系动词、助动词或情态动词之前），当有多个助动词时，副词一般放在第一个助动词后。副词可以位于句尾（方式副词 well、badly、hard 等只放在句尾），但宾语过长时，副词可以提前，以使句子平衡。例如：

- We could *see* very *clearly* a strange light ahead of us.（副词在动词后）

- I *go to bed late* in the evening everyday.（副词在动词短语后）
- She *is in*, let's *be out*.（她进来了，我们出去吧。）（副词在系动词后）
- Food *here is* hard to get.（副词在系动词前）
- San Francisco *is always* cool in summer, but Los Angeles *rarely is*.（副词在系动词后、前）
- My daughter *speaks* English *well*.（副词在句尾）

（2）修饰形容词

- The room is not *too big* but *very quiet*.
- The car bought last month is *quite conspicuous*.

（3）修饰副词

程度副词 almost、fairly、hard、much、nearly、quite、very、well 通常放在所修饰词的前面。例如：

- She is *nearly always* in the wrong.
- My daughter speaks English *quite well*.
- The train goes *very fast*.
- I don't know him *well enough*.

（4）修饰介词（短语）

- I often work or read *deep into* the night in oder to make great progress.
- The office is almost at the end of the street, *just before* the traffic lights.

（5）修饰全句

一些频度副词如 actually、briefly、certainly、（un）fortunately、importantly、surely 等对全句进行修饰时，常放在句首，并用逗号与后面的陈述句隔开。例如：

- *Fortunately*, you have the power to change all that.
- More *importantly*, we can share the information with all of them.

1.6.3　副词的形式

副词由形容词转变来的形式大致有以下四种情况。

1）在形容词词尾直接加 ly，以 l 结尾的加 ly，以 ll 结尾的加 y，以 e 结尾的多数直接加 ly。例如：bad→badly、quick→quickly、loud→loudly、sudden→suddenly、right→rightly、glad→gladly；usual→usually、real→really、medical→medically；full→fully；definite→definitely、complete→completely、late→lately。

2）以“辅音字母 + y”结尾的形容词，要变 y 为 i 再加 ly。例如：angry→angrily、heavy→heavily、happy→happily、lucky→luckily、noisy→noisily。

3）有些以 ble 或 le 结尾的形容词，要去掉 e 再加 y。例如：possible→possibly、terrible→terribly。

4）少数以 e 结尾的形容词，要去掉 e 再加 ly。例如：true→truly。

1.6.4　副词的顺序

多个副词并列出现时，应按一定的顺序先后排列，大致有以下两种情况。

1）几个同类副词或副词短语并列出现时，一般遵循“小在前、大在后”的原则，并用 and 或 but 等连词连接。例如：

- An important report should be written *slowly and carefully*.

2）通常按程度副词→方式副词→地点副词→时间副词的规律来排列。但对于 drive/go/head/leave/move/walk/run 等表示位置移动的动词，修饰它们的副词通常按地点副词→程度副词→方式副词→时间副词的规律来排列。例如：

- My daughter read *very quietly over there* all afternoon.
- I and my friends went *there very happily*

yesterday.

1.6.5 副词的级的变化

副词的级的变化与形容词的基本上一样，也有原级、比较级和最高级三个等级。原级是副词的原形，比较级和最高级是在原级的基础上变化产生的，有规则变化和不规则变化两种。

（1）规则变化

1）单音节词和少数双音节词，结尾加 er、est 来构成比较级和最高级，但以后缀 ly 结尾的除外。例如：fast、faster、fastest，hard、harder、hardest，late、later、latest，long、longer、longest，loud、louder、loudest，soon、sooner、soonest，wide、wider、widest，early[⊖]、earlier、earliest。

2）多音节词及以后缀 ly 结尾的副词（early 除外），前面须加 more 或 most。例如：carefully、more carefully、most carefully，happily、more happily、most happily，quickly、more quickly、most quickly，quietly、more quietly、most quietly。

（2）不规则变化

不规则变化没有变化规律，只能记住。例如：badly、worse、worst，far、farther（further）、farthest（furthest），little、less、least，much、more、most，well、better、best。

1.6.6 副词的级的用法

（1）单独使用

副词不加任何修饰语，独立出现。例如：

- I will try to do *better* in the future.
- She'll come back *sooner* or *later*.
- We have determined not to invest *farther* next year.
- It's necessary to speak *slowly* and *clearly*.
- I should have made the speech *earlier*.

（2）与 than 一起使用

副词与 than 连用，即在副词后加 than。例如：

- Can you do any *better than* that?
- He arrived *later than* usual.
- My daughter swims *faster than* I do.
- Xiao Li works *less than* he used to.
- She studied the problem *further than* he did.

（3）比较级前有状语修饰

副词以比较级的形式出现时，其前面加有状语。例如：

- You should do it *much faster* and *more effectively*.
- Tom came late, but his brother came *still later*.
- He walked *no further* because of his ill leg.
- Can you come over *a bit more quickly*?
- She could dance *even more gracefully* than a dancer.

（4）在 as…as 和 so…as 结构中使用

1）"as + 副词 + as"结构用在肯定句中，表示"像……一样"，其中副词用原级。例如：

- I love my daughter *as much as* you do.
- He can run *as fast as* a horse.
- They work *as hard as* she does.

在 as…as 结构中有名词时可用：as + 形容词 + a（an）+ 单数名词 + as；as + many/much + 名词 + as。例如：

- This is *as good an example as* the other is.

⊖ early 中的 ly 不是后缀，其比较级和最高级是把 y 变 i，再加 er 和 est。

- I can carry *as many books as* you can.

用表倍数的词或其他程度副词做修饰语时，应将其放在 as（as…as 结构中的首个 as）的前面。例如：

- This room is *two times as big as* that one.
- Your desk is twice *the size as* mine.

2）“as/so + 副词 + as”结构用在否定或疑问句中时，表示“不像（如）……那样”，其中副词用原级。例如：

- I *don't* go to swimming *as much as* I used to.
- She *didn't* do *as/so well as* she should.
- I *don't* like it *so/as* your other works.
- He *cannot* run *so/as fast as* you.

3）以上结构 2）中可以带一个表程度的状语。例如：

- I don't speak *half as* (*so*) *well as* you.（*half* 为表程度的状语）
- She can read *twice as fast as* he does.（*twice* 为表程度的状语）
- The substance reacts *three times as fast as* the other one.（*three times* 为表程度的状语）

1.7 介词

介词（prepositions）又称前置词，常放在名词之前。它是一种虚词，在句中不能单独做句子成分，只表示其后的宾语（名词或名词短语）与句子其他成分间的关系。

1.7.1 简单、复杂介词

1）简单介词为一个单词，如 after、between、for、in、of 等；复杂介词又叫多词介词，由几个单词组成，如 according to、at the expense of、at variance with、because of、out of 等。

2）有些词兼作介词和副词，形式无区别，但用法不同，介词带宾语，副词不带。例如：

- Do you have a plan to travel to Tibet *before* July this year?（*before* 为介词）
- She said that she had a true feeling that she had been here *before*.（*before* 为副词）

1.7.2 介词的用法

（1）表示时间或一段时间

这类介词常见的有：at、after、by、before、for、in、on、since 等。

1）at 表示在某具体时刻之前，如 at zero o'clock，at 6：50 等。含 at 的固定短语较为常见，如 at noon、at night、at that time、at the age of、at the weekend、at Christmas 等。

2）in 表示在某年、某月、某季节，或表示在一段时间之后，如 in 2019、in October、in Spring、in two hours、in a few days、in the next three years 等。

3）on 表示“在……天”，如 on Sunday、on May lst、on Mid-Autumn Festival 等。

注意：笼统地表示在上午、下午、晚上时，用 in the morning、in the afternoon、in the evening；但表示在某一天的上午、下午、晚上时，用 on 不用 in，如 on Monday morning、on the morning of Children's Day、on New Year's Eve 等。

另外，在一些词之前，如 all、every、each、last、next、one、this、these、today、tomorrow、yesterday，一般不用 at、in、on。但在某些名词词组前，如（on）that day、（in）the year before last，前面的介词可以省略，也可以不省。

4）for 后面接时间段，since 之后接时间点，如 for three years、since last year 等。

5）before、by 都可表示“在……之前”，但 by 含“不迟于……”“到……为止”之意，如 before 1945、before going to bed、by the end of this year 等。（by 后跟表

将来的时间时，与一般将来时连用；跟表过去的时间时，与过去完成时连用。）

6）after 表示“在……之后”，如 after 2019、after this meeting、after a meal 等。

（2）表示地点

这类介词常见的有：aboard、about、above、ahead of、along、amidst、among、around、at、before、behind、below、beneath、beside、between、by、in、inside、near、on、opposite、under、underneath、up、within 等。

1）at 表示“在……地点”，通常是空间某一点位置。例如：

- They arrived *at New York Airport* at three o'clock in the afternoon.
- My car is always parked *at the corner of the park*.

2）in 表示“在……内”“在……地点”。例如：

- There are two millennium trees *in the park*.
- They will arrived *in New York* tomorrow morning.

at、in、on 的区别：at 常指小地方，in 多指大地方，但当把大地方仅看作地图上的一个点时，就可以用 at（用 at 还是 in 取决于说话人是如何看待该地方的）；at 所指范围通常不太明确，而 in 表示“在……里”，所指范围较为明确；in 指“在内部”，on 指“在……之上”。

（3）表示方位、方向

这类介词常见的有：at、away、aboard、above、alongside、below、beside、from、for、in、inside、into、onto、over、to、toward、towards、under、up 等。

1）at 表示“向”或“对准……（目标）”。例如：

- You should shoot the arrow *at the target*.

表示“向（某方向移动）”或“朝着……”，用 towards（英式英语）和 toward（美式英语）。例如：

- He will be criticized for his intentions to move *towards* a market economy.

2）above、over 都表示“在……的上面”：above 只表示“在上方”，但不一定在“正上方”；而 over 表示“正上方”。above 还可表温度、水位等“高于”，over 还可表“越过……”。above 的反义词是 below，而 over 的反义词是 under。例如：

- Several planes are flying *above the thick clouds*.
- There is a new and beautiful bridge *over the river*.

3）for 表示“目的地”（常与表离开、出发的动词或短语如 depart、leave、make、start、sail、set out 连用）。例如：

- She *departed for* the office to take away her bag yesterday.
- Professor Liang will *leave for* the Athlete's Apartment of Laoshan on business.

4）to 表示位置“在……面（不属该地区）”“去向”（常与表示运动或移动的名词、动词连用）或“朝……（方向）”，但在名词 direction 前面常用介词 in 表示方向。例如：

- Mongolia lies *to the north of* China.
- Last year I went *to Xi'an* to visit the famous Huangdi Mausoleum.
- The thief hurried away *in the opposite direction*.

（4）表示使用的工具、方法、手段等

这类介词常见的有：by、in、into、on、off、to、with 等。

含 by 的短语常见的有：by car、by taxi、by train、by air、by plane、by foot、by ship、by bus、by bicycle、by road、by rail、by sea、by underground 等。

含 in 的短语常见的有：in my car、in Jack's car、in an elevator、in a car、in a taxi、 in a van、in the train、to get in（a car or a taxi）等。

含on的短语常见的有：on my bicycle、on the bus、on the 8：00 train、on a big ship、on the plane、on a boat、on foot、to get on（a bicycle、a bus or a train）等。

with多表示使用的工具、身体部位或器官，by表示使用的方法、手段，in指使用某种语言，如with a knife、with our eyes、by bike、in English等。

1.7.3 介词搭配短语

由介词与名词、动词或形容词搭配所组成的固定短语常见的有以下三类。

1）介词＋名词或名词＋介词。例如：at last、at first、after class、at home、for example、in surprise、on time、key to、visit to等。

2）动词＋介词。例如：ask for、get on、get to、hear from、knock at、listen to、laugh at、learn from、look for、look at、operate on、put on、shout at、turn on、take off、turn off、wait for、worry about、write to等。

3）形容词＋介词。例如：afraid of、angry with、busy with、famous for、far from、full of、good at、good/bad for、late for、polite to、ready for、sorry for、strict with等。

1.8 连接词

连接词（conjunction）就是在句中起连接作用（连接词与词、短语与短语、句与句）的词，将相关联语句连接起来，起过渡作用。连接词包括连词和连词以外的连接词：连词是虚词，不能独立做句子中的任何成分，只起连接作用；后一种包括有连接作用的副词、介词和短语。不少连词兼做副词，但语义有差别。

1.8.1 连接词的一般分类

（1）连接词通常分为并列、转折、选择和因果四类

1）并列连接词连接平行的词、短语和句子（分句），如and、or、but、for、because、besides、so、yet、nor、both…and、either…or、neither … nor、not only … but also、in the meantime等。例如：

- Time *and* tide wait for no man.
- They sat down *but* talked about nothing.
- We must work hard，*or* we'll fail to finish the given target.
- Professor Shi is *both* a good leader *and* a good friend of mine.
- Professor Shi is *not only* a good leader *but also* a good friend of mine.
- He will be back in fifteen minutes. *In the meantime*，let's wait outside.

2）转折连接词表示转折或对比，如but（转折）、while（对比）、not…but等。例如：

- I'd like to come to dinner tonight，*but* I'm too busy.
- Some people love dogs，*while* others hate them.
- They were *not* the bones of an animal，*but*（the bones）of a human being.

3）选择连接词表示让步，如however意为"不管怎样；无论如何"。例如：

- *However* we take any measure，it isn't going to save major amounts of money.

however做副词时，表示转折，意为"然而；不过；仍然"，也可表示让步，意为"不管怎样；无论如何"。例如：

- My leader Wang，*however*，did not agree to attend this meeting.
- *However* cold it is，I always go swimming./No matter how cold it is，I always

go swimming.

4）因果连接词表因果关系，如 so、for、because、therefore 等。例如：

- Mr. Wang is absent today, *for* he is ill.
- My daughter hurt her one finger, *so* she couldn't play the piano last month.
- I cannot go to bed early *because* I have important thing to do in the evening.

注意：because 和 for 意思相近，常可互换使用，前者可引导从句，后者引导从句可能受到某些限制。

（2）连接词还可从另一角度分为等立和从属连接词

1）等立连接词就是以上介绍的并列连接词。

2）从属连接词用于引导名词从句和状语从句，如 that、whether、when、although、because、since、as far as 等。例如：

- I set off half an hour earlier *that* I might get there in time.
- *Whether* they go *or whether* they stay, the result is the same.
- *Because* the plane was delayed, we were late for attending the meeting.

1.8.2 连接词的语义分类

连接词按在句中的语义作用或关联性大体上有以下类别。

1）表承接。例如：again、also、and、another、first、furthermore、moreover、second、third、too、and then、in addition、what's more、on top of that 等。

2）表时间顺序。例如：after、afterwards、before、earlier、finally、gradually、immediately、later、now、next、soon、suddenly、then、in a few days 等。

3）表空间顺序。例如：above、around、below、behind、beside、beyond、far (from)、near (to)、outside、in front of、to the right/left 等。

4）表比较。例如：in the same way、just like、just as 等。

5）表转折。例如：although、but、still、yet、however、nevertheless、nonetheless、even though、on the contrary、despite/despite the fact that、in spite of/in spite of the fact that 等。

6）表并存。例如：on the one hand、on the other hand、for one thing、for another 等。

7）表结果、原因。例如：because、furthermore、otherwise、therefore、then、so、since、as a result 等。

8）表目的。例如：for this reason、for this purpose、so that 等。

9）表强调。例如：certainly、indeed、necessarily、surely、truly、above all、in fact、most important、to repeat、without any doubt 等。

10）表解释、说明。例如：for example、for actually、for instance、in fact、in this case 等。

11）表总结。例如：above all、all in all、finally、at last、in a word、in conclusion、as I have shown、in another word、in brief、in short、in general、on the whole、as has been stated、last but、not least 等。

第 2 章　英语语法：动词

动词表示动作或状态，按作用和功能主要分为谓语动词和非谓语动词两大类。这一章讲述谓语动词的时态、语态，不规则动词、情态动词和助动词，虚拟语气，以及非谓语动词的几种类别，如不定式、动名词、现在分词、过去分词和分词独立结构，并对谓语动词和非谓语动词进行比较。

2.1　时态

时态（tense）是表示各种时间和动作方面的动词形式，常见的有以下 16 种，见表 2-1。

表 2-1　动词时态类别

类别	一般时	进行时	完成时	完成进行时
现在	do/does	am/is/are doing	has/have done	has/have been doing
过去	did	was/were doing	had done	had been doing
将来	will/shall do	will/shall be doing	will/shall have done	will/shall have been doing
过去将来	would/should do	would/should be doing	would/should have done	would/should have been doing

（1）一般现在时

一般现在时（simple present tense）表示通常性、规律性、习惯性、真理性的动作或事件有时间规律发生的一种状态，也可表示普遍真理或谚语。常与 often、usually、sometimes、every day/week、twice a week 等时间状语连用。有出发、到达等含义的动词（如 come、go、arrive、get、leave、move、sail、start）的一般现在时也可表示将来。在条件、时间、让步状语从句中还可用一般现在时替代一般将来时而表示将来。例如：

• In this image-conscious society, people *are* less afraid of losing wealth than of losing face.（在这个注重形象的社会里，人们更害怕丢脸而不是失去财富）

• The remainder of this paper *is organized* as follows.

• The train *starts* at 11：30 p. m.

• You will succeed if you *try* your best.

（2）一般过去时

一般过去时（simple past tense）表示过去某时间里发生的动作、状态，过去习惯性、经常性的动作、行为，过去主语所具备的能力和性格，包括经常或偶尔发生的动作。常与 just now、last year、in those days 等时间状语连用。一些暂时性、一次性动作常用一般过去时。例如：

• In this case, because the subject *had* a lower sweat rate at the wrist the sensors *were activated* at a later time.

• He *died* and *left* his family a large sum of money and several houses.

• The algorithms *were developed* with Visual C++, and the correctness of these algorithms *was verified* through examples test.

（3）一般将来时

一般将来时（simple future tense）表示将来某一时刻发生或某一时间段内经常发生

的动作、状态。常和表示将来的时间状语如 tomorrow、next week、in three days 等连用。例如：

• The book of *Standard Writing of Academic Papers will be published* in 2022.

• We everyone *shall be punished* if we break the rule.

（4）过去将来时

过去将来时（past future tense）表示从过去某时看将要发生的动作或情况，常用于宾语从句。例如：

• My wife told me that she *would go* to travel this Labour Day.

• Whenever something went wrong with my computer，I *should turn to* her for help.

（5）现在进行时

现在进行时（present continuous tense）表示正在进行的动作或存在的状态，动作发生的时间是"现在"，动作目前的状态是"正在进行中"。常与 now、at present、at this moment 等时间状语连用。现在进行时与频率副词连用可以表示某种感情色彩。具有"出发、到达"等含义的动词如 come、go、arrive、get、move、sail、leave、start 等，可用现在进行时表示计划将要发生的动作。例如：

• I see that many children *are swimming* in the river.

• The book of *Grammar and Rhetoric of Language Style for Science and Technology*, 2nd ed. *is being written* by several undergraduates.

• We'*re leaving* for London.

（6）过去进行时

过去进行时（past continuous tense）表示在过去某一时刻或时间段正在发生或进行的动作或状态。例如：

• My heart *was jumping* with the squeal.

• I *were wondering* if she can give me a help.

• A high-speed railway station *was being built* there last time I went back to my hometown.

过去进行时侧重动作持续时间的长度，表示持续的动作，一般过去时侧重说明动作的事实。

（7）将来进行时

将来进行时（future continuous tense）表示将来某时正在发生或持续发生的动作。例如：

• Professor Li *will be having* an project launch meeting this time the day after tomorrow.

• I *will be writing* my new book during next year.

（8）过去将来进行时

过去将来进行时（past future continuous tense）表示就过去某一时间来说将来某一时刻或时间段正在进行的动作。例如：

• She said she *would be taking care of* my old mother.

• Mr. Wang said he could not come because he *would be having* a party.

（9）现在完成时

现在完成时（present perfect tense）表示过去发生的动作或状态对现在造成的影响、结果或持续到现在，或表示过去某一时期发生情况的总和。常与 already，yet，these days，recently，lately，so far，by now，the past/last two years，for…，since…等时间状语连用。例如：

• Every year computer-based systems become ever more complex，enormous，and powerful，now deciding whether a system can be trusted *has become* a considerable problem.

• In this field，many testing and scripting languages *have been designed* year by year.

• It *has been* 23 years since I worked in this institute. /I *have been working* in this institute *for* 23 years.

• This（It）is the first time that I *have found myself in a kind of joy of success*.

（10）现在完成进行时

现在完成进行时（present perfect continuous tense）表示从过去某一时刻开始的动作一直延续到现时刻还在进行或可能继续下去，常与表某一段时间的状语连用。例如：

• My son *has been working* in America for ten years.

• The satellite *has been sending* back signals and photos ever since it landed on the moon.

（11）过去完成时

过去完成时（past perfect tense）表示过去某一时间或动作以前已发生或完成的动作或事件，对过去的某一点造成某种影响或结果。它表示动作发生的时间是"过去的过去"，侧重事情的结果。常与 by + 过去时间或短语、从句（如 by 2019、by the end of last year、by the time/when + 过去发生情况）等时间状语连用，也可与 before、never 连用，但不能与 ago 连用。有些动词如 plan、think、intend、hope、want、mean 常用过去完成时表示打算做而没有做的事。例如：

• I *had intended* to finish writing this grammar and vocabulary book by 2019.

• By the end of last year, I *had written* eight high impact creative monographs.

• The mean squared error between the predicted value and the actual game outcome is plotted against the stage of the game（how many moves *had been played* in the given position）.

（12）过去完成进行时

过去完成进行时（past perfect continuous tense）表示动作从过去某一时刻开始一直延续到过去某一时间而继续下去。例如：

• When I returned home, my daughter *had been playing* piano for two hours.

• Li told me that the employees *had been discussing* the project since 8:30 AM in the morning.

（13）将来完成时

将来完成时（future perfect tense）表示在将来某一时刻或动作发生前另一动作将要结束或完成，有时也表示一种推测。常与 by + 将来时间或短语（如 by 2050、by the end of next year）等时间状语连用。例如：

• I *will have worked* here for 28 years by the end of 2028.

• By next month we *shall have finished* the engineering project funded by NSFC.

（14）将来完成进行时

将来完成进行时（future perfect continuous tense）表示动作从某一时间开始一直延续到将来某一时间。例如：

• I *will have been working* here for 28 years by the end of 2028.

• The bookstore *will have been working* before we get there at 8 o'clock.

（15）过去将来完成时

过去将来完成时（past future perfect tense）表示从过去某一时间看将来某一时间之前已完成的动作。例如：

• They said that they *would have finished* the writing the report before the leader came back.

• I felt Mrs. Zhang *should have graduated* from Tsinghua University by then.

（16）过去将来完成进行时

过去将来完成进行时（past future perfect continuous tense）表示动作从过去某一时间开始一直延续到过去将来某一时间（从过去来看的将来，尽管从现在看此将来已是过去）。动作是否继续下去取决于上下文。通常使用较少，多用于转述方面，即间接引语中。例如：

• I heard by the end of this year she

would have been working here for 18 years.

2.2 语态

语态（voice）是通过动词形式的变化来表现的。分为主动语态（主动态，active voice）和被动语态（被动态，passive voice）两类，前者表示主语是动作的执行者，即施动者，后者表示主语是动作的承受者，即受动者（动作的对象）。被动语态常见的有以下几种，见表 2-2。

表 2-2 被动语态类别

类别	一般时	进行时	完成时
现在	am/is/are done	am/is/are being done	has/have been done
过去	was/were done	was/were being done	had been done
将来	will/shall be done		will/shall have been done
过去将来	would/should be done		would/should have been done

（1）被动语态的使用情况

下列情况宜用被动语态：①作者对动作的承受者更感兴趣，注意力在于对客观事物和过程的描述，不在于动作的发出者或执行者；②动作的执行者不明确，不必、不易、无从说出；③出于谦虚、礼貌、简练、微妙等感情方面的考虑或为使语气婉转、和谐，避免提及动作的执行者；④为使句子结构更加紧凑、连贯。例如：

- The degradation of permafrost stability in China over the past 30 years *is evaluated* using a new, high-resolution near-surface air temperature reanalysis dataset.
- Some important equipment in that factory *were stolen* several days ago.
- A more robust physical model *should be used* to evaluate the permafrost thermal stability at finer resolution in the future.
- I arrived at Los Angeles Airport, where I *was met* by old classmate Mrs. Liu.

（2）主动语态表被动的情况

1）某些以主动语态的形式表示某种状态或情况（如拥有、容纳、缺少、明白等）而本身有被动意味的动词或动词短语，如 befall、benefit、break out、catch、cost、contain、consist of、equal、fail、fill、fit、hold、happen、lack、look like、last、mean、occur、rise、resemble、take place 等。例如：

- Her skirt *caught* on the nail.
- Happiness perhaps *consist of* health, wealth and enterprise etc. one owns.
- Her mother's eyes *filled* with tears.
- The new designed meeting room *holds* about 100 persons.
- My daughter *resembles* her mother and *doesn't resemble* me at all.

2）某些做感官动词（即系动词）的不及物动词，如 feel、look、prove、smell、taste、sound 等。例如：

- This new material *feels* hard.
- The garden *looks* very beautiful.
- Your idea *proved* to be wrong.

3）某些表示主语某种属性的不及物动词，如 carry、clean、close、cook、cut、lock、move、open、peel、play、read、record、sell、shut、wash、wear、write 等。例如：

- The new bought glass *cleans* easily.
- The online course recorded by me *sells* quickly.
- The window *won't shut*.

4）be + 表被动意义的形容词、如 acceptable、eatable、drinkable、practicable、reliable、visible 等。例如：

- Most stars are not *visible* to the naked eye.
- His plan seems good，however it is not *practicable*.

5）某些做表语的不定式短语，如 to blame、to let 等。例如：

- Who *is to blame* for that thing?
- The old house *is to let*.

6）某些后面接不定式的表语形容词，如 convenient、dangerous、difficult、easy、hard、impossible、interesting、nice、pleasant、safe、tough、tricky、unpleasant 等，形成"be + 形容词 + to do"结构，句子主语就是此不定式的逻辑宾语。例如：

- This paper *is difficult to read* because of its bad writing.
- Teacher Liu *seems easy to approach*.
- This new building *is interesting to look at*.
- The music *isn't pleasant to listen to*.

7）某些后面带有宾语及修饰该宾语的后置不定式定语的动词，如 have、have got、get、want、need 等，不定式的逻辑主语就是句子的主语。例如：

- I *have* enough time *to finish* the record of this online course.
- I *want* something *to drink*.

但是，不定式的逻辑主语不是句子的主语时，要用被动式。例如：

- I *have* something *to be written*.（我有些东西要别人来写）（不定式的逻辑主语不是句子的主语）
- I *have* something *to write*.（我有些东西要自己写）（不定式的逻辑主语是句子的主语）

8）too…to do sth 和…enough to do sth 结构中的动词不定式。句子主语与其后不定式为被动关系时，不定式常用主动形式表示被动意义，有时也可直接用被动式。例如：

- The writing is *too faint to read*.
- These boxes are not strong *enough to use* as a platform.（不定式以主动形式表被动意义）
- These boxes are not strong *enough to be used* as a platform.（不定式用被动形式）

9）be worth 后的动名词。例如：

- This book *is worth reading*.
- She'*s not worth getting* angry with.（犯不上跟她生气）

但 be worthy 后面的动词用被动式。例如：

- This book is worthy *to be read*.
- This book is worthy *of being read*.

10）后接动名词表示被动之意的某些动词，如 deserve、need、require、want 等。例如：

- The new rent house *needs decorating*.（The new rent house *needs to be decorated*.）
- This activity does not *deserve doing*.（This activity does not *deserve to be done*.）
- These children *require looking* after.（These children *require to be looked after*.）

（3）被动语态与系表结构比较

1）时态方面：被动语态强调动作，其主语是动作的对象，可有各种时态。系表结构说明主语的状态和特征。做表语的过去分词相当于形容词，不强调动作，多用于一般现在时和一般过去时，表现在或过去的性质，没有进行时。用进行时结构的多是被动语态。例如：

- The book *was written* in 2010.（被动语态，表示过去的动作。）
- The book *is being written*.（被动语态，表示目前正在进行的动作。）
- The book *is well written*.（系表结构，表示状态和特征。）

2）状语方面：句中带有时间状语、方式状语或用by短语做状语的多是被动语态。例如：

• The handbook for Academic Writing was finished in 2019.（被动语态）

• The handbook for Academic Writing was finished.（系表结构）

• All photographs in this paper *were taken* by the authors.（被动语态）

3）动词性质方面：可做表语的动词通常有三类。

① 表示事物状态或人的感觉、情绪，具有形容词的特征，如broken、closed、excited、interested、lost、shut、worried等。

② 表示智力活动结果或行为结果的少数动词，如known、unknown、learned、deserted、injured、killed、mistaken、saved、wounded等。

③ 少数不及物动词的过去分词用于系动词后时一定是表语，多表运动或转变，如arrive、change、come、fall、go、grow、melt、pass、rise、return、set、turn等。

4）被动语态中的过去分词不可用程度副词如very、well修饰，但系表结构可以。例如：

• The book for Writing *is finished* in 2019.（被动语态侧重"已完成"动作）

• The book for Writing *is well finished* in 2019.（系表结构，侧重"完成得好"这一状态）

5）被动语态中的过去分词没有固定的介词搭配，但系表结构中的常常有。例如：

• The little boy *was* completely *absorbed* in the story.（被动语态）

• The little boy *was not accustomed to* this kind of sugar.（系表结构）

（4）不定式做宾补的结构变为被动语态

特殊动词（如feel、have、hear、let、make、notice、observe、see、watch）后面的不定式做宾语补足语时省略to，但此类结构变为被动语态时要补上to。例如：

• Teacher Liu *made* the little boy *read* the textbook.

• The little boy *was made to read* the textbook by teacher Liu.

（5）主动语态改成被动语态

主动语态改成被动语态时，把原句的宾语提前，做主语，原句的主语后置，做宾语。不及物动词没有被动语态，但与介词构成介词短语时，可以有被动语态。例如：

• The authors *designed* all figures and tables in this paper.（主动语态）

• All figures and tables in this paper *were designed* by the authors.（被动语态）

• The fire *had been put out* before the fireman arrived.（put一词在此是不及物动词，但短语put out相当于及物动词）

2.3　不规则动词

动词按其过去式和过去分词的构成方式分为规则（regular verb）和不规则（irregular verb）两类。规则动词的过去式和过去分词由词尾加ed构成，结尾是e时只加d，结尾是y时将y改为ied；不规则动词的构成变化没有什么规律，需要记住。英语中大部分动词是规则动词，少量是不规则动词。

英语中新生成的动词都归入ed的规则变化。例如：park→parked（停车，1864年）；fax→faxed（以传真传送，1979年）；e-mail→e-mailed（以电子邮件传送，1982年）[⊖]。可以说，不规则动词多是因古英语动词的不规则变化常用度很高而一直沿用到今天。

⊖ 括号中的年份为最早用例出现的年份，参见O. E. D.（The Oxford English Dictionary，牛津英语辞典）。

从英语的演变来看，不规则动词是强势动词，其词形变化依靠其自身的语音（尤其是元音）变化来进行而无须借助词尾变化，如 buy、bought、bought 和 speak、spoke、spoken 等。另外，借助 ed 词尾进行词形变化的规则动词叫弱势动词。学习不规则动词的有效方法是，了解词形变化现象，弄懂词形变化形式，发出声音念会、背会，自然融入英语日常和书面表达中。

不规则动词的构成大致有 AAA、AAB、ABA、ABB、ABC 几种形式。

（1）AAA 式

AAA 式是指现在式（原形）、过去式和过去分词同形（完全相同）。例如：cost、cost、cost，cut、cut、cut，hit、hit、hit，hurt、hurt、hurt，let、let、let，put、put、put，read、read、read（read 的过去式和过去分词的读音与动词原形不同）。

（2）AAB 式

AAB 式是指现在式和过去式同形。例如：beat、beat、beaten。

（3）ABA 式

ABA 式是指现在式和过去分词同形。例如：come、came、come，become、became、become，run、ran、run，overcome、overcame、overcome。

（4）ABB 式

ABB 式是指过去式和过去分词同形，分为三种情况：

1）在动词原形后加一个辅音字母 d 或 t。例如：hear、heard、heard，burn、burnt、burnt，learn、learned/learnt、learned/learnt，mean、meant、meant。

2）把动词原形的最后一个辅音字母 d 改为 t。例如：build、built、built，lend、lent、lent，send、sent、sent，spend、spent、spent。

3）其他不规则变化，数量较多。例如：bring、brought、brought，buy、bought、bought，have、had、had，get、got、got，keep、kept、kept，make、made、made，meet、met、met，pay、paid、paid，say、said、said，teach、taught、taught，think、thought、thought，understand、understood、understood，win、won、won。

（5）ABC 式

ABC 式是指现在式、过去式和过去分词不同形（完全不同），分为三种情况。

1）在动词原形后加 n 或 en 构成过去分词。例如：blow、blew、blown，draw、drew、drawn，drive、drove、driven，grow、grew、grown，eat、ate、eaten，fall、fell、fallen，give、gave、given，know、knew、known，ride、rode、ridden（双写 d），see、saw、seen，show、showed、shown，take、took、taken，throw、threw、thrown，write、wrote、written（双写 t）。

2）在过去式后加 n 或 en 构成过去分词。例如：break、broke、broken，choose、chose、chosen，forget、forgot、forgotten（双写 t），freeze、froze、frozen，speak、spoke、spoken。

3）将单词在重读音节中的元音字母 i 变成 a（过去式）和 u（过去分词）。例如：begin、began、begun，drink、drank、drunk，ring、rang、rung，sing、sang、sung，sink、sank、sunk，swim、swam、swum。

2.4　情态动词

情态动词（modal verb）属于动词，但不同于一般动词。一般动词是行为动词，表示一种动作或状态，而情态动词用在一般动词前，表达对动作或状态的某种观点、设想、想法，或某种语气、情绪、态度（如需要、可能、意愿或怀疑）等。情态动词本身虽有一定意义，但更多用来表达一种情态义。情态动词有以下八个特点：

1）有一定语义，但不能独立做谓语，只能和动词原形连用构成谓语；

2）两个情态动词不能连用；

3）无人称和数的变化；

4）时态性不强，与表意常不一致，现在时可表将来，过去时可表现在、将来；

5）有的有过去式变化，可表达更加客气、委婉的语气；

6）没有不定式、现在分词和过去分词形式；

7）属于不及物动词范畴，没有被动语态；

8）在后面加 not（'t）可构成否定式。

情态动词（含短语）用途很广，但数量不多，常见的有 can（could）、may（might）、must、have to、need（needed）、shall（should）、will（would）、ought to、dare（dared）等。下面简要介绍十个情态动词的用法。

1. can，could

1）表能力，如体力、知识、技能，可用 be able to 替代。can 只有一般现在时和一般过去式，be able to 还有其他时态。例如：

- *Can* you help me to design the inner cover of this new book?
- Mary *can* speak several foreign languages such as English, Germany and Japanese, etc.
- My daughter *could* play piano and cello skillfully when she was ten years old.
- I'll not *be able to* come this afternoon.

当表示经过努力或克服困难才得以做成某事时，应该用 be able to，而不用 can。例如：

- She *was able to* go to the party yesterday evening in spite of the heavy rain.

2）表请求、许可，可与 may 互换。在疑问句中还可用过去形式的 could、might 替代，但它们不是过去式，只用来表示更加委婉的语气，不能用在肯定句和答语中（不管疑问句由 can 还是 could 开头，回答总是用 can）。例如：

- *Can* I take away these research materials, please?
- The pupils asked whether they *could* go to swimming.
- "*Could* you help me writing the project report this weekend?"

"Yes, I can. /No, I can't. (No, I'm afraid not.)"

3）表示可能，即客观可能性（客观原因形成的能力）。例如：

- Some prisoners escaped the day before yesterday. They *can/could* be anywhere by now.
- They've changed the timetable, so we *can* go by air instead.
- This hall *can* hold 1000 people at least.

4）表示为别人提供帮助、服务等。例如：

- "*Can* I help you?" "No, thanks. I'm being served."

5）表示将来可能发生的动作、建议，或表示推测（惊讶、怀疑、不相信的态度），用于疑问句、否定句和感叹句中。例如：

- There *could* be a continued rise in building materials price in Beijing recently.
- *Can* this be true?
- This *can't* be done by him.
- How *can* this be true!

2. may，might

1）表请求和允许。might 比 may 语气更委婉，但不是过去式。否定回答时可用 can't 或 mustn't，表不可以、禁止。用 May I 征询对方许可比较正式和客气，而 Can I 在口语中更常见。例如：

● “*Might/May* I have a rest in this room” “Yes, you *can*. (No, you *can't/mustn't*.)”

2）表祝愿，用于祈使句。例如：

● *May* you succeed!

3）表推测、可能性（不用于疑问句）。might 所表示的可能性比 may 小。例如：

● He *may/might* be very busy now.

● Your girl friend *may/might* not know the truth.

3. must，have to

1）表责任、义务，必须、必要。否定式为 needn't、don't need、don't have to，不用 mustn't。例如：

● Citizens *must* obey orders.

● You *needn't/don't need* sleep so early everyday if you want to succeed.

● We *don't have to* do thing so rough. Details determine success or failure.

● “*Must* we hand in the files today?” “Yes, you *must*. (No, you *don't have to*/ you *needn't*.)”

must 表必须之意时，常可与 have to 互换，但在意义上可能有区别：must 是从说话人的角度出发来表示个人意愿和主观看法的，而 have to 强调客观需要，表示客观、事实上的必须之意。must 只有一般现在时，而 have to 有更多的时态形式。例如：

● It will perhaps rain. I really *must* go now.

● I *had to* work when I was your age.

2）表示推测、可能性，只用于肯定的陈述句。例如：

● You *must* be very hungry when you get off work in the afternoon.

● Professor Du *must* be waiting for you now.

4. should，ought to

1）表示责任、义务。should 常表示一种主观和道义上的责任、义务，而不是一种外来、强加的法律或规则所应承担的责任，但在语气上比 ought to 弱一些，比 must 弱更多。例如：

● Everyone *should* wash his hands before dinner.

● You *ought to* take care of your baby.

2）表示劝告、建议和命令。should 和 ought to 可通用，但在疑问句中常用 should。例如：

● You *should/ought to* exercise more if you wants to lose weight.

● She *ought not to* go swimming, because she is too tired.

● *Should* I open the window now?

3）表示客观推测，有含蓄、不太肯定之意。注意与 must 的区别：must 表示主观推测，有直爽、断定之意。例如：

● It's time to leave off work. Mr. Wang *should/ought to* be here in a few minutes.

● He *should/ought to* be home by now. (He *must* be home by now.)

● This is where the oil *should/ought to* be.（这里应该就是石油存在的地方）

● This is where the oil *must* be.（这里肯定就是石油存在的地方）

4）表示的时间概念可以是过去或现在。无论是表达“推测和可能性”，还是表达“虚拟”，只要是对过去已经发生的事情进行描述，就用“情态动词 + have done”这一结构，而对现在或将来的事情进行描述，则用“情态动词 + 动词或系动词原形”。例如：

● Mrs. Qin *should/ought to have arrived* at ten, but she didn't turn up.

● Mrs. Qin *should/ought to arrive* at ten, but she may not turn up.

5. need

1）表示必须、需要。常用于疑问句、否定句，否定形式为 needn't，在肯定句中一般用 must、have to、should、ought to 替代，见本节标题3、4。例如：

- You *needn't* come so early today.
- "*Need* I finish the work today" "Yes, you *must*. /No, you *needn't*."

2）常做普通（实义）动词，与其做情态动词的意思相同，后面接带 to 的不定式，或名词及名词短语。例如：

- My daughter *needs to* finish her piano test this semester
- You didn't *need to* tell her the truth, it just made her bad.
- I *need* lots of time *to* finish the writing of the new book.

6. shall

1）用于第一人称，表征求对方意见。例如：

- What *shall* we do in this 2021 International Conference of Design Innovation?

2）用于第二、三人称，表说话人给对方的命令、警告、允诺或威胁。例如：

- You *shall* help to check homework for my daughter when I am on a business trip.
- You *shall* fail if you don't work hard.
- She *shall* have the book when I finish it. （允诺）
- He *shall* be punished.

7. will, would

1）表请求、建议等。would 比 will 更委婉。例如：

- *Will/Would* you pass me the book, please?

2）表意志、愿望和决心。例如：

- I *will* never do that again.
- They asked him if he *would* go abroad next year.

3）Would 表过去习惯或某种倾向，比 used to 正式，而且没有现已无此习惯之意。例如：

- During the whole training period, she *would* visit me every other day.
- The wound *would* not heal.

4）表估计、猜想。例如：

- It *would* be about two o'clock in the morning when I finished the daily writing.

8. had better

这是一个固定习语，意为"还是……好""最好还是……"，表建议、忠告，可用于表现在或将来，但不能表过去。其否定形式是 had better not。例如：

- As requested, I *had better* send my PPT to you at the 15th of this month.
- There is a great rise in building materials price, we *had better not* buy now.

9. dare

1）表敢、竟敢，主要用于否定句、疑问句或条件从句，一般不用于肯定陈述句。例如：

- *Dare* you swim cross the river?
- I *dare not* give a report in such environment because I don't want to make trouble.
- If we *dared not* go there before Wednesday, we couldn't get the good books.

2）I dare say/I daresay 是固定习语，相当于 I think probably 或 I suppose，做插入语，表我想、大概之意。例如：

- There's something wrong with the car, *I dare say*.
- It's worth a few pounds, *I daresay*, but no more.

3）dare 常做普通（实义）动词，尤其在美式英语里，有人称、时态和数的变化。在肯定句中，dare 后面常接带 to 的不定式；在疑问句和否定句中，dare 后面可接带或不带 to 的不定式，过去式为 dared。用在否定句中较为普遍。例如：

- Some brave girls usually *dare* to go out

at night.

• I wonder how she *dared* （to） say such words.

4）dare 也可做及物动词，表挑战别人做什么危险之事，常用于表示互相争斗、挑战的情形。例如：

• I *dare* you to do it. （量你也不敢）

• He *dare* me to jump over the river.

10. 情态动词 + 不定式完成式

1）can/could + 动词完成式：在肯定句中表示本来可以做而实际上未能做某事，是虚拟语气；在疑问句或否定句中表对过去行为的推测、怀疑或不肯定。例如：

• I *could have done* better，but I didn't try my best.

• *Can* they *have got* the money？

• She *can't have been* to that city.

2）may/might + 不定式完成式：表示对过去行为的推测，意为可能、或许。不能用于疑问句，没有虚拟语气的用法。might 所表示的可能性比 may 小。例如：

• My friend *may not have finished* the work.

• If I had taken the other road，I *might have arrived* earlier.

3）must + 不定式完成式：用于肯定句中，表示对过去行为的推测，意为一定、想必。疑问、否定形式用 can、can't（could、could't）替代。例如：

• Her father *must have seen* the film Titanic several years ago.

• Could she *have read* the great classical novel "Dream of Red Mansion"？

• He could't *have gone* to foreign countries，such as Canada，America and France，etc.

4）should + 不定式完成式：用于肯定句中，表示对过去行为的推测，意为应该、理应。例如：

• You *should have finished* the file writing by now.

当表示本应该做而实际上未做某事时，可与"ought to + 不定式完成式"互换，其否定式表示某种行为本不该发生却发生了。例如：

• You *should* （*ought to*） *have helped* her. （But you didn't.）

• She *shouldn't have taken away* my English dictionary，for I wanted to consult it.

5）needn't + 不定式完成式：表示本来不必做而实际上做了某事。例如：

• You *needn't have watered* the flowers，for they have been watered by my daughter.

6）will + 不定式完成式：主要用于第二、三人称，表示对已完成动作或事态的推测。例如：

• Academician Tan *will have arrived* by now，it's time for him to make a keynote speech.

2.5 助动词

助动词（auxiliary verb）用来协助主要动词（实义动词）构成谓语。其本质作用是形成句型结构，但本身没有语义，不可单独做谓语，只能与其他动词一起构成谓语，表示时态、语态、语气等，或表示疑问或否定。例如"I *doesn't* like flattery."中，doesn't 是助动词，无词义；like 是主要动词，有词义。助动词的作用有以下七个方面。

1）表示时态。例如：

• I *am writing* a book about scientific paper writing.

• I *have got married* for more than twenty years.

2）表示语态。例如：

• My daughter *was sent* to England for learning English.

• The new book *was awarded* the best work of mine in 2019.

3）构成疑问句。例如：

• *Do* you like college life in America?

• *Did* you study engineering physics before you came here?

4）与否定副词 not 合用，构成否定句。例如：

• I *don't* like wasting time.

5）加强语气。例如：

• *Do come* to the tea party tomorrow evening.（明天晚上一定来参加茶话会）

• I *do love* you.（我真的喜欢你）

• They *did know* that.（他们确实知道那件事）

6）构成倒装句。例如：

• Never *did* I hear of such a bad thing.

• Never *will* I waste time.

7）替代动词。例如：

• She works hard than you *do*.（do 替代 work）

常用的助动词有：be（am、is、are、was、were），have、has、had，do、does、did，will、would、shall、should 等。下面简要介绍几个助动词的用法。

1. have

1）表状态（如拥有、患病等）或必须（指 have to）。不能用于进行时，没有被动语态；在否定句或疑问句中，可不用助动词 do（在非正式文体中可在 have 后加 got，有时还可将 have 省去）。例如：

• They *haven't*（*got*）a house of their own.

• He *doesn't have* a house of his own.

• I（*'ve*）*got* a problem.

在英式英语中，指经常性现象时，常用助动词 do 来构成疑问式和否定式，且不与 got 搭配；指暂时现象时，不用 do 构成疑问式和否定式，且可与 got 连用。但在美式英语中无此区别，一般都用 do 构成疑问式和否定式。例如：

• Do you often *have* colds?

• *Have* you（*got*）a cold now?

• I don't usually *have to* work on Sundays.

• I *haven't got* to work tomorrow.

表示拥有时，构成否定式有两种情形：其后名词没有限定词修饰时，常用 have no；有限定词（如 a、any、much、many、enough 等）修饰时，常用 have not。例如：

• I *have no* friends./I *haven't any* friends.

• The new employee *hasn't much* money.

2）表动作，做实意动词，如表示吃（相当于 eat）、喝（相当于 drink）、拿（相当于 take）、收到（相当于 receive）、度过（相当于 spend）等，可用进行时，但一般没有被动语态；不与 got 连用；一般不用缩略式；构成否定式和疑问式时，必须用助动词 do。例如：

• He *is having* breakfast.（不能说 Breakfast *is had* by him.）

• I'd like to *have* an invitation letter.（不能说 I'd like to *have got* an invitation letter.）

• *We had* a good time at travelling in Guilin.（不能说 *We'd* a good time at travelling in Guilin.）

• When *did* you *have* breakfast?（不能说 When *had* you breakfast?）

2. will, would

will 和 would 表习惯或规律，前者表现在的习惯或一般规律，后者表过去的习惯。例如：

• I *will get pleasant* over my cumulative results.

• My daughter *will keep leaving* things all over the room.（我的女儿总是把东西扔得满

屋都是）

• Joy and pain *will mix together*.

• In the process of thinking, I *would walk* many hours without having a break.

3. shall, should

1）shall用于陈述句，主语为第一人称时，表示单纯将来，意指将要、会，意志、意愿；主语为第二、三人称时，表示说话者的意图、警告、命令、威胁、许诺、规定等。例如：

• I *shall* be fifty next July.

• I *shall* never forget your help.

• We *shall* start off to Guilin on February 3, 2021.

• You *shall* not catch me so easily next time.

• The pupils *shall* wear uniforms.

2）shall用在疑问句中，用于征求对方意见。例如：

• *Shall* we all take a walk in Beihai Park next Sunday?

• *Shall* she go to Shanghai instead of Beijing?

• Let's go on to the speech, *shall* we?

3）should表示义务、责任，或建议、劝告，意为应该；还可表示委婉而郑重的说法，用于第一人称，表"想"之意。例如：

• Everyone *should* care for his parents.（义务、责任）

• You *should* make enough preparations before writing this paper.（建议、劝告）

• I *should* like to make a phone call to my friend Mrs. Li.（想）

2.6 虚拟语气

英语中用不同的谓语动词形式即语气表示说话者的不同意图，分为陈述、祈使和虚拟三类。虚拟语气表示的动作或状态不是客观存在的事实，而是说话人的主观想法、愿望、假设、推测、怀疑或建议等，是在一种假设成立时所发生的某种动作或呈现的某种状态，是由句中谓语动词的特殊形式表示出来的。

2.6.1 非真实条件句中的虚拟语气

非真实条件句使用虚拟语气表达假设或实际发生的可能性极小甚至不可能。此条件句主要是if引导的虚拟条件状语从句，简称if条件句，表示一种假设的情况或主观的愿望，其主句、从句的谓语形式见表2-3。

表2-3　If条件句的谓语形式

事实的时间属性	从句时态	主句时态
与现在事实相反	did/were（不能用was）	would/should/could/might do
与过去事实相反	had done	would/should/could/might have done
与将来事实相反	did/were to do/should do	would/should/could/might do

（1）一致条件句（主、从句时间一致）的虚拟语气

从句中的动作与主句中的动作是同时发生（发生时间一致）的情况下，主句用过去时（一般过去时或过去完成时），从句用过去时（动词原形前加would/should/could/might）。

1）与现在事实相反。

从句：If + 主语 + did（be用were）。主句：主语 + would/should/could/might + do。例如：

• If I *knew* the password of that computer,

I *should try* to test the new developed software.

● If I *had* enough money these years, I *could buy* a house in Beijing.

● If I *were* you, I *would take* an action at once.

● If *there were* no air or water, *there might be* no living things on the earth.

2）与过去事实相反。

从句：If + 主语 + had + done。主句：主语 + would/should/could/might + have done。例如：

● If we *had got* there earlier, we *should/would have met* teacher Feng.

● If you *had got up* earlier, you *could have caught* the plane.

● If I *had taken* your advice, I *would not have made* such a big mistake.

3）与将来事实相反。

从句：①If + 主语 + did（be 用 were）；②If + 主语 + were to do；③If + 主语 + should do。

主句：主语 + would/should/could/might do。例如：

● If I *were* a leader, I *would take* an important measure to develop the new media.

● If *there were* a heavy snow next Sunday, we *should go skating*.

● If she *were* to be here next Monday, I *might* tell her about the matter.

● If she *should come* here tomorrow, I *could talk* to her.

（2）混合条件句（主、从句时间不一致）的虚拟语气

从句中的动作与主句中的动作不是同时发生（发生时间不一致）的情况下，动词的形式应根据其所表示的时间加以调整。这种条件句也叫错综条件句。

1）从句与过去事实相反，主句与将来事实相反。例如：

● If she *had taken* my advice, she *would not have* any trouble in next meeting.

● If we *hadn't made* adequate preparation, we *shouldn't dare* to report the work tomorrow.

2）从句与过去事实相反，主句与现在事实相反。例如：

● If he *had studied* hard at college, he *would be* a senior engineer now.

● If the leader *had informed* him, he *would not come* here now.

3）从句与现在事实相反，主句与过去事实相反。例如：

● If he *were* free today, we *would have sent* him to Shanghai.

● If he *knew* her, he *would have greeted* her.

4）从句与将来事实相反，主句与过去事实相反。例如：

● If I *were not to make* an experiment this afternoon, I *would have gone to* Beijing yesterday.

● If she *were* to start tomorrow, she would have arranged everything well.

5）从句与将来事实相反，主句与现在事实相反。例如：

● If I *shouldn't make* an experiment this afternoon, I *would go to* visit Tom now.

● If I *could go to* a better college in the further, I would't study hard now.

（3）注意事项

1）if 条件句中的谓语动词含有 were、should、had 等词时，if 可以省略，这时从句用倒装语序，即把 were、should、had 等提到主语前。例如：

● *Were* she *here*, she *would agree* with us.

● *Should* he *agree* to go to London, we *would send* him there.

● *Had* he learnt about engineering, we

would have hired him to this position.

但是，if 条件句为否定句时，否定词 not 应置于主语之后，而不能与 were、should、had 等缩略成 Weren't、Shouldn't、Hadn't 而置于句首。例如：

- *Had* he *not learnt* about engineering, we wouldn't have hired him to this position.

有时省略 if 后提前的 had 不是助动词，而是实体动词。例如：

- *Had* I time, I *would come to* Beijing. (If I *had* time, I *would come* to Beijing.)

2）if 条件句中的谓语动词如果是 be，其过去形式一般用 were。

- If I *were* you, I would vigorously improve the publishing quality of popular science books.

- If there *were* no air and water, people would die or would never come to the world.

3）从句有时可以不表达出来，而是暗含在介词短语、副词、不定式短语、上下文中或以其他方式表示出来。这种句子称作含蓄条件句，实际上是一种省略句，通常借助一定的上下文语境能够明白此省略的部分的意思。例如：

- *Without* my wife's help, I *would not have made* such great progress in English writing.（介词短语 Without…暗含从句"If my wife hadn't helped me"。）

- *But for* the bad weather, we *should have arrived* earlier.（介词短语 But for…暗含从句"If it had not been for the bad weather"。）

- We didn't know her telephone number, *otherwise* we *would have telephoned* her.（otherwise 是副词，暗含从句"If we knew her telephone number"。）

- It *would cause* great trouble *not to* lubricate the bearing immediately.（不定式短语 not to…可能暗含从句"if you（we）were not to lubricate the bearing immediately"。）

- I *would not have done* it that way.（上下文可能暗含从句"If I were you"。）

- You *might come to* join us in the discussion.（上下文可能暗含从句"If you wanted to join us"。）

- I *would have bought* the new style of car.（上下文可能暗含从句"If I had enough money"。）

4）主、从句主语一致且从句谓语部分包含动词 be 时，通常可将从句的主语和动词 be 省略。例如：

- If *repaired* earlier, the tractor *would not have broken down*.

- If necessary, we *would send* more students to help you.

前一句的从句省略了主语 it（指 the tractor）和动词 be（had been），其完整表达应是"If it had been repaired earlier"；后一句的完整表达应是"If it was（were）necessary"。

2.6.2　名词性从句中的虚拟语气

（1）主语从句中的虚拟语气

主语从句中的虚拟语气结构：It is（was）+ 形容词/过去分词/名词 + that…，其中 that 部分的谓语表达形式为 should + do，而 should 可以省略，直接用动词原形。

1）常用的形容词：advisable、appropriate、better、crucial、desirable、essential、funny、important、impossible、imperative、natural、necessary、probable、possible、preferable、incredible、right、strange、surprising、urgent、vital、wrong 等。

2）常用的过去分词：demanded、desired、ordered、required、suggested、requested、recommended 等。

3）常用的名词：advice、decision、desire、demand、idea、order、pity、proposal、 requirement、recommendation、shame、suggestion、surprise、with、wonder等。

例如：

- It is *necessary* that I (should) read or write something everyday.
- It is *required* that nobody (should) park here.
- It is my wife's *proposal* that my daughter (should) be sent abroad for further study.

(2) 宾语从句中的虚拟语气

1) 动词wish后接宾语从句时的基本结构：主语 + wish (ed) + that + 主语 + did (were)/had done。

表示与现在事实相反时，从句谓语用过去时（did、were、were doing）；与过去事实相反时，从句谓语用would/could/might + do或had done（注意这里不能用should）；将来没有把握或不太可能实现时，用would (could) + 动词原形。如果将wish改为过去式wished，其后that从句中的动词形式不变。例如：

- I *wish* (that) I *could be* of some use.
- I *wish* prices *would come down*.
- I *wish* she *had listened to* my advice.
- I *wished* that she *had listened to* my advice.

2) 下列动词后接宾语从句时，从句谓语用should + do（美式英语中可省略should）。

① insist（坚持）。

- She *insisted* that I (*should*) *go* with them.
- The policeman *insisted* that she *should have* a look.
- Rose *insisted* that they *be* present.

insist后接宾语从句时，还可用陈述语气：从句谓语动词所表示的情况尚未发生、没有成为事实时用虚拟语气，已经发生、成为事实时（坚持认为、坚决宣称）用陈述语气。例如：

- He *insisted* that I *should read* his letter. （他坚持要我看他的信。）（虚拟语气）
- He *insisted* that I *had read* his letter. （他坚持说我看过他的信。）（陈述语气）

② command, order（命令）。

- The officer *commanded* that *soldiers* (*should*) *attack* at once.
- The King *ordered* that those prisoners *be released*.
- He *ordered* that the parcel (*should*) *be sent back*.

③ advise, propose, recommend, suggest（建议）。

- The doctor *advised* (*suggested*) that he (*should*) *not smoke*.
- The committee *proposed* (that) Mr. Liang *be elected*.
- The people *recommend* that this tax *be abolished*.
- She *suggested* that I (*should*) *be responsible for* the arrangements.

suggest后接宾语从句时，还可用陈述语气：从句谓语动词所表示的情况尚未成为事实时用虚拟语气，suggest意为“建议”；已成事实时用陈述语气，suggest意为“表明”“认为”。例如：

- He *suggested* that we (*should*) *stay* for dinner.
- What he said *suggested* that he was a cheat.

④ ask, demand, require, request（要求）。

- He *asked* that the message *be given to* Madame immediately.
- They *demanded* that the right to vote *be given to* every adult man.
- She *requires* that I (*should*) *appear*.

● I *request* that Mrs. Lu（should）go there at once.

⑤ move，vote（提议、投票）。

● She *moves* that I *accept* her proposal.

● Mrs. Yao *moved* that Mr. Liang *should be made* a member of the editorial committee.

● Congress *has voted* that the present immigrant law *be maintained*.

⑥ urge（敦促）。

● I *urged* that the matter *should go to* arbitration.

● They *urged* that relief work *should be given* priority.

⑦ arrange（安排）。

● He *arranged* that I *should go abroad* for further English study.

● It *was arranged* that the president *should be met* at the airport.

⑧ desire，intend（希望、打算）。

● The doctor *desire* the patient *should have a calm* mentality *every day*.

● They *intended* that the news（*should*）*be suppressed*.

⑨ direct（指示）。

● The new emperor *directed* that the prisoners *should be set free*.

had rather、would rather、would sooner 等后面的宾语从句常表示与客观事实不相符的一种愿望，故用虚拟语气。此虚拟语气的结构为：一般过去时 did（与现在事实相反，be 用 were）；过去完成时 had + done（与过去事实相反）；一般过去时 did（与将来事实相反，be 用 were）。例如：

● I'*d rather* they *were* here now.

● I'*d rather* you *had gone to* America last year.

● We'*d rather* you *went to* Shanghai tomorrow.

3）表语或同位语从句中的虚拟语气

某些表示建议、请求、命令等主观意向的名词做主语或宾语时，其后表语或同位语从句用虚拟语气，句型为 should + do（should 可省略）。这类名词常见的有：advice、demand、desire、idea、necessity、order、plan、preference、proposal、requirement、recommendation、suggestion 等。例如：

● The *plan* is that we（*should*）*work* at home during office decorating.（表语从句）

● They make a *proposal* that we（*should*）*hold* a meeting next Friday.（同位语从句）

● They are all for her *proposal* that the driving test（*should*）*be canceled*.（同位语从句）

● The *suggestion* that the leader（*should*）*present* the prizes was accepted by us.（同位语从句）

2.6.3　其他从句中的虚拟语气

（1）It is time 后的定语从句

这种虚拟语气的结构为 It is/was（high/about）time that + 主语 + did（should + do），从句谓语常用过去式表示现在或将来，有时也用过去进行时或“should + 动词原形”（较少见，且 should 不能省略），表示“（早）该干某事了”之意。time 前有时有 high 和 about 修饰；与其他虚拟结构不同，该用 was 时不能用 were。例如：

● *It's high time* that she *made up her mind* to go to New York.

● *It's time* we *went*（*were going/should go*）.（我们该走了）

● *It's about time* I *was* in bed.（was 不能改为 were）

● *It is high time* you *should write* the scientific report.

如果该句型调整为“It is/was the first time that + 从句”，那么 that 后面的从句的谓语动词不再用上述形式：主句用 is 时，从句谓语动词用现在完成时（have/has done）；主句用 was 时，从句谓语动词用过去完成时（had done）。例如：

● *It is the first time* that I *have obtained* such a high prize.

● *It was the first time* that I *had obtained* such a high prize before 2018.

（2）一些目的状语从句

1）在 for fear that、lest、in case 引导的目的状语从句中，从句谓语为（should）+ do。用 for fear that、lest 引导时，should 能省略；用 in case 引导时，should 不能省略。例如：

● Her mother examined the door again *for fear that* a thief（*should*）*come* in.

● I and Mrs. Qin started out earlier *lest* we（*should*）*be* late.

● I and Mrs. Qin started out earlier in case we *should be* late.

2）在 so that、in order that 引导的目的状语从句中，从句谓语为 can/could/may/might/will/would/should + do。例如：

● He goes closer to the speaker *so that* he *can hear* him clearer.

● I read the letter carefully *in order that* I *should not miss* a word.

（3）一些让步状语从句

1）在 even if、even though 引导的让步状语从句中，主句谓语用 may/might + 动词原形，从句用过去时，表示与现在或过去事实相反。类似的词语有 though、even though、whatever、however、so long as 等。例如：

● *Even if* he were here himself, he *may not know* what to do.

● Nobody *might save* him *even though* Hua Tuo *Should come* here.

2）在 whatever、whichever、whenever、whoever、wherever、however、no matter wh-等引导的让步状语从句中，从句谓语为 may + do（指现在或将来）、may + have done（指过去）。例如：

● We will find him *wherever* he *may be*.

● She will finish the paper writing on time *no matter what may happen*.

● They mustn't be proud *whatever* great progress they *may have made*.

● You must respect doctor Liang *no matter what* mistakes he *may have made*.

（4）一些方式状语从句

由 as if、as though（好像）引导的方式状语从句中，从句结构为 as if/as though + 主语 + were（did 或 had done）（与现在或过去事实相反）。例如：

● Li and Wang talk *as if* they *were* old friends.

● Li and Wang talked *as if* they *had been* old friends.

2.6.4　简单句中的虚拟语气

1）表示客气、谦虚、委婉、礼貌，常用"would/could/might/should + do"。例如：

● *Would* you *mind* my closing the windows?

● You *should* always *learn* this lesson by heart.

● She *should agree* with you.（她本该同意你。）

2）表示祝愿，常用"may + 主语 + do"。例如：

● *May* you *have* a good journey!（祝你一路顺风。）

● *May* your youth *last* forever!（祝你青春永驻。）

3）表示强烈愿望、祝愿，常用"do"。例如：

● God *bless* us.（上帝保佑）

4）一些习惯表达。

① 提出请求或邀请。例如：

- *Would* you like to have a talk with us this evening?
- *Could* I use your car to go to Shanghai now?

② 陈述自己的观点或看法。例如：

- I *should* be glad to meet you and your parents.
- She *would* try her best to finish the task at end of this year.

③ 提出劝告或建议。例如：

- You '*d better* ask your friend this question first.
- The policemen *should* make a full investigation of them first.

④ 提出问题。例如：

- Do you think the buyer *could* buy the computer on time?
- Do you expect Mrs. Li *would* tell us the truth?

⑤ 指责过去情况，常用"情态动词 + have done"。例如：

- You *should* have arrived here earlier.
- They *should* have finished the report.

2.7 非谓语动词

非谓语动词（non-finite verbs）又称非限定动词，即动词的非谓语形式或非限定形式，是指在句子中充当除谓语以外的其他句子成分的动词形式，主要包括不定式、动名词和分词（现在分词和过去分词）。非谓语动词属于动词的一种，具有动词的特点，但起名词、形容词或副词的作用，做主语、宾语、状语等多种句子成分。

2.7.1 不定式

不定式（infinitive）是动词中的一种不带词形变化从而不指示人称、数量、时态的一种形式，因其不被限定或者不被词形变化所限而得名。

（1）不定式的构成与形式

不定式具有名词、形容词、副词的特征，构成是"（to）+ do"，否定式是"not +（to） do"，包括一般式、进行式、完成式和完成进行式，有主动语态和被动语态之分，见表 2-4。

表 2-4　不定式的形式

形式类别	主动语态	被动语态
一般式	to do	to be done
进行式	to be doing	
完成式	to have done	to have been done
完成进行式	to have been doing	

不定式不能在句中单独做谓语，没有语法上的主语，但由于可表示一种动作或状态，因此应有动作的执行者和体现者。它的这种逻辑或意义上的主语叫逻辑主语，在句中往往可以找到或按语境可以确定出来。

（2）不定式的用法

不定式在句中可做主语、表语、宾语、宾补、定语、状语和独立成分。

1）做主语。例如：

- *To finish* the writing of this book in one year is very hard.
- *To be writing* this book for such long time is really boring.
- *To have known* you is a privilege.（认识你真是荣幸。）

主语较长、谓语较短时，常用 it 做形式主语，不定式短语放在谓语后面。常用句式有：①It + be + 名词或名词短语 + to do；②It takes sb. + sometime + to do；③It + be + 形容词 + for（of）sb. + to do。例如：

- *It is* her *pleasure to look into* the matter objectively.
- *It took me three years to bring* this

research to an end.

• *It is very hard for* me *to finish* the writing of this book in one year.

• *It is courageous of* you *to jump* into the water and *save* the drowning child.

2）做表语。例如：

• My job is *to edit and publish* scientific papers and journals.

• His son appears *to have* caught a heavy cold.

• The next step is *to make sure* that you know exactly what is required.

不定式做表语通常说明或解释主语的内容，表示目的或将来的动作。句子主语用 aim、ambition、duty、function、goal、idea、intention、objective、plan、purpose、reason、suggestion、wish 等抽象名词充当时，常用不定式做表语。

不定式做表语有时可用主动语态表示被动意义。例如：

• Mr. Zhang and Mrs. Liu are *to blame*.（张先生和刘女士应受到责备。）（主动表被动）

• Something is still *to find out*.（有些事还有待查明。）（主动表被动）

3）做宾语。例如：

• They *determine to start* the project next year.（动词宾语）

• I have no choice *but to stay* here.（介宾）

• He gave us some advices on *how to learn* English.（不定式与疑问词连用）

常用不定式做宾语的动词或动词短语有：ask、apply、agree、arrange、afford、aim、attempt、begin、beg、bother、care、claim、consent、choose、continue、dare、decide、desire、demand、determine、expect、endeavor、fail、forget、guarantee、know、hope、help、intend、like、learn、long、mean、manage、need、offer、plan、pledge、prefer、promise、pretend、prepare、proceed、prove、refuse、request、resolve、seek、start、teach、tend、threaten、volunteer、try、wait、wish、want、would like 等。

常用省略 to 的不定式做宾语的动词短语有：do nothing but、have no choice but、had better/best（not）、would rather、rather than、would sooner/would just as soon、may/might as well、why not 等。例如：

• He *did nothing* last year *but* learn English well.

• Since you've finished recording of the course，we *might as well sell* it.

在 believe、consider、deem、find、make、prove、show、take、think 等动词后用不定式做宾语且有宾补时，用 it 做形式宾语替代不定式，不定式后置，放在宾补后面。例如：

• Marx found *it important to study* the situation in Russia.（it 做形式宾语，替代 to study）

• This has made *it* necessary for me *to apply for* a new licence.（it 做形式宾语，替代 *to apply*）

4）做宾补。例如：

• Teacher Liang encouraged *the students to ask* more questions in class.

• Mrs. Wang remind *me not to be* late for the meeting.

• I consider *her*（*to be*）my best friend.（谓语动词 think、consider、find 后可省略 to be）

• With *a lot of work to do*，professor Liang didn't take part in that conference.（介宾的补语）

常用不定式做宾补的动词或动词短语有：ask、advise、afford、agree、allow、apply、bear、beg、cause、choose、command、compel、convince、call on、decide、desire、 determine、demand、enable、encourage、expect、fail、forbid、force、get、happen、help、hope、inform、intend、instruct、intend、invite、learn、like、manage、mean、offer、order、persuade、permit、plan、prepare、prefer、press、pretend、promise、refuse、remind、request、require、select、tell、trouble、urge、wait for、want、warn、wish 等。

不定式前的谓语动词是使役动词（如 make、let、have）和感官动词或短语（如 see、watch、look at、observe、notice、feel、hear、listen to）时，不定式符号 to 可以省略，但改为被动语态时，不能省略。例如：

- He saw the thief *steal* a lady's cellphone.
- The thief was seen *to steal* a lady's cellphone.

5）做定语。例如：

- I have *a meeting to attend* next month.（动宾关系，attend + meeting）
- She is always *the last* （*person*） *to speak* at the meeting.（主谓关系，person + speak）
- I'm going to the post office, for I have *a letter to post*.（动宾关系，post + letter，不定式的逻辑主语是 I，不是 letter）
- I have no *letters to be posted* now.（主谓关系，letters + be posted，不定式的逻辑主语是 letters，不是 I）
- We have made *a plan to finish* the work.（不定式说明被修饰名词 plan 的内容）

不定式定语放在被修饰名词或代词后形成某种关系，或说明被修饰名词的内容。

常用不定式做定语的名词有：ability、agreement、ambition、anxiety、attempt、campaign、claim、chance、curiosity、decision、desire、determination、disposition、eagerness、effort、failure、freedom、inclination、intention、impatience、mind、movement、need、offer、obligation、opportunity、plan、pressure、promise、readiness、reason、refuse、reluctance、right、temptation、tendency、time、way、willingness、wish、yearning 等。

一些短语如 the first、the second、the last、the best、the only thing 等后面也常用不定式做定语。例如：

- He is always *the first to come to* the bookstore.
- A solicitor is *the best person to advise* her about solving the marriage.

被修饰名词或代词后面的不定式的动词为不及物动词时，该动词后通常应有必要的介词，但被修饰名词是 time、place、way 时，通常省略介词。例如：

- Please lend me *a pen to write with*.
- The old man has *nothing to worry about*.
- This is *the best way to work out* this problem.
- They have *no money and no place to live* （*in*）.
- I think *the best way to travel* （*by*） is on foot .

6）做状语。例如：

- *To learn* English well, I need a dictionary.（表目的）
- *To learn* English well, a dictionary is needed.（错句，不定式逻辑主语与句子主语不一致）
- The academic book is too hard for children *to read*.（表结果）
- We shall be very happy *to cooperate with* professor Qin in the project.（表原因）
- It's too dark for us *to see* anything.（表程度）
- How can you catch the air plane *to get up* so late?（表条件）

不定式做状语可以表示目的、结果、原

因、程度、条件等意义。表示原因时常放在形容词的后面，这类形容词常见的有：angry、anxious、clever、delighted、disappointed、eager、foolish、fortunate、frightened、glad、happy、lucky、pleased、proud、ready、surprised、sorry、worthy 等。

不定式并列时，第一个不定式后面的其他不定式可省略 to。例如：

- I wish *to study* physics and（*to*）*become* a scientist.

7）做独立成分。例如：

- *To tell the truth*, I don't like the way he talked.
- *To make a long story short*, this film is rather fantastic.

不定式短语做独立成分相当于做状语。这类不定式短语常见的有：to begin with、to make matters worse（更糟糕的是）、to put it straight（直截了当地说）、to put it in another way（换句话说）、to be honest（老实说）、to sum up 等。

（3）不定式的时态与语态

1）一般式：表示的动作与谓语动词动作同时发生或在谓语动词动作之后发生。例如：

- It's very glad *to meet* Mrs. Qing.（同时发生）
- She *seems to know* a lot about Chinese art.（同时发生）
- I *want to be* a scientist or a writer.（之后发生）
- The patient *asked to be operated* on at once.（之后发生）
- The chairman *ordered* the work *to be done*.（之后发生）

2）进行式：表示的动作与谓语动词动作同时发生，即谓语动作发生时不定式动作正在进行，且不定式多由动态动词充当。例如：

- His son *seems to be reading* in his room.
- These young men *pretended to be working* hard.

3）完成式：表示的动作在谓语动词动作之前发生。例如：

- She *regretted to have said* such a bad word.
- I *happened to have met* Mr. Wang in London.
- They *are pleased to have studied* the ancient physics.

4）完成进行式：表示的动作在谓语动词动作之前发生，且该动作一直持续进行，即截至谓语动作发生之时不定式动作从过去某一时刻开始一直在进行着。例如：

- I *was happy to have been writing* so many books of academic paper writing.
- She was still young to have been working for thirty years.

5）不定式的逻辑主语是不定式动作的执行者时，不定式常用主动语态，是承受者时多用被动语态，也可用主动语态。例如：

- *They* have many, many words *to say* in the meeting.

此句中 They 为不定式 to say 的逻辑主语，也就是动作的执行者，因此不定式用主动语态。

- There is still *an important role to be played* by people in a world by science and technology.

此句中 an important role 为不定式 to be played 的逻辑主语，也是不定式动作的承受者，因此不定式为被动语态。

- *I* prefer *to be given* difficult work *to do*.

此句有两处不定式：第一处的逻辑主语 I 是不定式 to be given 所表示动作的承受者，不定式用被动语态；第二处的逻辑主语 I 是不定式 to do 所表示动作的执行者，不定式用主动语态。

● It's a great honor for *me to be invited to speak* in the academic conference.

此句中两处不定式的逻辑主语均为 me：它是第一处不定式 to be invited 动作的承受者，因此，此不定式用被动语态；同时，它又是第二处不定式 to speak 动作的执行者，因此，此不定式用主动语态。

● Have you got *anything to send*?

● Have you got *anything to be sent*?

此两句中不定式所修饰的代词是不定式动作的承受者，不定式分别用主动、被动语态。

6）在某些形容词如 easy、difficult、good、fit、hard 等后面或在某些结构及习惯用法中，可用不定式的主动语态表被动含义。例如：

● This language is very *difficult* for me *to speak*.

● This kind of book is *not fit to read* for children.

● There is *so much work to do*.

（4）不带 to 的不定式

1）不定式做一些感知、使役动词或动词短语（如 feel、have、hear、listen to、look at、let、make、observe、perceive、see、watch 等）的宾语时一般不带 to，但这类动词用于被动语态时，不定式带 to。例如：

● I *let* Mrs. Qin *design* the structure of the room.

● Mrs. Qin made Tom *answer* the phone in her absence.

● Tom was made *to answer* the phone by Mrs. Qin in her absence.

2）不定式在 do nothing/anything/everything but，can/could not but，do nothing (else) than，would/had sooner ... than，would/had rather...than 等习惯用法中 but 或 than 后面的不定式不带 to。例如：

● Last year I *did nothing but write* a book.

● I *would rather walk home than take* my car in such a jammed road.

3）句子主语含有实意动词 do 时，做表语的不定式可不带 to。例如：

● The first important thing this year I *did* was (*to*) *have* a speech about scientific writing in 21th Century Hotel of Beijing.

4）充当 help 的宾语或宾补的不定式可不带 to。例如：

● Enough sleep *helps bring* a good mood.

● Mrs. Qing and Zhang can *help me* (*to*) *write* the new book.

（5）不定式的其他要点

1）在"It + be + 形容词 + for（of）sb. + to do"句型中，for（of）引导的是不定式的逻辑主语，引导词常用 for，但句型中的形容词是表示人的性格品质、行为特征或表示赞扬、批评时，引导词常用 of。这类形容词有：brave、cruel、careless、careful、clever、courageous、considerate、foolish、good、greedy、generous、honest、kind、lazy、modest、negative、nice、polite、rash、rude、right、silly、selfish、stupid、thoughtful、wise、wrong、wicked 等。例如：

● It's *negative for me to make* such a big mistake.

● It's *careless of me to make* such a big mistake.

2）在动词原形之前插入修饰性副词的不定式称为分裂不定式，这个副词是用来修饰不定式而不是做其他成分的。例如：

● He wishes *to completely change* his bad habit.

3）不定式的否定式是 not + to do，其中否定词还可为 never、seldom 等词。例如：

● I appear *never to say* no to anyone.

4）不定式前面可加上 how、what、

which、where、whether、when、whom 等疑问词而形成“疑问词 + 不定式”结构，做主语、宾语、表语和同位语等。例如：

- The difficulty was *how to improve* my oral English level.

5）“介词 + 关系代词 + 不定式”结构一般做名词的定语，类似于定语从句。例如：

- My daughter and her grandmother will go to the countryside so that they will have a garden *in which to play* this summer vacation.

2.7.2 动名词

动名词（gerund）是由动词变化而来的，仍保留着动词的某些特征和变化形式，用来表达一般名词所不能表达的较为复杂的语义，同时在句中有着与名词类似的用法和功能，可以做多种句子成分，还可以被副词修饰或用来支配宾语。

（1）动名词的构成与形式

动名词的构成形式是“do + ing”，否定式是“not + doing”。它包括一般式、完成式，有主动语态和被动语态之分，见表 2-5。动名词还可以和其逻辑主语构成复合结构式，逻辑主语是代词时要用形容词性的物主代词，表示有生命的名词时要用名词所有格。

表 2-5　动名词的构成形式

形式类别	主动语态	被动语态
一般式	doing	being done
完成式	having done	having been done

动名词形式示例：

- *Reading* English aloud is very helpful to learn English.（动名词的一般式）
- She apologized for *having broken* her promise.（动名词的完成式）
- I regret *not following* his advice.（动名词的否定式）
- The president suggested *our trying* it once again.（动名词的复合结构式）
- *My mother's not knowing* medicine troubled her a lot.（动名词的复合结构式）

（2）动名词的用法

动名词在句中可以做主语、表语、宾语、定语和同位语。例如：

- *Making* speech of scientific writing is interesting for me.（做主语）
- *It*'s no use *quarrelling*.（it 替代动名词，做形式主语）
- In the ant city, the queen's job is *laying* eggs.（做表语）
- The workers haven't finished *building* the dam.（做动宾）
- We everyone have to prevent the air from *being polluted*.（做介宾）
- We found it no good *making* fun of others.（it 替代动名词，做形式宾语，后跟有宾补）
- There is four *swimming* pools near my *living* house.（做定语）
- My habit, *listening* to the news on the mobile phone remains unchanged.（做同位语）

以下动词或动词短语只跟动名词做宾语：admit、anticipate、attach importance to、avoid、appreciate、add to、be engaged in、be busy、be worth、be accustomed to、be contrary to、be fond of、be opposed to、be used to、consider、can't help、can't stand、conform to、contribute to、delay、deny、devote to、dislike、dream of、enjoy、escape、excuse、favor、fall to、feel like、finish、forbid、feel like、give up、imagine、in addition to、insist on、involve、keep、keep… from、keep on、lead to、leave off、look forward to、mind、miss、object to、permit、propose、practise、protect…from、postpone、pay attention to、prevent…(from)、put off、 resist、risk、resent、resort to、stop ...（from）、set about、stick to、suggest、spend...（in）、succeed in、thanks to、think of、the way to、the approach to、the solution to、the key to 等。

注意：在动词短语后跟动名词的条件是，动词短语在句中做谓语。例如：

● I'm *looking forward to receiving* your email from London.（looking forward to 做谓语）

● The date I'm *looking forward to* came years ago.（looking forward to 做定语）

动名词或其短语直接位于句首时，谓语动词用单数；做主语时常用 it 替代它做形式主语，被替代的动名词或其短语是真正的主语。例如：

● *Discussing this* is a mere waste of time. /*It*'s a mere waste of time *discussing this*.

● *It*'s no use *crying over spilt milk*. /*It*'s no point in *crying over spilt milk*.（覆水难收）

含做前置定语的动名词短语有：driving licence、fishing rod、sleeping car、swimming pool、waiting room、working language 等。

（3）动名词的时态与语态

动名词一般式表示的动作与谓语动词动作同时发生或在其后发生，完成式表示的动作发生在谓语动词动作之前。动名词的主动语态表示主动行为，被动语态表示被动行为。动名词表示的动作的逻辑主语一般可以从句中或按语境找到。例如：

● *Seeing* is believing.（眼见为实）（一般式 + 主动语态）

● He came to the party without *being invited*.（一般式 + 被动语态）

● We remembered *having seen* her.（完成式 + 主动语态）

● Tom forgot *having been taken* to Guangxi Province when he was six years old.（完成式 + 被动语态）

2.7.3　现在分词

现在分词（present participle）是分词的一种（分词分为现在分词和过去分词），在句中不能单独做谓语，但能做其他成分，且具有动词的性质。

（1）现在分词的构成与形式

现在分词的构成是"do + ing"，否定式是 not + doing。它包括一般式、完成式，也有主动和被动语态之分，见表 2-6。

表 2-6　现在分词的形式

形式类别	主动语态	被动语态
一般式	doing	being done
完成式	having done	having been done

（2）现在分词的用法

现在分词有动词、形容词和副词的特征，在句中可以做定语、表语、宾补和状语，还可作独立结构。

1）做定语。现在分词单独做定语时放在中心词（被修饰语）前，现在分词短语做定语时放在中心词后，与中心词之间是主动关系，其作用相当于一个定语从句。这个定语从句既可以是限定性（限制性）的，现在分词对中心词直接修饰或限制，不用逗号与中心词隔开；也可以是非限定性的，现在分词对中心词作进一步的附加或解释说明，用逗号与中心词隔开。例如：

● In the *following years* I worked even harder.（现在分词做 years 的前置定语，此定语相当于一个限定性定语从句：In the years *that followed* I worked even harder.）

● *The man speaking to teacher Liang* is our old leader.（现在分词短语做 The man 的后置定语，此定语相当于一个限定性定语从句：The man *who is speaking to teacher Liang* is our old leader.）

● This book，*dealing with simple Japanese*，

is suitable for beginners.（现在分词短语做 This book 的非限定性修饰语，相当于一个非限定性定语从句：This book，*which is dealing with simple Japanese*，is suitable for beginners.）

2）做表语。现在分词做表语一般表示主动或主语的性质和特征，含有“令人……”的意思，主语多数情况下表示物。例如：

- The cartoon being shown on the TV is *amusing and exciting*.
- The present situation *is inspiring* to everyone.

常用现在分词做表语的系动词有：appear、be、become、feel、get、grow、look、keep、remain、seem、sound 等。做表语的现在分词多是由能表示人们某种感情或情绪的动词变化而来的，常见的有 amusing、astonishing、boring、exciting、encouraging、interesting、inspiring、moving、puzzling、missing、promising、surprising 等。

注意：be + doing 可能表示现在分词做表语（强调特征）或现在进行时（动作正在进行）。

3）做宾补。现在分词只能在感觉、感官和使役、致使动词后做宾补，表示正在进行的主动意义。例如：

- Can you *hear* Mrs. Liang *singing* the song in her room?
- He *kept* the car *waiting* at the gate.

常用现在分词做宾补的使役、致使动词有 catch、get、have、keep、leave、set、start 等，感觉、感官动词或短语有 feel、find、hear、listen to、look at、notice、observe、see、smell、watch 等。

4）做状语。现在分词或其短语做状语时，句子的主语是现在分词的逻辑主语，即现在分词的逻辑主语和主句的主语一致，常用逗号将现在分词状语同其他成分隔开。现在分词状语包括时间、原因、方式、条件、结果、目的、让步、伴随状语与独立成分（独立成分可做伴随状语，但伴随状语并不一定都能做独立成分），除方式和伴随状语不能用状语从句替换（但可以写成并列句）外，其他均可用相应的状语从句（如时间状语从句、原因状语从句、条件状语从句、结果状语从句、目的状语从句和让步状语从句）替换。例如：

- （While）*Working* in the editorial office，he was an advanced editor.（时间状语）
- *Being a Party member*，he is always helping others.（原因状语）
- He stayed at home，*reading and writing*.（方式状语）
- （If）*Playing all day*，you will waste your valuable time.（条件状语）
- He dropped the glass，*breaking it into pieces*.（结果状语）
- I and my daughter often go *swimming at weekends*.（目的状语）
- *Though raining heavily*，it cleared up very soon.（让步状语）
- She laid on the river bank，*feeling the warmth of the sun against her body*.（伴随状语）
- Love is a beautiful song，*bringing me great happiness and motivation*.（伴随状语）
- *Judging from*（*by*）*her appearance*，she must be an actress.（独立成分）
- *Generally speaking*，girls are more careful than boys.（独立成分）

现在分词短语做状语时，有时前面可用一个连词或连词短语表示强调或出于某种表达的需要。这样的连词或短语有 after、as if、before、if、though、unless、while、whether…or 等，如以上例句前面括号中有 While 和 If 的句子。

（3）现在分词的时态与语态

1）一般式：现在分词所表示的动作与谓语动词动作同时发生，或在谓语动词动作之后发生。例如：

- They *went to* the meeting room, *singing and talking*.（主动式）
- The problem *being discussed is* very important.（被动式）

2）完成式：现在分词所表示的动作在谓语动词动作之前发生。例如：

- *Having finished* the writing, I *went to* the Beihai Park of Beijing.（主动式）
- *Having been told* many times, those naughty boys still *made* the same mistake.（被动式）

2.7.4 过去分词

过去分词（past participle）是分词的另一种形式，兼有动词、副词和形容词的特征，可以带宾语（如介宾）或受状语的修饰，和宾语或状语一起构成过去分词短语。规则动词的过去分词一般是由动词加 ed 构成，不规则动词的过去分词没有统一规则。

过去分词在句中可以充当定语、表语、宾补和状语。

1）做定语。过去分词与中心词之间有被动意义，其作用相当于一个被动语态的定语从句。做后置定语时，既有限定性的，也有非限定性的，非限定性的对中心词作进一步的附加说明，相当于一个非限定性定语从句，用逗号与中心词隔开。例如：

- Our group went on an *organized* trip last Monday.（过去分词做 trip 的前置定语，此定语相当于一个限定性定语从句：Our group went on a trip which was *organized* last Monday.）
- Those *selected as committee members* will attend the meeting.（过去分词短语做 Those 的后置定语，此定语相当于一个限定性定语从句：Those *who were selected as committee members* will attend the meeting.）
- These ten books in the bookshelf, *written by me*, are confluence of my writing ideas of academic paper.（过去分词短语做非限定性修饰语，相当于一个非限定性定语从句：These ten books in the bookshelf, *which were written by me*, are confluence of my writing ideas of academic paper.）

过去分词是单词时，一般位于所修饰的中心词前，而对于过去分词短语，则要放在中心词的后面。

2）做表语。过去分词做表语一般表示被动或主语所处的状态，含有“感到……”的意思，主语多数情况下是人，也可以是物。例如：

- Those children *were frightened* at the sad sight.
- I have *been married* for 30 years.

做表语的过去分词多是由能表示人们某种感情或情绪的动词变化来的，常见的有：amused、astonished、completed、crowded、bored、delighted、discouraged、excited、frighten、hurt、interested、inspired、moved、married、missed、pleased、satisfied、surprised、tired、worried 等。

be + 过去分词，表示状态时是表语结构，表示被动的动作是被动语态。例如：

- The door *was broken*, and it should be repaired at once.（broken 表损坏，过去分词做表语）
- The door *was broken*, and something *were stolen*.（broken 表被破坏，谓语被动语态）

有些过去分词由不及物动词构成，常见的有：arrived、boiled、changed、come、fallen、gone、passed、risen、returned 等。这类过去分词不表示被动，只表示完成，如 boiled water（开水）、fallen leaves（落叶）、

newly arrived goods（新到的货）、the risen sun（升起的太阳）、the changed world（变了的世界）等。

3）做宾补。过去分词在感觉、感官和使役、致使动词后做宾补，表示已经完成的被动意义。做宾补的分词表示正在发生的被动动作时，需要在其前面加 being。例如：

- I *heard the song sung* several times last week.
- My friend Mr. Wang still could not *make himself understood* in English.
- Mr. Yang *saw the wounded girl being carried into* the nearby hospital.

4）做状语。过去分词或其短语做状语同现在分词，做状语时句子的主语也是分词的逻辑主语，即过去分词的逻辑主语和主句的主语一致。例如：

- *Praised* by the teachers and the classmates, he became the pride of his parents.（表示原因）
- Once *seen*, this beautiful scenery can never be forgotten.（表示时间）
- *Given more time*, I'll be able to do it better.（表示条件）
- *Though told of the danger*, he still risked his life to save the boy.（表示让步）
- The president went to the office, *followed* by some soldiers.（做方式或伴随状语）
- When *compared with the whole universe*, the earth is too small.（过去分词短语前加连词表强调）

2.7.5 分词独立结构

分词或其短语做状语时有意义上的逻辑主语，既可以与句子的主语一致（二者相同），也可以不一致（二者不同）而有自己的主语来构成分词的独立结构，也称独立主格。例如：

- *Time permitting*, *we*'ll do another two projects.（逻辑主语 Time + 现在分词）
- *I waiting for the bus*, *a bird* fell on my head.（逻辑主语 I + 现在分词短语）
- *This done*, *I* went to Yichang.（逻辑主语 This + 过去分词）
- *All taken into consideration*, *my plan* seems to be practicable.（逻辑主语 All + 过去分词短语）
- *The sun having risen*, *they* started to do the farm work.（逻辑主语 The sun + 现在分词完成主动式）
- *All the tickets having been sold out*, they went away disappointedly.（逻辑主语 All the tickets + 现在分词完成被动式）

独立结构中的逻辑主语前面有时可以加 with 或 without（with 通常可以省略，而 without 不能），做伴随状语。例如：

- *With the lights burning*, *he* fell asleep.（With + 逻辑主语 the lights + 现在分词）
- *With the work done*, *they* went out to play.（With + 逻辑主语 the work + 过去分词）
- *I* started on the trip to Xinjiang *without anyone accompanying me*.（without + 逻辑主语 anyone + 现在分词）
- *Without a word more spoken*, *Xiao Li* left the office depressingly.（Without + 逻辑主语 a word more + 过去分词）

分词没有逻辑主语时称为悬垂分词，可表示说话人的一种态度或看问题的角度。悬垂分词通常为现在分词，已被看作一种固定词组，也列入独立结构之列。例如：

- *Strictly speaking*, the quantity in this equation should be called modulus of elasticity not Young modulus.（悬垂现在分词）
- *Taken as a whole*, there is nothing important with this new project.（悬垂过去分词）

悬垂分词结构常见的有：allowing for…,

frankly（broadly、generally、properly、strictly）speaking，judging from（by），taking ... into consideration 等。

2.8　谓语、非谓语动词的比较

谓语动词做谓语，构成句子的主体。非谓语动词只作为一种修饰语存在，本身并非一个真正的动词，故得名非谓语动词。

（1）相同点

1）做及物动词时，均可与宾语连用。例如：

- They will *build a training school of academic paper*.
- They want *to build a training school of academic paper*.

2）可被状语修饰。例如：

- The new shoes *fit* me *very well*.
- The old shoes used *to fit* me *very well*.

3）有语态和时态变化。例如：

- The naughty boy *was punished* by his parents.（谓语动词被动语态）
- The naughty boy avoided *being punished* by his parents.（动名词被动式）
- We *have written* the scientific report.（谓语动词完成时）
- *Having written* the scientific report，we handed it in.（现在分词完成式）

4）可有逻辑主语。例如：

- *They started* the work at once.
- The boss ordered *them to start* the work.（不定式有逻辑主语）
- *We being* Party members，the work was well done.（现在分词有逻辑主语）

5）可用 not 构成否定式。例如：

- I *don't waste* time and money.（谓语动词的否定式）
- The teacher tell students *not to waste* time and money.（不定式的否定式）
- The real problem is *not having* enough time and money.（动名词的否定式）
- *Not wasting* time and money，I feel proud.（现在分词的否定式）
- *Not given* enough time and money，I couldn't finish the research well.（过去分词的否定式）

（2）不同点

谓语动词在句中做谓语，受主语的人称和数的限制；非谓语动词在句中不能单独做谓语，不受主语的人称和数的限制。

非谓语动词可有名词作用，如不定式、动名词，在句中做主语、宾语、表语；也可有形容词作用，如不定式、分词，在句中做定语、表语、宾补；还可有副词作用，如不定式和分词，在句中做状语。

第3章 英语语法：句法

句子是一个语言单位，由词按语法规则构成，表示一个相对完整独立的思想。本章主要讲述基础句法，包括句子的成分、基本句型、句子的分类、各类从句的用法、倒装句与强调句、否定句、主谓一致。

3.1 句子的成分

一个句子一般由主语部分（subject group）和谓语部分（predicate group）组成。这两部分由各种句子成分构成，句子成分是句中起一定功用的组成部分，包括主语、谓语、宾语、定语、状语、补语、表语、同位语、独立成分九种。通常，每个句子都应有主语、谓语和宾语。例如：

- China International Piano Competition Committee elected officially excellent Miss Liang the winner of the piano competition.
- Excellent Miss Liang was officially elected the winner of the piano competition by China International Piano Competition Committee.

此两句分别为同一意思的主动语态和被动语态两种表达形式。

句一中，China International Piano Competition Committee 是主语，elected 是谓语，Miss Liang 是宾语，the winner of the piano competition 是宾补，其中 China International Piano Competition 是 Committee 的定语，excellent 是 Miss Liang 的定语，officially 是状语，the piano competition 是 the winner 的后置定语。

句二中，Miss Liang 是主语，was 是系动词，与 elected 一起做谓语，the winner of the piano competition 是主补，officially 是状语，excellent 是定语。

（1）主语

主语（subject）是句子叙述的主体，是陈述或叙说的对象，说明是谁或是什么，表示句子说的是什么人、什么事、什么东西、什么地方等等，一般置于句首，可由名词、代词、数词、名词化的形容词、不定式、动名词和主语从句等来充当。

（2）谓语

谓语（predicate）说明主语所发出的动作或具有的特征、状态，一般由动词来充当，旨在对主语动作或状态进行陈述或说明，指出“做什么”“是什么”“怎么样”，一般在主语之后。

（3）宾语

宾语（object）又称受词，是一个动作的对象或接受者，常位于及物动词或介词后面。宾语分为直接宾语和间接宾语两类，直接宾语指动作的直接对象，间接宾语（宾补）说明动作的非直接性，但受动作的影响。一般而言，及物动词后面最少要有一个宾语，该宾语常为直接宾语。有些及物动词要求两个宾语，通常一个为直接宾语，另一个为间接宾语。名词、代词、数词、动名词、不定式、句子都可充当宾语。

（4）定语

定语（attributive）用来修饰、限定、说明名词（名词词组）或代词的品质与特征，常为形容词，还可以是名词、代词、数词、介词短语、不定式及其短语、分词、定语从句，以及相当于形容词的词、短语或句子。定语和中心语之间是修饰和被修饰、限制和被限制的关系。

（5）状语

状语（adverbial）用来修饰动词、形容词、副词或全句，说明地点、时间、原因、目的、结果、条件、方向、程度、方式和伴随状况等。状语一般由副词、介词短语、分词及其短语、不定式及其短语，以及相当于副词的词、短语来充当，一般放在句末，也可放在句首或句中。

（6）补语

补语（complement）用来补充说明主语或宾语，相应地有主语补语（主补）和宾语补语（宾补），有鲜明的定语性描写或限定性功能，在句法上通常是不可缺少的。名词、动名词、形容词、副词、不定式、现在分词、过去分词都可以做补语。

（7）表语

表语（predicative）充当主语补语，用来说明主语的身份、性质、品性、特征和状态。表语常由名词、形容词、副词、介词短语、不定式、分词、从句来充当，多位于系动词之后。如果句子的表语也由一个句子充当，那么此充当表语的句子称为表语从句。

（8）同位语

同位语（appositive）是当两个指称同一事物的句子成分处于同等位置时，其中后一个句子成分可用来解释、说明前一个句子成分，那么这后一个句子成分就称为这前一个成分的同位语，即同位语常位于其用来解释、说明的成分之后。这两个成分多由名词或代词充当。同位语和补语的区别在于：补语不能缺少，而同位语可以。

（9）独立成分

独立成分（independent element）是指句中与其他句子成分只有意义关联而无语法关系的一个词、短语或从句，常见的有呼吁、惊叹语，答语，插入语，介词短语，非谓语动词所构成的短语，形容词、副词短语等。

3.2　基本句型

英语基本句型有以下六类。

（1）主语 + 谓语 + 状语（S. + predicate + adverbial）

此句型为主谓结构，谓语一般是不及物动词。例如：

- All things change, my thought also changes.（两个主谓结构并列）
- The sun is rising in the east, and people are starting to work in the field.（两个主谓结构并列）
- “Did you go to Changsha by train?” “No, we flew”.

（2）主语 + 谓语 + 宾语（S. + predicate + O.）

此句型为主谓宾结构，谓语一般是及物动词或动词短语。例如：

- We never beat children.
- My daughter can play the piano skillfully.
- The students are looking after their sick teacher.
- She will never forget those romantic, joyous and glorious days in her college days.

（3）主语 + 系动词 + 表语（S. + link V. + predicative）

此句型称主系表结构，系动词在形式上也是一种谓语动词，与表语一起构成复合谓语。系动词常见的有：appear、be、become、feel、go、get、look、seem、smell、sound、turn、taste 等。例如：

- I am a scientific editor of *Nature*.
- It seems to be a good day tomorrow.
- The lake often smells awful in summer.
- She turned a power engineer after her graduation from Tsinghua University.

（4）主语 + 谓语 + 间接宾语 + 直接宾

语（S. + predicate + OI. + OD.）

此句型为主谓双宾结构，其谓语是可以有双宾语的及物动词，双宾语分别是间接宾语、直接宾语。能加间接宾语的动词常见的有：allow、buy、do、find、give、keep、lend、make、offer、pay、pass、promise、save、send、show、sing、tell、teach、wish 等。例如：

- Please do *me a favour*.
- He gave *the book* to *a good friend* of his.
- She lent *me* her new *car*.
- I send *her a postcard* on her birthday.

（5）主语 + 谓语 + 宾语 + 宾补（S. + predicate + O. + C.）

此句型为主谓宾补结构，谓语通常是及物动词或动词短语，其补语是宾补，与宾语一起构成复合宾语。能做宾补的词有名词、形容词、分词、不定式和介词短语。例如：

- He made London *the base* of his revolution.（名词宾补）
- I found the task *easy*.（形容词宾补）
- I heard her *playing* the piano happily.（分词宾补）
- The leader let us *go home* at once.（不定式宾补）
- The leader allowed us *to go home* at once.（不定式宾补）
- I found the key *on the table*.（介词短语宾补）

（6）there be 句型

此句型中，be 的单复数取决于其后面的名词是否可数、单复数，并采取邻近原则。例如：

- There *are two bottles* on the table, and *there is some beer* in each bottle.
- There *is a computer*, a clock and three books and other things in the desk.
- There *are a lot of students* watching the football match on TV.

3.3 句子的分类

英语句子可从结构和功能两个层面进行分类。

（1）按结构分类

句子按结构可分为简单句、并列句和复合句。

1）简单句：只包含一套主、谓结构。例如：

- Think it over.
- He went out of the room, moved towards his car.

一个句子有一个共同的谓语时，不论主语、谓语分别并列几个要素，仍属简单句。例如：

- *China and other countries* in the east Asia are developing rapidly.（并列主语的简单句）
- *Mr. Wu and I* often *work* together and *help* each other.（并列两个主语和两个谓语的简单句）

2）并列句：包括几个互不依存的主谓结构，用并列连词连接。基本结构是“简单句 + 并列连词 + 简单句”，并列连词有 and、but、or、so 等，各个简单句的意义同等重要，相互之间是平行并列关系，没有从属关系。例如：

- Honey is sweet but the bee stings.
- Every employee must work hard, or he will miss his chance.

3）复合句：包括几个主谓结构，其中充当句子主体的主谓结构为主句，充当句子某个成分的为从句，从句对主句的某个成分进行限定、修饰、补充或说明，因此又称主从复合句。主句是全句的主体，可以独立存在；从句充当句子成分，不能独立存在，但有主语和谓语，如同一个句子，但需要由关联词（引导词、连词）来引导，关联词将主句和从句连接起来，表明主句和从句之间的关系。

按照句子成分，从句分为名词性从句、形容词性从句（定语从句）、副词从句（状语从句）三大类，每类下还可以继续分类。

（2）按功能分类

句子按功能（使用目的）可分为陈述句、疑问句、祈使句和感叹句。

1）陈述句：对事物依照其实在情形进行述说，句末使用句号，以表示陈述结束。分为肯定句和否定句，基本结构为“主语＋谓语＋其他成分”。

2）疑问句：表示疑问，分为一般、特殊、选择和反意疑问句四类。

一般疑问句：基本结构为“系动词/助动词/情态动词＋主语＋其他成分”，一般读升调，相当于汉语的“……吗?”，用Yes/No回答。例如：

- “Have you ever been to England?” “Yes, I have/No, I haven't.”

特殊疑问句：以疑问词开头，对句中某一成分提问，一般读降调。常用疑问词或短语有how、how many、how much、how old、how soon、how often、what、who、why、whose、which、when、where等。如果疑问词做主语或主语的定语，即对主语或主语的定语提问，其语序是“疑问词（＋主语）＋谓语＋其他成分”；如果疑问词做其他成分，即对其他成分提问，其语序是“疑问词＋一般疑问句语序”。例如：

- “How often do you go swimming recent years?” “About once a week.”

选择疑问句：用or连接询问的两部分，以供选择，回答是完整的句子或其省略形式，不能用Yes/No来回答。其结构可用一般疑问句或特殊疑问句。供选择的两部分，前一部分读升调，后一部分读降调；后一部分与前一部分相同的地方应该省略，但如果名词前面有不定冠词或省略会影响句义的完整性，则不要省略。例如：

- “Which phone do you like best, Apple or Huawei?” “Huawei, of course.”

反意疑问句：又叫附加疑问句，是指提问人对前面的事实不能肯定时，需要向对方加以证实时所提出的问句，用Yes/No来回答。其基本结构为：陈述句＋反意疑问句。陈述部分用肯定语气，反意部分则用否定语气；陈述部分用否定语气，反意部分则用肯定语气。例如：

- Mrs. Qin likes math, doesn't she?
- Mrs. Qin doesn't like math, does she?

3）祈使句：表示请求、命令、劝告等，用动词原形，否定式用don't或never，有时也可用let引起祈使句（如Let's read and write.）。

4）感叹句：表示说话时的惊喜、愤怒等情绪，用what、how引导。

3.4　名词性从句

名词性从句（noun clauses）是指在句子中起名词作用的句子。它相当于名词短语，在复合句中具有主语、宾语、表语、同位语、介宾等语法功能，根据这些不同的语法功能，名词性从句可分为主语从句、宾语从句、表语从句和同位语从句。能充当名词性从句的句子有陈述句、一般疑问句和特殊疑问句三类。用陈述句时，需要在其句首用that来引导；用一般疑问句时，需要把疑问倒装语序变为陈述语序，再用whether或if来引导（if只用来引导宾语从句）；用特殊疑问句时，则把特殊疑问语序变成陈述语序。例如：

- I think *that my daughter should be admitted to Tsinghua University.*（陈述句充当名词性从句）

My daughter should be admitted to Tsinghua University.（陈述句）

- I don't know *whether/if my daughter*

could be admitted to Tsinghua University.（一般疑问句充当名词性从句）

Could my daughter be admitted to Tsinghua University?（一般疑问句）

• I don't know *when my daughter could obtain the U. S. visa.*（特殊疑问句充当名词性从句）

When could my daughter obtain the U. S. visa?（特殊疑问句）

3.4.1　主语从句

主语从句在复合句中做主语，应保持陈述语序。当单个主语从句做主语时，主句谓语动词用单数。

（1）主语从句由从属连词、连接代词和连接副词引导

主语从句常由从属连词 that、whether、if，连接代词 who、whom、whose、what、which、whatever、whichever、whoever，连接副词 how、when、where、why 等引导。that 在句中无实在意义，只起连接作用；连接代词和连接副词既保留自身疑问含义起连接作用，又充当从句的成分。if 引导的主语从句不能位于复合句句首。例如：

• *That the new media is becoming more and more popular* is known to us all.

• *Whether my daughter comes* (*or not*) on my birthday is a secret.

• *Who will be our new leade*r has not been decided.

• *What she wants to tell us* is not clear.

• *Whatever is worth doing* should be done at once.

• *When we can meet next time* depends on *when the conference will be hold.*

如果 what 引导的主语从句本身明显表复数意义，特别是它的谓语动词和补语都是复数时，主句谓语动词可以用复数。例如：

• *What I say and think* are none of your business.

（2）常用 it 做形式主语

常用 it 代替主语从句做形式主语放于句首，而把真正的主语（主语从句）置于句末，这样有时还可以避免头重脚轻式的句子。例如：

• *It* is a wonderful achievement *how I should succeed in learning swimming.*

• *It* is not sure *whether I can run for the Chief Editor of the journal.*

（3）it 做形式主语的主语从句句型

it 做形式主语的主语从句的常用句型有：it + be + 名词 + that 从句；it + be + 形容词 + that 从句；it + be + 过去分词 + that 从句；it + 不及物动词 + that 从句。具体结构常见的有以下十几种，谓语多用单数形式：

1）It is a fact that…

2）It is a pity that…

3）It is a shame that…

4）It is a fortunate that…

5）It is important that…

6）It is strange that…

7）It seems that…

8）It is said that…

9）It is reported that…

10）It has been proved that…

11）It is believed that…

另外，主语从句用来表示惊奇、惋惜、不相信、理应如此等语气时，谓语动词宜用虚拟语气“(should) + do”，常用句型有：

1）It is necessary (important, natural, strange, etc.) that…

2）It is suggested (requested, proposed, desired, etc.) that…

注意：It is said/reported that…中的主语从句不可提前。

3.4.2　宾语从句

宾语从句在复合句中做宾语，包括及物

动词的宾语和介词的宾语，应保持陈述语序。

（1）宾语从句常由连接词引导

宾语从句常由连接词 that、whether/if、who、whom、whose、what、which、how、when、where、why 引导。that 引导时，不做句子成分（可省略）；what 引导时，做句子成分。例如：

- He think *that we could finish the work at the end of this month.*
- The old man asked *if/whether the boy had the right to drive so fast.*
- I don't know *what I should do for my wife on our 20th wedding anniversary.*
- She didn't explain *why she came so early.*

（2）用 it 做形式宾语

常用 it 代替宾语从句做形式宾语放于谓语之后，把真正的宾语（宾语从句）置于句末。例如：

- They considered *it* impossible *that we could finish the work before* 2021.
- We think *it* necessary *that we should widen our field of observation to the world.*

（3）（should）+动词原形式宾语从句

在 command、decide、desire、doubt、demand、insist、order、request、suggest 等表示要求、命令、建议、决定等意义的动词后面，宾语从句常用“（should）+动词原形”。例如：

- I insist *that she* （*should*） *do her work alone.*
- The commander ordered *that troops* （*should*） *set off at once.*

（4）特殊疑问句式宾语从句

用 who、whom、which、whose、what、when、where、why、how、whoever、whatever、whichever 等词引导的宾语从句相当于特殊疑问句，要注意用陈述句语序。

（5）用 whether/if 引导的宾语从句

用 whether/if 引导的宾语从句，保持陈述句语序，即主语和谓语的顺序不能颠倒。另外，下列情况一般只能用 whether，而不用 if：①引导的从句在句首；②引导表语从句；③引导的从句做介词宾语；④从句后有 or not；⑤后接动词不定式。例如：

- *Whether there is life on the moon* is an interesting question.
- *Whether we can succeed or not* depends on how well we cooperate.

（6）主句、从句谓语时态的匹配性

主句谓语是一般现在时、一般将来时或祈使句时，从句谓语应按自身需要而使用相应时态。主句是一般过去时（could、would 除外），从句多用过去时（一般过去时、过去进行时、过去将来时等）；主句表示自然现象、客观真理、科学原理和格言时，从句仍用一般现在时。例如：

- I *know* （that） he *will study* English next year.
- I *know* （that） he *has studied* English since 1998.
- She *said* that she *would come to take part in* the meeting.
- The teacher *told* us that Jason *had left* us for America.
- The teacher *told* us that light *travels* much faster than sound.

（7）宾语从句的否定转移

在 believe、expect、imagine、suppose、think 等动词引起的否定式宾语从句中，要用否定转移，就是把主句的谓语动词变为否定式，即将从句的否定形式移到主句中。例如：

- We *don't think* she *has* the ability to do the job. （We *think* she *has't* the ability to do the job.）
- I *don't believe* he *will* do so in right

way.（I *believe* he *won't do* so in right way.）

（8）"谓语+间接宾语+直接宾语"式宾语从句

主句谓语动词是ask、tell、show、teach、reminded等词且后面带双宾语时，宾语从句可做其中的直接宾语，句子结构为"谓语+间接宾语+直接宾语（宾语从句）"，而且从句的主语应与间接宾语相一致。例如：

- He has *asked me whether I can take part in the conference.*
- My wife *reminded me what I had done for my old mother.*

（9）"引导词+do you think+陈述句语序"式宾语从句

当do you think后面接特殊疑问句而转化成宾语从句时，do you think位于引导词和陈述句之间，句式结构为"引导词+do you think+陈述句"。例如：

- How many times *do you think* she has read the book?
- When *do you think* she'll come back?
- Who *do you think* Jason is waiting for?

3.4.3 表语从句

表语从句在复合句中做表语，应保持陈述语序，位于系动词后，基本结构为"主语+系动词+连接词+从句"。引导主语从句的连接词都可以引导表语从句，应用原则与主语从句一致。

（1）表语从句的引导词语

引导表语从句的连接词或短语有：that、who、when、where、which、what、why、whether、if、how、because、as if等，系动词有be（am、is、are、was、were）、become、feel、look、remain、smell、seem、sound、taste、turn等。例如：

- The fact is *that we have lost the chance to make a great progress.*
- The years of peace are *when everyone can lead a free and happy life.*
- This is *where our problem lies.*
- That's just *what I don't want and say.*
- That is *why I can write such many books in recent years.*
- It looks *as if it is going to snow.*

（2）与that is the reason why…和that is because…的区别

表语从句的that is why…与that is the reason why…同义，不过后者是定语从句；that is why…与that is because…语义有差别，前者指（由于某原因或理由所造成的）后果，后者指（造成某后果的）原因或理由。例如：

- I had to write a report in two hours. That is *why I didn't attend the meeting.*
- I didn't attend the meeting. That is *because I had to write a report in two hours.*

（3）表语从句的主语是reason时的引导词

句子的主语是reason时，表语从句用that而不是because引导。例如：

- The reason why she was late was *that she missed the train by one minute that morning.*

（4）两个引导词whether和if的区别

注意区分whether和if，前者可引导表语从句，但后者通常不可以。例如：

- The question is *whether we have enough time to do it.*

3.4.4 同位语从句

同位语从句在复合句中修饰其前面的某个名词或代词（先行词），它与先行词是等同关系，但内容往往更详尽具体，对先行词所表示的内容作进一步的解释和说明。同位语从句多由that引导，但that只起连接作用，在从句中不做任何句法成分。

（1）同位语从句的引导词

同位语从句通常由 that、whether、how、when、where 等连接词引导（if、which 不能引导）。可以跟同位语从句的先行词通常有：advice、answer、belive、conclusion、demand、doubt、evidence、fact、hope、idea、information、likelihood、message、news、order、plan、proof、proposal、problem、possibility、promise、question、request、suggestion、suspicion、thought、truth、wish、words（消息）、worry 等。例如：

- The news *that Chinese team won the game* is exciting.
- I have no idea *when she will come back from Canada*.

（2）同位语从句不一定紧跟在名词后面

同位语从句有时可以不紧跟在名词后面，而是被别的词语隔开。例如：

- *The thought* came to him *that his mother had probably fallen ill*.
- I got a *message* from the official news *that I would have obtained a S&T award*.

（3）同位语从句和定语从句有差别

同位语从句中的连接词只起连词的作用，连接主句和从句，没有实际意义，不充当句子成分，一般不能省略；而定语从句中的关系词既代替先行词，又充当句子成分（做宾语时可以省略）。例如：

- I had no idea *that you were coming here*.（同位语从句）
- The biological book (*that/which*) *a famous historian wrote lost year* gives you of life in ancient Greece.（定语从句）

3.5 形容词性从句

形容词性从句就是定语从句（attributive clauses），充当修饰语，修饰某个名词、代词或整个句子。被定语从句修饰的部分叫先行词，引导定语从句的词叫关系词，定语从句通常出现在先行词之后。关系词分为关系代词或关系副词，前者有 that、which、who、whom、whose，后者有 when、where、why。

3.5.1 关系代词的用法

关系词在定语从句中做主语、宾语或定语时，用关系代词引导定语从句。

（1）先行词是人时的关系代词

先行词是人时，关系代词用 that、who、whom（宾格）引导定语从句。例如：

- *Some people who* have made lots of money do not know what to do with their money.
- She is *the girl*（*who/whom/that*）I'm looking for.
- She is *the girl with whom* I want to have a talk.

（2）先行词是物时的关系代词

先行词是物时，关系代词用 which、that 引导定语从句。例如：

- A country is a *nation which/that* has its own government, land and population.
- *My grandmother is out of danger*, *which* excites me very much.（先行词是前面整个句子）

（3）关系代词的省略

关系代词在定语从句中可做主语、宾语，做宾语时可以省略。例如：

- Mr. Zhang is *the editor who/that* is in charge of mechanical engineering major.（做主语）
- *The book*（*that/which*）I bought from Beijing Books Building is very important for me.（做宾语，省略）

（4）关系代词 whose 可以指人或物

关系代词 whose 引导定语从句时，既可以指人，也可以指物。例如：

- *The editor whose name* is Zhang Xiaoli is in charge of mechanical engineering major.（指人）
- *These books whose covers* are interesting are best-selling.（指物）

（5）关系代词用 that 不用 which

用 that 不用 which 引导定语从句的情况有以下多种。

1）先行词为不定代词或被不定代词所修饰，这类不定代词有 all、anything、everything、little、much、none、something、those。例如：

- Is there *anything*（that）we can write?
- There is *little money*（that）we can use.

2）先行词被形容词最高级所修饰。例如：

- This is *the most biggest and highest building*（that）I have ever seen.

3）先行词被序数词所修饰。例如：

- *The* 8*th book*（that）I wrote is *Writing and Submission of SCI Papers*.

4）先行词被 all、any、first、last、next、only、very 所修饰。例如：

- *All the books* that I bought online last week were sent to my office in two hours.
- *The first step*（that）we are to take is very important.
- He was *the only person* that joined the army in his village that year.
- This is *the very computer* that I want to buy.

5）先行词前面已有 who、which 等疑问代词。例如：

- *Who* is the *man* that is speaking loudly there?
- *Which* is *the coat* that fits me most?

6）在 there be 句型中。例如：

- *There are* two choices that I must choose in this year.

（6）关系代词用 which（whom）不用 that

用 which（whom）不用 that 引导定语从句的情况有以下两种。

1）在“介词 + 关系代词”结构中。例如：

- She is a good teacher *from whom* we should learn well.
- This is a good book *from which* we have learnt a lot.

2）用逗号分隔先行词与其后的非限制定语从句。例如：

- I have written *ten books*, *which* are all very popular to college students.

（7）关系代词用 as

先行词前有修饰词语 such 或 the same 时，关系代词用 as。例如：

- I have *such a car as* she has.
- He *has the same idea as* I have.

（8）关系代词用 the one

关系代词前缺少先行词时，需要用 the one 来充当。例如：

- This university is *the one*（that）I once visited before.

（9）关系代词用 which、who

先行词是集体名词时，关系代词用 which、who。which 强调整体，who 强调整体中的个体。例如：

- *The football team which is* playing very well will come out first.
- *The football team who are* having a rest will begin another match in 20 min.

（10）关系代词用 who/whom

先行词是 anyone、anybody、everybody、everyone、someone、somebody 等不定代词时，关系代词用 who/whom。例如：

- Is there *anyone* here *who* can speak Chinese?
- He saw his friend talking with *someone*

whom he doesn't know.

3.5.2 关系副词的用法

关系词在定语从句中做状语时，用关系副词引导定语从句，表示时间用 when，表示地点用 where，表示原因用 why。

（1）关系副词 when

when 引导的定语从句用来修饰表示时间的名词（如 time、day、month、year、century），在从句中做时间状语。例如：

● There comes *a time when* I have to make a choice.

● Gone are *the days when* they could do what they liked.

● We'll put off the meeting until *next month*, *when* the time may be sufficient.

注意，不要一看到先行词是时间名词，就用 when 来引导定语从句，还要看它在从句中做何成分，做时间状语用 when，做主语或宾语不用 when 而用 that 和 which 等。例如：

● I will never forget *the day when* I met Mrs. Meng for the first time.（先行词在定语从句中做时间状语，用 when）

● Don't forget *the time*（*that*, *which*）the leader has told you.（先行词在定语从句中做直接宾语，用 that 或 which）

● I will never forget *the day on which* I met Mrs. Meng for the first time.（先行词在定语从句中做介词宾语，用 which）

（2）关系副词 where

where 引导的定语从句用来修饰表示地点的名词（如 city、country、room、house、office），在从句中做地点状语。例如：

● This is *the Guancun village where* I was born in 1968.

● That's *the scenic spot where* we traveled last summer.

● Mr. Cheng was living in *London*, *where* he went to work by bus.

要注意，不要一看到先行词是地点名词，就用 where 来引导定语从句，还要看它在从句中做何成分，做地点状语用 where，做主语或宾语用 that 和 which 等而不用 where。例如：

● This is *the Guancun village where* I used to live.（先行词在定语从句中做地点状语，用 where）

● He works in *a factory that/which* makes TV sets.（先行词在定语从句中做主语，用 that 或 which）

● This is *the Guancun village in which* I used to live.（先行词在定语从句中做介词宾语，用 which）

另外，where 有时还可用于抽象名词后引导定语从句。例如：

● We have reached *a point where* a change is needed.

● There are *cases where* the word "mighty" is used as an adverb.

● I got into *a situation where* it is hard to decide what is right and wrong.

● She doesn't want *a job where* she is chained to a desk all day.

（3）关系副词 why

why 引导的定语从句用来修饰表示原因的名词，在从句中做原因状语。例如：

● We don't know *the reason why* she didn't show up.

● My friend Mr. Zhang didn't tell me *the reason why* he refused the help.

● The *reason why*（*for which*）he refused the help is still not clear.

注意，与 when 和 where 不同，why 可以换成 that 或省略。例如：

● That's one of the reasons（*why*, *that*）I asked you to come.（用 why 或 that 均可，或省略）

另外，与 when 和 where 可引导非限定性定语从句不同，why 只能引导限定性定语从句而不能引导非限定性定语从句。例如：

- The main reason why he lost his way was that he drank.

- The main reason, why he lost his way, was that he drank.（错句）

3.5.3　限定性、非限定性定语从句

（1）限定性定语从句

限定性定语从句与主句的关系较为紧密，对先行词起限定、修饰作用，如果将其去掉，会影响句意的完整性，还可能会引起误解。例如：

- I walk into *the canteen where* students are eating.

- She has *a daughter who* works in America.（她有一个在美国工作的女儿。）

- This is *the computer*（*which*）I bought from Jingdong last month.

此三句均为限定性定语从句。第一句由 where 引导，说明说话者走进的不是别的食堂，而是一个学生正在就餐的食堂。第二句由 who 引导，专门指出她有一个在美国工作的女儿而不是指别的女儿，折射出“她也许还有在别处工作的女儿”之意。第三句省略了关系代词 which，其后的定语从句限定该电脑不是别的电脑，而是上月从京东购买的电脑。

（2）非限定性定语从句

非限定性定语从句与主句的关系较为松散，其间用逗号分隔；对先行词没有限定、修饰作用，只做补充、说明。这种补充、说明可以针对整个主句或主句的一部分，并用 which 或 as 引导（不能用 that 引导）。这种从句去掉后，并不影响句子的本意，因此将其看作一个“并列”的分句而非定语从句可能更容易理解。例如：

- We went *home*, *where* we had dinner.（We went home, we had dinner at home.）

- She has *a daughter*, *who* works in America.（她有一个女儿，在美国工作。）

- I tried to *get out of business*, *which* I found difficult.

- *As* is known to us, *Taiwan belongs to China*.

此四句均为非限定性定语从句。第一句的从句由 where（相当于 at home）引导，去掉后并不影响“我们回家了”这一本意。第二句的从句由 who 引导，去掉后并不影响“她只有一个女儿”这一本意。第三句的从句由 which 引导，补充、说明主句的一部分（get out of business），去掉后并不影响主句的意思。第四句的从句位于主句的前面，由 as 引导，对整个主句进行补充、说明，去掉后并不影响主句的意思。

3.6　副词性从句

副词性从句就是状语从句（adverbial clause），由从句担任状语，在句中可修饰谓语动词或其他动词、形容词、副词。状语从句位于主句前时，常用逗号与主句分开；位于主句后时，一般不用逗号分开。状语从句按意义和作用可分为时间、地点、原因、条件、目的、让步、比较、方式、结果等类别。

3.6.1　时间状语从句

时间状语从句在复合句中表示主句谓语动词发生的时间属性。常用引导词语有：as、as soon as、after、before、not ... until、once、since、till、until、when、whenever、while 等。特殊引导词语有：any time、all the time、by the time、directly、each time、everytime、hardly（scarcely, rarely）...when、immediately、next time、no sooner ... than、the day、the minute、the moment（that）、 the second、the first time、the instant 等。

（1）when，whenever，while，as

when 表示具体时间，从句谓语动词表示瞬时或持续性动作，与主句动作同时发生或先于主句动作发生，可以是时间点或时间段。例如：

- *When* she came in，I stopped watching TV.
- I used to swim in the river，*when* I lived in the countryside.

whenever 表示不具体时间。例如：

- *Whenever* you think you know nothing，then you begin to know something.
- *Whenever* that boy says "To tell the truth"，I suspect that he's about to tell a lie.

while 只表示持续性动作或状态，强调主从句动作同时发生或相对应，或表示对比。例如：

- Strike *while* the iron is hot.（趁热打铁。）
- *While* my wife was reading the newspaper，I was writing a paper.（同时发生）
- I like playing football *while* you like playing basketball.（对比）

as 多表示延续性动作，与主句动作同时发生（一边……一边），或强调一先一后。例如：

- *As* we were going out，it began to snow.
- You can feel the air moving *as* your hand pushes through it.

（2）before，after

before 引导的时间状语从句，主句动作发生在从句动作前，从句多位于主句后，且不用否定式谓语。主句用将来时，从句用现在时；主句用过去时，从句用过去时；主句用过去完成时，从句多用一般过去时。after 引导的时间状语从句，主句动作发生在从句动作后。例如：

- It will be two years *before* I come back from London.
- Jason almost knocked me down *before* he saw me.
- My wife had left for Canada just *before* the letter arrived.
- *After* you think it over，please let me know what you decide.
- *After* I finished the writing，I had a pleasant holiday
- They had not been married two months *before* they were divorced.

（3）till，until

主句谓语为瞬时动词时用否定形式，是延续性动词时用肯定、否定均可，但表意有所不同。until 可放在句首，till 不可，not until 放在句首时用倒装。二者常可以互换，强调句中多用 until。例如：

- She didn't realize how great her mother was *until*（*till*）she became a mother.
- I worked *until* she came back.（我工作到她回来为止。）
- I didn't work *until* she came back.（她回来时我才开始工作。）
- *Not until* last year did I know I was wrong.
- *It* was *not until* the meeting was over *that* she began to see me.

（4）since

since 引导的时间状语从句，谓语动词是瞬时或持续性动词。从句谓语用一般过去时，主句用现在完成时，但在 It is + 时间 + since 从句中，主句多用一般现在时。例如：

- I have been in Beijing *since* I left my hometown.
- *Since* she went to London，I have not heard from her.
- Where have you been *since* we last met

in Beihang University?

● It is near twenty-four years *since* I lived in Beijing.

(5) as soon as

as soon as 引导的时间状语从句，表示主句动作紧接着从句动作发生（一……就），从句用一般现在时，主句用一般将来时。例如：

● *As soon as* I reach Canada, I will send you a message.

● I will go to China Three Gorges University directly *as soon as* I get ready for the PPT.

(6) no sooner...than, hardly (scarcely, rarely)... when/before

这几个短语引导的时间状语从句，从句用一般过去时，主句用过去完成时，表意相当于 as soon as。当 no sooner、hardly、scarcely、rarely 位于句首时，主句应该部分倒装。例如：

● I had *no sooner* finished a lecture on writing *than* I was asked to start on another lecture.

● *No sooner* had the sun shown itself above the horizon *than* the *farmer* started to work.

● He had *hardly* fallen asleep *when* he felt a soft touch on his shoulder.

● *Hardly* (*scarcely*, *rarely*) had I got to the station *when* the train left.

(7) by the time

by the time 引导的时间状语从句，通常从句用一般过去时，主句用过去完成时；从句用一般现在时，主句用将来完成时。例如：

● *By the time* you came back yesterday, I had finished this work.

● *By the time* you come here tomorrow, I will have finished this work.

(8) each time, every time

这两个短语的意思相同，引导的时间状语从句，通常主句、从句都用一般过去时，或从句用一般现在时代替一般将来时，主句仍用一般现在时。例如：

● *Each time* I went to Harbin, I would call on her.

● *Every time* I listen to your advice, I always can avoid getting into trouble.

● She grows younger *every time* I see her.

(9) as long as, so long as

两词意思相同，引导的时间状语从句，通常从句用一般现在时，主句用一般现在时或一般将来时。例如：

● You can go where you like *as long as* you get back before dark.

● I will fight against these conditions *so long as* there is a breath in my body!

(10) the moment

the moment 引导的时间状语从句，通常主句、从句都用一般过去时。例如：

● The children ran away from the orchard, *the moment* they saw the guard.

● *The moment* I heard the news, I hastened to the spot.

(11) once

once 引导的时间状语从句，从句用一般现在时表一般将来时，主句用一般将来时。例如：

● *Once* you understand this rule, you will have no difficulty in understanding this.

3.6.2 地点状语从句

地点状语从句在复合句中表示地点、方位，一般位于主句之后，但有时为了强调，也可放在句首。引导词有 where、wherever、anywhere、everywhere 等。例如：

● We all should work hard *wherever* (*anywhere*, *everywhere*) *we go*.

● *Where* the Communist Party of China

goes, there the people are liberated.

where 引导的地点状语从句、定语从句的区别：前者从句前不需先行词，相当于修饰范围较大的副词，后者从句前应有一个表示地点的词做先行词，相当于修饰范围较大的形容词。例如：

- Go back *where* you came from. （where 引导地点状语从句）
- Go back to the place *where* you came. （where 引导定语从句，place 为先行词）

3.6.3 原因状语从句

原因状语从句在复合句中表示缘由、起因。常用引导词有：because、since、as、for 等。特殊引导词语有：considering（that）、now（that）、in that、seeing（that）、seeing as、given that、for the reason that、when（既然）等。

（1）because

because 引导的原因状语从句，从句一般在主句前面，和主句间用逗号分隔，但从句在主句后面时，不用逗号分隔。because 表示直接原因，语气最强，最适合回答 why 提出的问题，且不能与 so 连用；because 还可和 of 组成短语 because of，表示原因，后面只接名词、代词或动名词，不接从句。强调原因状语时，句型为 It is/was because… that…。例如：

- Some persons dislike me *because* I'm handsome and successful.
- *Because* it was cheaper, we went by train. （It was cheaper so we went by bus.）
- My daughter can't go to school *because of* her illness.
- *It was because* Mrs. Wang didn't feel well *that* she didn't go to work.

（2）since

since 引导的原因状语从句，从句一般放于主句之前，表示一种已知、显然或附带的原因，相当于 now that（既然），较为正式，语气比 because 弱。例如：

- *Since* you are free today, you had better help me with my homework.
- *Since*（*Now that*）you are grown up, you should rely on yourself.

（3）as

as 引导的原因状语从句，从句表示附带说明的双方已知的原因，所表示的理由最弱，含有对比说明的意味，语气比 since 弱，较为正式，位置较为灵活，常放于主句之前。例如：

- *As* you are tired, you had better have a good rest.
- My wife went to bed early yesterday, *as* she was exhausted.

（4）for

for 引导的原因状语从句，只起补充说明的作用，虽表示主句行为发生的原因，但并不说明直接原因，只提供一些辅助性的补充说明，而且只能放于主句之后，并用逗号隔开。例如：

- Mr. Wang seldom goes out now, *for* he is very old.
- It must have rained heavily last night, *for* the ground is still wet this morning.

（5）not…because，not because

这两词语等义，用来构成 because 从句的否定结构。not…because 结构中的 not 既可否定主句也可否定从句，需按表意或语境做出正确或合乎逻辑的理解，必要时可在 because 前使用逗号，以免引起歧义。但是，because 前有 just 修饰时，一般认为 not 是否定从句而不是主句的。例如：

- I like her *not because* she is beautiful, but because she is active.
- The country is *not* strong *because* it is larger. （国强不在大。/国不强在大。）
- I *didn't* go *because* I was afraid. （我

不是因为怕才去的。/我没有去是因为怕。）

• I *shouldn't* get angry *just because* they speak ill of me.（我不会因他们说我坏话而生气。）

（6）now（that），in that，seeing（that），when

还有一些词语如 now（that）、in that、seeing（that）、when，可以引导原因状语从句。例如：

• *Now that* everybody has come，let's begin our conference.

• The higher income tax is harmful *in that* it may discourage people from trying to earn more.

• *Seeing that* it's difficult for me to do that，I'd better go on studying paper writing.

• I won't tell you *when* you won't listen.（既然你不想听，那我就不告诉你了。）

3.6.4　条件状语从句

条件状语从句在复合句中充当表示条件的状语，分为真实条件从句和虚拟条件从句。条件是指某事完成（状语从句动作）后，其他事情才能发生（主句动作）。常用引导词有：if、unless（if…not）。特殊引导词语有：assuming that、but that、given（that）、in case、if only（if）、on condition（that）、providing/provided（that）、suppose/supposing（that）、so（as）far as、so（as）long as 等。

（1）if

If 引导的条件状语从句表示在某种条件下某事可能发生。此条件既可以实现，主句用一般将来时，从句用一般现在时，也可能不能实现或根本不可能存在，即一种虚拟的条件或假设，从句多用一般过去时或过去完成时。例如：

• *If* you *fail* in the test，you *will let* your parent down.

• We *will stop* going out and running *if* it *rains*.

• *If* I *were* you，I *would invite* her to the party.

• I *would have arrived* much earlier *if* I *had not been caught* in the traffic.

（2）unless

unless 引导的条件状语从句，意思是"如果不；除非"，相当于 if…not…。例如：

• You *will fail to* arrive there in time *unless* you *start* earlier.（You *will fail to* arrive there in time *if* you *don't start* earlier.）

• *Unless* it *snows*，the athletic meeting *will be held* on schedule.（*If* it *doesn't snow*，the athletic meeting *will be held* on schedule.）

（3）on condition（that）

on condition（that）用于强调某人采取行动前必须事先认可的条件，意为"如果；只要；条件是"等，常位于主句之后，表示从句是主句事件发生的前提或唯一条件。例如：

• I can tell you the truth *on condition that* you promise to keep a secret.

• You can go swimming *on condition*（*that*）you don't go too far from the river bank.

（4）providing/provided（that）

providing/provided（that）引导的条件状语从句，意为"只要"（as long as）、"如果"（if）等，从句表示一种假设条件。这类从句置于主句前或后均可。例如：

• *Providing/provided* you promise not to tell anyone else，I'll explain the secret.

• I will sign the contract *providing/provided* they offer more favorable terms.

• Mrs. Li won't be against him in the meeting *providing/provided* that he asks for her advice in advance.

（5）suppose/supposing（that）

suppose/supposing 引导的条件状语从句，表示一种假设条件，意为“假若；如果”等，主句通常为疑问句。suppose 引导条件状语从句必须置于主句之前，但 supposing 引导条件状语从句可置于主句前或后。例如：

- *Suppose* you were given a chance to study in Japan, would you accept?
- *Supposing* it rains, shall we continue the field operation in the village?
- What would you do then, *supposing* something should go wrong?

有些句子虽未出现条件引导词，却隐含着条件关系，常用 but for（若非，要不是）、without 等引出一个表示条件的介词结构，条件应是虚拟的或与事实相反的假设，但不属从句。例如：

- *But for* the strong wind and rain, we should have a pleasant journey.
- *But for* your help, I should not have finished studying for my doctor's degree in time.
- *Without air*, there would be no living things on the earth.

3.6.5 目的状语从句

目的状语从句在复合句中充当表示目的的状语，补充说明主句中谓语动作发生的目的。常用引导词语有：that、so that、in order that 等。特殊引导词有：lest、in case、for fear that、in the hope that、on（for）the purpose that、so much so that、to the end that 等。此类从句谓语常含 may、might、can、could、should、will、would 等情态动词，表示未来行为或可能性。

（1）that, so that

that、so that 引导的目的状语从句，相当于 in order that，往往带有情态动词，表示的目的往往非常明确（还可以引导结果状语从句）。例如：

- Please say it louder（*so*）*that* everyone can hear you.
- She worked hard *so that* everything would be ready by 5 o'clock.
- The boss asked his secretary to prepare the letters *so that*（*in order that*）he could sign them.

（2）in order（that），to the end that, on（for）the purpose that

这几个短语引导的目的状语从句，意为“以便；为……起见；为……目的”，从句多用 may/shall/can/will（might/should/could/would）+动词原形（shall 很少用）。例如：

- My wife and I work hard *in order that* our family may be richer.
- He shouted at the top of his voice, *to the end that*（*in order that*）he might be heard.
- Teacher Liang raised his voice *on the purpose that* the students in the back could hear more clearly.

注意 in order that 与 in order to 的区别：in order that + 从句，in order to + 动词原形，后者不是目的状语从句，而是表示目的的短语。

（3）lest

lest 引导的目的状语从句，常用虚拟语气，表示“以防，以免”等意思，其谓语动词多为 should/may/might/would + 动词原形（should 可省略）。例如：

- I am teaching my daughter patiently *lest* she *should* make the same mistake.
- *Lest* the wall（*should*）collapse, they evacuated the building.
- We took our umbrellas and rain shoes *lest* it *may* rain.

（4）in case

in case 引导的目的状语从句，表示“万一；假使；以免；以备；以防”之意。例如：

• Please take more clothes *in case* the weather becomes cold.

• Record the telephone number in your mobile phone *in case* you forget it.

（5） for fear（that）

for fear（that）引导的目的状语从句，表示“以免；生怕；唯恐”，相当于lest，用should + 动词原形，should可以省略。例如：

• He wrote the name down *for fear that*（*lest*） he should forget it.

• He was worried *for fear*（*that*） the child might hurt himself.

3.6.6 让步状语从句

让步状语从句表示退一步说，以便是主句表达的递进一步，属于一种“以退为进”的表达策略，意为“尽管……”或“即使……”，相当于日常生活用语“退一步说……”。常用引导词语有：although、though、as、while；even if、even though；whether，whether…or…；no matter + 疑问词（who、what、where、when等），疑问词 + ever等。

（1） although，though

这两个词同义，表示“虽然、纵然、尽管、即使”，通常可以互换使用，但although比though要正式，语气较重，though显得比较随意。它们都可与yet、still、nevertheless连用，但不能与but连用。although引导的让步状语从句多位于主句前，though引导的位于主句前后均可。例如：

• *Although/Though* she was very old，she（still） kept on studying and exercising.

• I insisted on writing books *although* my friend Mr. Zheng warned me not to.

• I will remain firm *though* I must face many real problems.

• *Though* I believe it，yet I must consider it again.

（2） as（though）

as（though）引导的让步状语从句，表示“虽然……但是；纵使”，而且必须以部分倒装的形式出现，倒装的部分可以是表语、状语或动词原形，名词提至句首时不加冠词。例如：

• Small *as*（*though*） atoms are，they are made up of still smaller units.（尽管原子很小，但它们由更小的单位构成。）（表语）

• Teacher *as* he is，he likes writing very much.（teacher前不加冠词）（表语）

• Hard *as*（*though*） he works，he makes little progress.（状语）

• Object *as*（*though*） you may，I'll do it.（动词原形）（纵使你反对，我也要去。）

（3） while

while引导的让步状语从句，表示“虽然”，一般位于句首。例如：

• *While* she likes the colour，she doesn't like the shape.

• *While* I am unlucky these years，I firmly believe that I will be successful in the future.

（4） even if（even though）

even if（even though）引导的让步状语从句，表示“即使”。例如：

• We won't be discouraged *even if*（*even though*） we fail one hundred times.

• *Even if* you saw him pick up the money，you can't be sure he was a thief.

1） even if引导的从句的内容往往是假设性的，相当于“即使、纵然、就算、哪怕”，有时还用于虚拟语气，这时与单独使用if比较接近。例如：

• *Even if* I had the money，I wouldn't buy the useless.

• She would have married him *even if* he had been penniless.

2）even though 引导的从句的内容往往是真实的，主要引出不利于主句所表意思的信息，相当于“尽管、虽然”，这时与 though 或 although 的意思比较接近，通常可以互换。例如：

• He went out *even though* it was raining.

• *Even though* I didn't know anybody at the party, I had a nice time.

• *Even though* (*though*, *although*) I felt sorry for her, I was secretly pleased that she was having difficulties.

3）even if 与 even though 有时可以不加区别地混用。例如：

• *Even if* (*Even though*) she laughs at him, she likes him.

• We thoroughly understand each other, *even if* (*even though*) we don't always agree.

（5）whether

whether 引导的让步状语从句，表示“不论；不管”，旨在说明正反两方面的可能性都不会影响主句的意向或结果，语气比较强烈，主句的内容得到加强。例如：

• You'll have to attend training class *whether* you're free or busy.

• *Whether* you believe it or not, it's really true.

• *Whether or not* they win this battle, they won't win the war.

（6）whether…or…

whether…or…引导的让步状语从句，表示“不论是否；不管是……还是……”。例如：

• *Whether* it's a friendly rivalry *or* a fight to the death, the end result should be the same.（但不管是友好竞争还是生死决斗，结果应该是相同的。）

• *Whether* I succeed *or* fail, I will keep on writing.

（7）no matter + 疑问词，疑问词 + ever

这两类引导词都可用来引导让步状语从句，表示“无论；不管”，通常可以互换。前者中的疑问词包括 what、who、which、when、how 等；后者包括 whatever、whoever、whichever、whenever、wherever 等，用途很广，还可用来引导主语从句、宾语从句、时间或地点状语从句等。例如：

• *No matter what* (*Whatever*) happened, you would not mind.

• *No matter who* (*Whoever*) you are, you must keep the law.

• When anyone does something for you, *no matter how* small and *no matter whether* he's a superior *or* servant, it's proper to say “Thank you”.

• *Whoever* comes will be welcome.（Whoever 引导主语从句）

• I'll eat *whatever* you give me.（whatever 引导宾语从句）（其中 whatever 不能替换为 no matter what）

注意：despite、in spite of 也可表示让步，但二者后面只能接名词或动名词，不能构成从句，因此不能引导让步状语从句，但能构成状语短语，表示让步之意。

3.6.7 比较状语从句

比较状语从句在复合句中起副词作用做状语，用来进行对比、比照。此类从句和大多数状语从句不同，不是用来修饰动词，而是修饰 as、so、less、more 等副词或比较级形容词。常用引导词有：as、than；特殊引导词语有：as…as…, not so/as…as…, the more…the more…, no…more than, not A so much as B, according as（取决于），in proportion as（与……成正比）。

（1）同级比较

同级比较是指比较事物双方无论是否相

似，都是在同一等级上。常用句式：as/so + 原级 + as（as/so 是副词，后一个 as 是连词）。否定句式：no more than（最多和……一样，只不过）；not more than（不多于）；no less than（不亚于，至少和……一样）；not less than（不少于）。例如：

• My hometown is *as/so beautiful as* yours.

• I have *no more than* ten RMB Yuan left in my pocket.（有钱少的含义）

• I have *not more than* ten RMB Yuan left in my pocket.（没有钱多钱少的含义）

• She has got *no less* presents *than* she did last time.（有收到礼物多少的含义）

• She has got *not less* presents *than* she did last time.（没有礼物多少的含义）

（2）优级比较

优级比较是指比较事物双方中一方比另一方更优越。常用句式：比较级 + than；the 比较级 + the 比较级。例如：

• My room is *bigger than* yours.

• Man developed *earlier than* people thought.（人类的出现比人们所想的要早。）

• I usually walk *more slowly than* my daughter does.

• *The more* I can do for this community, *the happier* I will be.

（3）差级比较

差级比较是指比较事物双方中一方比另一方要差。常用句式：less + 原级 + than。否定句式：not so + 原级 + as。例如：

• This kind of machine is *less expensive than* that one.

• The activity this year is *not so interesting as* the one last year.

一些特殊引导词也可以引导比较状语从句。例如：

• You may go or stay, *according as* you decide.

• People are happy *in proportion as* they are virtuous.（人越是有德行就越幸福。）

3.6.8　方式状语从句

方式状语从句在复合句中用来表述一个人的行为或者做某事的方式。常用引导词语有：as，（just）as，（just）as…so…，as if（as though），by，with，（in）the way（that）。例如：

（1）as，（just）as…so…

as 和（just）as…so…表示“就像；正如”之意。此类从句通常位于主句后，但在（just）as…so…结构中位于句首，这时 as 从句有比喻义，多用于正式文体。例如：

• Always do to the others *as* you would be done by（the others）.（永远用你想让别人待你的方式待别人，类似于“己所不欲，勿施于人”。）

• *As* water is to fish, *so* air is to man.（水之于鱼就像空气之于人。）

• *Just as* we sweep our rooms, *so* we should sweep backward ideas from our minds.

（2）as if，as though

as if 和 as though 表示“仿佛（好像、犹如）……似的”之意，常用在 act、appear、be、behave、feel、look、seem、smell、sound、taste 及其他描写行为举止的动词之后，多用虚拟语气，表示与事实相反；有时也用陈述语气，表示所说情况是事实或实现的可能性较大。也可用来引导分词短语、不定式短语、形容词或介词短语。例如：

• I completely ignore these facts *as if*（*as though*）they never existed.（虚拟语气）

• When the spaceship leaves the earth at tremendous speed, the astronauts feel *as if* they are being crushed against the spaceship floor.（陈述语气）

• She stared at me *as if seeing me* for the first time.（引导分词短语）

● Mr. Li cleared his throat *as if to say something*.（引导不定式短语）

● The waves dashed on the rocks *as if in anger*.（引导介词短语）

（3）（in）the way（that）

（in）the way（that）表示“按照；像”的意思。例如：

● Please do this thing *the way* it does.

● We didn't do it *the way* they do now.

3.6.9 结果状语从句

结果状语从句在复合句中常位于主句之后，补充说明主句谓语动作发生的结果。结果状语从句中通常不用情态助动词，但must、can、could除外。常用引导词语有：so that、so ... that、such ... that、so much so that。such是形容词，修饰名词或名词词组，so是副词，只能修饰形容词或副词。

（1）so that

so that既可引导结果状语从句（表示结果的意思很明确），又可引导目的状语从句（表示目的意思很明确）。例如：

● He worried too much *so that* he wanted to see a psychologist.（结果状语从句）

● Jason is badly ill *so that* he has to rest for several days.（结果状语从句）

● I came to the office early *so that* I *could* have a breakfast in staff restaurant.（目的状语从句）

（2）so...that（so + adj./adv. + that；so + adj.（+ a/an）+ n. + that）

表示“如此……以致”之意，so是副词，修饰其后的形容词或副词，说明程度。例如；

1）so + adj. + that

● Her classmates were *so moved* by the sight *that* they began to cry.

● The wind was *so strong that* I could hardly move forward.

● Shicha Lake in Beijing is *so beautiful* a place *that* I often visit it.

2）so + adv. + that

● The ball struck him *so hard that* he nearly fell into the water.

● I ran *so fast that* I caught the early bus and came to the meeting room firstly.

3）so + adj. + a/an + 单数名词 + that

● It was *so hot a day that* I wanted to go swimming or to stay at home.

● There is *so rapid an increase* in population *that* a food shortage is caused in that country.

4）so + many/few（+ 复数名词）+ that

● There are *so many picture-story books* in the books building *that* the girl won't leave.

● I have so few good friends that I often feel lonely.

5）so + much/little（+ 单数不可数名词）+ that

● There is *so much* contradictory *advice* about exercising *that* I become confused.

● He gave me *so little time that* it was impossible for me to finish the work on time.

so与表示数量的代词many、few、much、little等连用已经形成一些固定搭配，对于这种搭配，不能用such替代其中的so。

（3）such ... that（such（+ a/an）（+ adj.）+ n. + that）

也表示“如此……以致”之意，such后用名词，名词前面可带形容词。例如：

● It is *such nice weather that* I would like to go to the beach.

● Tom is *such a fine teacher that* we all hold him in great respect.

● Shicha Lake in Beijing is *such a beautiful place* that I often visit it.

（4）so much so that

so much so that用于形容词或副词之后，

表示“达到如此程度以至于”之意。例如：

● Mr. Zhang is very *ill*, *so much so that* he could not go to work these days.

● I *long* to visit the famous professor Bai, *so much so that* I dream about her every night.

结果状语从句的主句、从句的主语一致时，可用结果状语 so（adj./adv.）as to 取代从句，或省去从句的主语，从句的谓语变成非谓语，主句的 so...换一种形式陈述（so 是中性程度副词，从句是肯定句时，so 可换为 enough，是否定句时，so 可换为 too）。例如：

● How could you be *so stupid that* you believe him?（结果状语从句）

● How could you be *so stupid as to* believe him?（结果状语）

● The light was *so weak that* it made a good photo hard to take.（结果状语从句）

● The light was *so weak*, *so as to* make a good photo hard to take.（结果状语）

● My daughter is *so courageous that* she can go to school alone.（结果状语从句）

● My daughter is *courageous enough* to go to school alone.（上句的省略式，简单句）

● The man is *so old that* he can't go to school.（结果状语从句）

● The man is *too old to* go to school.（上句的省略式，简单句）

3.7　倒装句与强调句

倒装就是主谓颠倒，即谓语置于主语之前。主语通常位于谓语之前，但有时由于结构或修辞（如为了强调）需要，而将谓语全部或部分移到主语前面，这样就形成倒装句。强调是通过某种方式有意对句中某个部分特别或着重提出，即将句子某个成分甚至整个句子用某种手段加以突出，以更好地表达意愿或情感。

3.7.1　完全倒装

完全倒装（full inversion）是将句子整个谓语放在主语之前或置于句首，常见的有以下两种情况。

（1）副词、介词短语类

1）表示地点的副词（如 here、there）置于句首，而且主语是名词（不是代词）。句型为：There/Here + 谓语 + 主语。其中谓语动词常见的有：be、go、come、exist、lie、follow、remain 等，用一般现在时。例如：

● *There goes* the last bus.

● *Here is* the address of Hotel of Fragrant Hills, Beijing.

● *Here you* are.（代词做主语，不倒装）

2）表示时间的副词（如 now、then 等）、运动方向的副词（如 in、out、up、down、away、ahead 等），以及表示地点的介词短语（如 beyond…，behind…）置于句首，而且主语是名词（不是代词）。句型为：副词或介词短语 + 谓语 + 主语。其中谓语动词常见的有：come、exist、fall、follow、go、lie、live、run、remain、rush、stand 等，用一般现在时或一般过去时。例如：

● *Now comes* your turn to decide the work discipline!

● *Then came* the order to start the key.

● *Out rushed* a missile from under the bomber.

● *Up went* the arrow into the air.

● *Ahead stood* a small boy and a beautiful girl.

● *Beyond the river lived* an old fisherman.

● *Behind* the river lived *Mrs. Qin* for a long time.（人名做主语，倒装）

● *Behind* the river *they* lived for a long time.（代词做主语，不倒装）

（2）表语类

为了保持句子平衡或强调表语部分等，

将做表语的形容词、现在分词、过去分词、介词短语、such 置于句首。句型为：形容词/现在分词/过去分词/介词短语/such + be + 主语。例如：

- *Happy are* those who are contented.（知足者常乐）（形容词）
- *Growing all over the mountain are* wild flowers.（现在分词）
- *Seated on the ground are* a group of young boys and girls.（过去分词）
- *Inside the parcel was* nine books written by myself.（介词短语）
- *Such was* her bad behavior.（*Such was* what she did badly.）（such）

3.7.2 部分倒装

部分倒装（partial inversion）也称半倒装，是将谓语的一部分如助动词、情态动词或系动词放在主语之前或置于句首。如果句子的谓语没有助动词或情态动词，则需要添加助动词如 do、does、did，并将其置于主语之前。部分倒装常见的有下列三种情况。

（1）only 修饰副词、介词短语或状语从句且放在句首

这种情况下谓语部分无助动词时，需要添加助动词；only 修饰状语从句时，从句不倒装，主句倒装；only 修饰主语时，句子不倒装。例如：

- *Only then did* people realize the importance of science and technology.（did 为添加的助动词）
- *Only in this way can* I achieve a great success.
- *Only after the war did* he learn the sad news.（did 为添加的助动词）
- *Only when you can find peace in your heart will* you keep good relationships with others.（only 修饰状语从句，从句不倒装，主句倒装）
- *Only I can* write such good writing books.（only 修饰主语，句子不倒装）

（2）否定词及表否定意义的介词短语置于句首

这种否定词有 nor、not、never、nowhere、hardly、little、rarely、seldom、scarcely 等，表否定意义的介词短语有 at no time、under/in no circumstances、in no case、in no way、by no means、on no condition、on no account、not on any account、not on consideration 等。例如：

- *Not a single mistake did* he make.
- *Never*（*before*）*have* I read such a interesting book.
- *Hardly do I think* it possible to finish the report writing before this week.
- *At no time did* they actually break the rules of the game. It was unfair to punish them.

（3）使用固定句型

1）so + do（be/助动词/情态动词）+ 主语（也是如此）。同义的不倒装结构为 it is/was the same with…或 so it is/was with…。例如：

- You are a scientific editor, *so am I.*（倒装）
- They love making friends, *so do those* with disabilities.（倒装）
- They love making friends, *it is the same with* those with disabilities.（不倒装）
- They love making friends, *so it is with* those with disabilities.（不倒装）

如果仅对前面的内容做肯定或附和，这时 so 相当于 indeed，则不用倒装。例如：

- A：I was afraid.（I 指 A）B：So was I.（I 指 B，意为 I was afraid, too.）
- A：I was afraid.（I 指 A）B：So you were.（you 也指 A，意为 Indeed you were afraid.）

2）neither/nor + do（be/助动词/情态

动词）+ 主语（也不这样）。同义的不倒装结构为 it is/was the same with…或 so it is/was with…。例如：

• You aren't a scientific editor, *neither/nor am I.*（倒装）

• She can't drive a car, *neither/nor can* he.（倒装）

• She can't drive a car, *it is the same with* he.（不倒装）

• She can't drive a car, *so it is with* he.（不倒装）

注意：neither/nor 不可用 so…not 替代，但可用 not…either 改写。例如：

• Jack has never been abroad. *Neither/Nor has* Lily.（倒装）

• Jack has never been abroad. *So hasn't* Lily.（so…not）（错误）

• Jack has never been abroad. Lily has *never/not* been abroad *either.*（not…either）

3）so + adj. /adv…. that…, such +（a/an +）adj. + n. …that…（如此……以至于）。so、such 后面的句子倒装，而 that 从句不倒装。例如：

• *So clearly does* she speak English that she can always make herself understood.

• *Such a valuable book does* Dr. Liang write that we all want to buy and read it.

4）neither…nor…（也不）。neither 和 nor 这两个词均为否定词，其后句子均需倒装。例如：

• *Neither do* I know it, *nor do* I care about it.

5）not only…but also…（不仅……而且）。也可写成 not only…but…或 not only…but…as well。not only 放在句首，后接句子时用倒装，而 but（also）后面的不倒装。例如：

• *Not only will* help be given to people to find jobs, but also medical treatment will be provided for people who need it.

6）not until…（直到……才）。not until 引导名词短语时，主句倒装；引导从句时，从句不倒装，主句倒装。例如：

• *Not until* my 30th birthday *could I* got married.

• *Not until* my wife returned *did we* make a decision to visit Guilin on this festival.

• *Not until the boy* left his home *did he* begin to know how important the family was for him.

3.7.3　形式倒装

形式倒装（formal inversion）语法上称为前置，是将强调的内容提至句首而其实主谓并不倒装。这种倒装常见的有以下四种类型。

（1）感叹句

对名词或名词短语感叹用 what 引出，对形容词或副词感叹用 how 引出。例如“I have done a sigrificant thing.”的感叹句为：

• *What a sigrificant thing* I have done!

• *How sigrificant* the thing I have done is!

（2）the more…, the more…句型

两个 the more…连用，前一个相当于条件状语从句，后一个相当于主句。more 代表形容词或副词的比较级。例如：

• *The more* you listen to English, *the easier* it becomes.（If you listen to English more, it will becomes easier.）

• *The harder* you work, *the greater* progress you will make.（If you work harder, you will make greater progress.）

（3）however、whatever 引导的让步状语从句

此类倒装的常用句型是 however + adj. /adv. 和 whatever + n. 。例如：

• *However severe* the situation may be, I must face it optimistically.

• *Whatever reasons* you have, you should carry out your promise.

（4）as、though 引导的让步状语从句

此类倒装主要有表语倒装、谓语动词和状语倒装。注意：单数名词或形容词最高级做表语时不再需要用冠词。行为动词前置，从句主语后用 may、might、can、could、will、would 等情态动词时，若没有情态动词，则加 do、does 或 did。行为动词是及物动词时，其宾语也随其一同提至前面。例如：

• *Tired as/though* I was, I still went on with my writing.（表语倒装）

• *Child as/though* my daughter is, she knows a lot of history and stories.（表语倒装）

• *Youngest as/though* she is in our class, she speaks English best.（表语倒装）

• *Search as* they *would* here and there, they could find nothing in the room.（谓语动词倒装）

• *Hard as* Tom worked he could not catch up with his colleagues.（状语倒装）

• *Strange as/though* it might sound, Li's idea was accepted by most young scholars.（状语倒装）

• *Change your mind as/though* you *do*, you will get no help from others.（行为动词及其宾语也随其一同提至前面）

3.7.4 强调句

强调句是为了表达某种意愿或情感而使用的一种具有某种修辞效果的特别句式，可通过某种方式对句中某个部分进行强调而实现。下面介绍强调句的常见类别。

（1）it 强调句

it 强调句有以下四种，其中陈述句为基本句型，其他是在基本句型基础上转化成的。被强调部分指人时用 who，指物时用 that，但 that 也可指人，美式英语中还常用 which 指物。

1）It is（was）+ 被强调部分（主语、宾语或状语）+ who（that）…（陈述句）。例如：

• It was *on December 24, 2018* that I finished writing the book.

2）Is（was）it + 被强调部分（主语、宾语或状语）+ who（that）…?（一般疑问句）。例如：

• Was it *on December 24, 2018* that I finished writing the book?

3）被强调部分（疑问代词或疑问副词）+ is/was + it + that/who…?（特殊疑问句）。例如：

• *When* was it that I finished writing the book?

4）It is（was）not until + 被强调部分 + that…（not…until…句型）。例如：

• It was *not until December 24, 2018 that* I finished writing the book.

• It was *not until her son came back* that she went to bed.（She didn't go to bed until/till her son came back.）

强调句中用 until 不用 till，普通句中 till 和 until 可通用。强调句中的 It is（was）not…已经是否定句了，其后 that 从句不再用否定句。

另外，被强调部分是原因状语从句时，用 because 引导，不用 since、as 或 why。例如：

• It was *because the water had risen* a lot that the army could not cross the river.

（2）助动词强调句

句型结构为“助动词（do、does、did）+ 动词原形”，重在强调谓语动词。例如：

• Some people *do believe* that nuclear power poses a threat to the world peace.

• Many people *do think* that H7N9 virulent mutants detected in chickens pose an increased threat to humans.

（3）形容词强调句

句型结构为“形容词（如 very、only、single、such）+ 名词或形容词”，重在强调名词或其修饰性成分。例如：

- This is the *very question* that deserves careful analysis.（强调做主语的名词 question）
- How dare you buy *such expensive* jewels?（强调做宾语的名词 jewels 的定语 expensive）

（4）副词强调句

句型结构为“副词（如 ever、never、very、just、badly、highly、really）+ 谓语动词”，重在强调谓语动词。例如：

- The train *just arrived* at 8 AM.
- I *really don't know* what to do next.

（5）介词短语强调句

句型结构为“插入语式介词短语（如 in the world、on earth、at all）”，疑问句常见。例如：

- Where *in the world* could she be?
- What *on earth* are you trying to say?

（6）双重否定句

双重否定句是通过两次否定来表示肯定，语气比肯定句更为强烈，句型结构为“否定词 + 具有否定意义的词”，否定词常见的有 no、not、never、nobody、few、little 等（多为否定副词），具有否定意义的词包括介词（如 without）、形容词（如 impossible、uncommon、unquestioning）、动词或短语（如 deny、discomfort、disagree、fail to）。例如：

- It is conflict and *not unquestioning* agreement that keeps freedom alive.（使自由保持活力的是冲突而不是绝对的一致。）
- Taking part-time jobs is *never without* drawbacks.（采取兼职工作从来都不是没有缺点。）

（7）what 引导的主语从句

这是由关系代词 what 引导的做句子主语的名词性从句，具有一定的强调作用。例如：

- *What really matters* is a close cooperation between you and me.
- *What is really important to a country* is that the development of its S&T and the talent training must be continuous.

（8）比较状语从句

比较状语从句用形容词和副词的原级、比较级和最高级三种形态，表示事物的等级差别，有比较作用，可以产生强调作用。例如：

- Nothing is *more imperative than* to learn from the failure.

（9）感叹句

感叹句表示喜怒哀乐等强烈的情感，能产生强调作用。例如：

- *How interesting* your story is!（*What an interesting story* it is!）

（10）强调式重复句

强调式重复句通过有意连用几个词语而产生强调作用。例如：

Why! Why! Why! My car in the small courtyard is lost!

（11）倒装句

倒装句使用非常规的句法结构，颠倒了句子成分之间的正常顺序，比如将谓语全部或部分放在主语前，具有明显的强调作用，可以说凡倒装句都可表示强调。例如：

- *Little do* people take into account the seriousness of this problem.
- *On the desk* were some new books of mine and a new computer.

（12）if 从句

if 从句往往是正话反说，反话正说，也有强调作用。

1）if 从句 + ①或 + ②或 + ③（① I don't know who/what, etc. does/is/has/can, etc. ② nobody does/is/has/can, etc. ③ everybody does/is/has/can，etc.）。例如：

● *If Jason can't do it*，I don't know who can.（强调只有Jason能做）

● *If Jason can't do it*，nobody can.（强调只有Jason能做）

● *If Jason can do it*，everybody can.（强调Jason不能做）

2）if从句+it be主句。把所强调的内容放在it be之后，其他放在if从句中。例如：

● If anyone knew the truth，it was *Newton*.

（13）用破折号、黑体（加粗）字体

在破折号后面给出强调语句，或使用某种字体形式如黑体（加粗）、斜体和放大或缩小等，均可达到强调的作用。例如：

● It's because of *hard work—twenty four years of hard work*.（破折号后面的部分强调艰苦工作的长期性）

● I began the writing of academic paper writing **in 2005**.（黑体或粗体强调时间）

3.7.5 强调句与定语、状语从句比较

（1）与定语从句比较

1）强调句中的it没有实际意义，it be与that可同时省略；定语从句中的it是主语，it be与that不可同时省略。

2）强调句中be的时态与后面句子的时态相一致；定语从句中主句谓语动词be的时态由主句的时间属性确定。

3）强调句中的that不能省略，即使前面的名词表事物，也不能将that换成which；定语从句中的that做宾语时可省略，而且先行词表事物时可用which代替。

4）对于定语从句，当it be后面的时间、地点名词做从句的主语、宾语或表语时，引导词可用that/which，做其他成分如状语时，引导词用when/which。

例如：

● It is a Huawei mobile phone（*that/which*）*I bought yesterday*.（that/which引导的定语从句）

● It was the Huawei mobile phone *that I bought yesterday*.（强调句，强调the Huawei mobile phone）

● It is a day *when we celebrate the liberation victory*.（when引导的定语从句）

● It was on that day *that we celebrated the liberation victory*.（强调句，强调on that day）

（2）与状语从句比较

1）句首的It，在强调句中不做任何成分，也无实际意义，在状语从句中是主语。

2）强调句中的that及句子前面的It be可以省略，而状语从句中的不能省略。

3）强调句中的that不能用其他词代替，而状语从句的引导词还可以是when/where。

例如：

● *It* is such an interesting story *that we all like it very much*.（结果状语从句，句首的It做主语）

● It is such an interesting story *that we all like very much*.（强调句，强调such an interesting story）

● It is such an interesting story *as we all like very much*.（定语从句，引导词as指代an interesting story）

● It was already morning *when I woke up*.（时间状语从句，引导词是when）

● It was the next morning *that I woke up*.（强调句，强调the next morning）

3.8 否定句

否定句是一种基本句型，用来对所述情况进行否认。它的形式繁杂、种类多样，就所用词语类型（词汇）来看，有副词、名

词、代词、动词、形容词、介词、连词，以及由这些词构成的短语和固定短语等，从否定方式（意义）上看，有全部否定、部分否定、双重否定等。

3.8.1　否定词语的类型

否定句的形成需要使用一定的词语，包括词、短语和固定短语，大体上分为以下七类。

（1）副词型

副词型是指否定副词或相当于副词的短语，按否定的程度，分为完全否定词（全否定词）和部分否定词（半否定词）。全否定词有 no、not、nor、none、never、neither、nobody、nothing、nowhere 等，半否定词有 barely、few、hardly、little、otherwise、rarely、scarcely、seldom 等。例如：

- I have *no*（*not* any）friends in New York.
- They have *never* been to Germany.
- *Hardly* had the train arrived at the station when we ran towards the exit looking for her.
- *Little* did I think that he could be back alive.（我没有想到他竟能活着回来。）

否定副词直接表示否定，表意是直接的，称为真正意义的否定词；而其他否定类词语的意义是暗含的，称为含否定意义的词语，如以下所述的各类。

（2）名词型

名词型是指含否定意义的名词，如 absence、denial、exclusion、failure、ignorance、lack、loss、negation、refusal、reluctance、shortage、want、zero（乌有）等。例如：

- George was also in *ignorance* of their whereabouts.
- My *reluctance* to think of an non-editorial career caught up with me.（我不愿考虑编辑以外的职业给我带来的不幸。）
- *Shortage* of man power is the chief cause of the delay at the publishing house.
- Our hopes were reduced to *zero*.（我们的希望化为乌有。）

（3）动词型

动词型是指含否定意义的动词或动词短语，如动词有 avoid、absent、baffle、cease、deny、defy、decline、doubt、escape（被……忘掉）、exclude、fail、forbid、ignore、lack、miss、negate、neglect、overlook、resist、reject、refuse、wonder 等，动词短语有 deprive…of、differ from、give up、get rid of、keep off、keep/prevent/stop/protect … from、keep sb. in the dark、live up to（不辜负）、lose sight of、make light of、prefer … to、refrain from、turn a deaf ear to 等。例如：

- Clinton *absented* himself from a meeting on some pretext.
- The president's name *escaped* me for a moment.（我一时记不起那位首长的名字了。）
- Sickness *deprived* him *of* the pleasure of meeting his mother.
- Why should you *prevent* me *from* attending the conference?

（4）形容词型

形容词型是指含否定意义的形容词或形容词短语，如 absent from、alien to、blind to、clear of/from、deaf to、different、different from、far from、free of/from、foreign to、ignorant of、new to、reluctant、safe from、short of 等。例如：

- We should be *free of*（*from*）arrogance and rashness.（我们应该不骄不躁。）
- They are *safe from* suspicion.（他们没有嫌疑。）
- I am *new to* recent year's situation.

（5）介词型

介词型是指含否定意义的介词或介词短

语，如介词有 above、against、beneath（不值得），before、beside、below、beyond、but、except、off、past、without 等，介词短语有 as opposed to（而非）、at a loss、at one's wits' end（智穷计尽，不知所措）、at the end of one's rope/at the end of one's row（山穷水尽，智穷力竭）、beneath one's notice（不屑一顾）、instead of、in spite of、in the dark（不知道）、in vain、out of the question、off one's guard（不提防）、out of the swim（不合潮流）、regardless of 等。例如：

- This problem is *beneath* (*our*) *notice*.
- The youngster of the present day are *beyond* our common comprehension.
- Her stupidity is *past* all belief.（她的愚蠢简直不可思议。）
- The doctor is *at a loss* for an explanation of her illness.
- I am completely *in the dark* concerning the development of the journal.
- I have so much writing task to finish that a holiday for me this year is *out of the question*.

（6）连词型

连词型是指含否定意义的连接词（含短语），如 before、unless、rather than 等。例如：

- I left my office *before* I finished the work.
- I *rather than* she am to blame. /I, *rather than* she, am to blame.（是我而不是她该受批评。）
- It is I, *rather than* she, that am to blame.（与上句同义）

（7）固定短语型

固定短语型是指含否定意义的习惯用语，如 aside from、anything but/by no means/in no way/in no case/on no account/under no circumstances、instead of、least of all、let alone/to say nothing/still less、not at all、not in the least、other than、rather than、still less 等。例如：

- The little bridge is *anything but* safe.（那座小桥一点也不安全。）
- The truth is quite *other than* what they think.（事实与他们所想的不一样。）
- Jake can't run a hundred yards, *still less* a mile.

3.8.2 否定方式的类型

以下为否定句的否定方式的八个类别。

（1）全部否定

全部否定就是完全否定，表示强烈的否定意味，使用完全否定词，表示“都不是”“全不是”“绝对不”“没有”“远不（非）”“一点也不”“根本不”“无论如何也没有”之意。例如：

- *No* reference books are allowed to be taken out of the library.
- *None* of us understand the theory of the machine.
- The book of *Writing and Submission of SCI Papers* is *nowhere* to be bought now.
- *Nothing* is difficult if you put your heart into it.（世上无难事，只怕有心人。）

（2）部分否定

部分否定就是不完全否定，表示一般的否定意味，使用部分否定词，表示“不都是”“不全是”“并非都是”“并不全是”“不总是”“不是每个都是”之意，主要有以下三种情况。

1）由总括性（全体）代词或副词与否定词 not 连用来构成，不能按字面理解为“所有……都不是”。这样的代词或副词有 absolutely、all、always、altogether、anything、anywhere、both、completely、each、every、everybody（everyone）、anyone（anybody）、 entirely、everywhere、everything、many、necessarily、often、quite、wholly 等。例如：

• *All* is *not* gold that glitters. /*Not all* is gold that glitters.（闪光的未必都是金子。）

• The rich are *not always* happy.

• *Both* knowledge and techniques of modern medicine *can not* bring him back to life.

• I *can't* remember *entirely* the tale happened here last year.

• *Every* instrument here is *not* good.

• Money is *not everything*.（金钱并非万能。）

2）all the time 的否定式表示"并非一直……；未必老是……"。例如：

• A foolish man *does't* make a mistake *all the time*, while a wise man *can't* be right *all the time*.

3）not…and…的否定式中，and 连接前后两个部分（被连接的两个部分可以是表语、谓语动词、状语、定语或宾语），被否定的通常是后一部分。例如：

• That book is *not* interesting *and valuable*.（那部书有趣但无价值。）

• *Don't* drink *and drive*.（不要酒后驾车。）

• The persons who *don't* say *and do* are unwelcome.（只说不做的人不受欢迎。）

（3）双重否定

双重否定即否定之否定，是指一个句子里出现了两个否定词（否定词连用），表示肯定，强调或委婉意味浓厚，语气很重。例如：

• I am *not unwilling* to follow your action.

• *Nothing* is *impossible* to a willing heart.（世上无难事，只怕有心人。/有志者，事竟成。）

• It's *never too* late *to* learn.（活到老，学到老。）

• There is *no* rule that has *no* exception.（凡是规则总有例外。）

• We *can't think* of China *without* thinking the Great Wall.

（4）几乎否定

几乎否定也叫半否定，由半否定词与谓语动词肯定式连用构成。其否定意义不像完全否定那么绝对，而是留有余地，语气也较弱。例如：

• This technique is *rarely used* in such an engineering project.

• I really *know little* about the history of Ancient Europe.

• These boys *will leave* college, though, with *little* education and *few* social skills.

（5）转移否定

转移否定是指在某些主从复合句中，否定在形式上出现在主句的谓语部分，而在意义上转移到了从句部分。

1）否定词加在表示判断、看法的动词如 assume、anticipate、believe、expect、fancy、figure、guess、imagine、reckon、suppose、think 等的前面，否定意义转移到其后的宾语从句。例如：

• The leader *didn't believe* that he *was quite suitable* for this position.（领导认为他不太胜任这个职位。）

• I *don't think* she *can finish* the work in time.（我认为她不能及时完成这项工作。）

2）否定词加在不及物动词如 happen、prove、pretend 等的前面，否定意义转移到其后的动词不定式。例如：

• Her answer *didn't prove to be* right.（她的回答证明不对。）

• I *didn't pretend to know* her.（我假装不认识她。）

3）否定词加在系动词如 appear、feel、seem、sound 或类似于系动词的动词短语如 look like 等的前面，否定意义转移到这些词的后面部分。例如：

• She *doesn't seem to be* happy these days.（她似乎这几天不太高兴。）

• It *doesn't look like* it's *going to stop* snowing today.（好像今天不会停止下雪。）

4）在原因状语从句如 not…because 句型中，有时形式上否定主句的谓语，实际上否定意义可能转移到从句，是否发生这种转移还需按上下文语境和逻辑关系来确定。例如：

• I *don't do* editing *because* editing is easy for me.（我不是因为编辑容易才当编辑。）

• The boy *didn't leave* home *because* he was afraid of his mother.（①那个男孩离开家并不是因为怕母亲；②那个男孩没有离开家，因为怕母亲。前者有否定转移，后者没有）

5）当 view、wish、opinion 等词做否定句主句的表语时，否定意义转移到从句。例如：

• It *is not my wish* that you should *break* your word.（希望你不要违背诺言。）

• It *is not my opinion* that she is the best person for the new position.（我认为她并不是担任此新职位的最佳人选。）

（6）持续否定

持续否定也叫追加否定，是指对前面的否定句作进一步的补充否定。例如：

• The old man *can't* fall asleep，*neither* at night *nor* in the daytime.

• They have *no* experience，*not* to speak of scholarship.

（7）特指否定

特指否定也称局部否定，是指对句子的非谓语成分进行否定。（常规的否定是对句子的谓语动词进行否定。）例如：

• The computer is working，but *not properly*.（否定状语副词 properly）

• I told Mrs. Li *not to worry* about that thing.（否定直接宾语 to worry）

（8）多余否定

多余否定是指在句中有可有可无的否定词，即去掉该否定词丝毫不影响句义。例如：

• I wonder if I *cannot* finish the task before Saturday.（cannot 可用 can 替换）

• I'm not sure if she *won't* come over for a visit.（won't 可用 will 替换）

3.8.3 否定句的固定结构

以下为否定句的五个固定结构类别。

（1）too、to、not 等组合结构

此类结构常见的有：too…to…，not…enough to…，too…not to（双重否定），not/never too … to …（怎么……也不过分，越……越……），but/only too … to …（非常）。例如：

• The man is *too* much of a coward *to* do such a thing.（那个男人太懦弱，不敢做这样的事。）

• The little girl is *too* young *to* go to school.（这个小女孩年龄太小了，不能上学。）

• The little girl is *not* old *enough to* go to school.（这个小女孩没到上学的年龄。）

• The woman is *too* brilliant *not to* be elected.（这个女人太优秀了，不能不当选。）

• You *can't* be *too* careful *to* drive a car.（你开车越小心越好。）

• He is *but too* glad *to* join our team.（他非常高兴加入我们的团队。）

（2）more、than 组合结构

此类结构常见的有：more A than B（B 不如 A），more than + 含 can 的从句（不能）。例如：

• He is *more* brave *than* wise.（他有勇无谋。）

• My gratitude for your help is *more than* I *can* express.（我对你的帮助的感激之情是

不能用语言表达的。)

(3) than 后不定式比较级结构

这是一种 than 后带动词不定式的比较级结构，即比较级 + than + 不定式（不至于做）。例如：

• You should know *better than to have* a smoke in the room.（你应懂得不该在房间里抽烟。)

• You were *wiser than to have done* such a thing.（你不至于愚蠢得竟然做出这样的事情。)

(4) before、unless 等引导的状语从句

以连词 before、unless 等引导的状语从句也能表示否定的意思。例如：

• They slipped out *before* the opening ceremony started.（开幕式还没开始，他们就开溜了。)

• *Unless* you follow our advice, you'll hit a brick wall.

(5) 虚拟语气

虚拟语气所陈述的是一个条件，往往不是事实，甚至与事实完全相反，其中蕴含着对现实或事实的否定意味。例如：

• *But for your help*, I should have failed.

• *If only you had worked with greater care*, you would made a greater progress.

3.9 主谓一致

一致是指句中各个成分之间或词语之间在人称、数、性等方面保持一定的语法上的协调关系。主谓一致是其中最主要的一种，是指句子的谓语动词在人称和数上必须同主语一致，分为语法一致（形式一致）、意义一致（语义一致）和就近一致。

3.9.1 语法一致

语法一致也称形式一致，是主语和谓语动词在语法形式上保持一致关系。主语是单数时，谓语动词用单数形式；主语是复数时，谓语动词用复数形式。

1) 代词 anyone、anybody、another、anything、each、either、everyone、everybody、nothing、no one、one（指 you、we）、someone、somebody、something、the other 等做主语时，谓语动词用第三人称单数形式。例如：

• If *anyone calls*, tell him I'll be back in a moment.

• It is necessary that *one does his* best to increase *his* personal quality.

• *Someone was* unhappy with the decision of *his* leader.

2) "each/every + 单数可数名词" 做主语时，谓语动词用单数形式。

3) "one of + 代词/复数名词" 做主语时，谓语动词用单数形式；"one of + 代词/复数名词" 后面跟有定语从句时，该从句的谓语动词用复数形式；"the one/the only one of + 代词/复数名词" 后面跟有定语从句时，该从句的谓语动词用单数形式。

4) 主语后面跟有 along with、as well as、besides、but、except、including、in addition to、like、rather than、together with、with 等起连接作用的词语及相应的单数或复数名词时，谓语动词应与主语的人称、数相一致。

5) 非谓语动词（动名词、不定式）做主语时，谓语动词用第三人称单数形式。

6) 名词性从句及 "疑问代词/副词 + 不定式" 做主语时，谓语动词用第三人称单数形式。

7) "one and a half + 复数名词" 做主语时，谓语动词用第三人称单数形式。

8) "more than one + 单数名词"，意义上表示许多，但谓语动词用单数形式。

9) 数词做主语时，谓语动词用单数形式；表示加减乘除时，谓语动词既可用单数形式，也可用复数形式。

10）连词 and 连接几个并列名词或代词做主语时，谓语动词用复数形式。

11）由 all、both、few、many、several 等修饰的可数名词的复数形式做主语时，谓语动词用复数形式。

12）倒装句中，谓语动词应根据后面的主语在人称、数上保持一致。

13）代词在数、性、人称等方面彼此保持一致。

3.9.2　意义一致

意义一致是主语和谓语动词在语义上保持一致关系。主语在形式上是单数而意义上是复数时，谓语动词用复数形式；主语在形式上是复数而意义上是单数时，谓语动词用单数形式。

1）集体名词（如 class、data、family、group、public、team）做主语时，作为一个整体看待时，谓语动词用单数形式；作为整体中的各个成员看待时，谓语动词用复数形式。例如：

- The *data shows* that a chemical change takes place.
- The *data* in Fig. 4 *have* five segments of turbulent information.

2）表总称意义的名词（如 people、clothes、police 等）做主语时，谓语动词用复数形式。

3）maths（mathematics）、news、physics、politics、works（工厂）等名词做主语时，谓语动词用单数形式。

4）“分数 + of + 复数名词”做主语时，谓语动词用复数形式，但“分数 + of + 不可数名词/单数可数名词”做主语时，谓语动词用单数形式。

5）由复数名词构成的国家、书刊、作品等的名称做主语时，谓语动词用单数形式。例如：

- *The United Nations was* formed in 1945.

6）表示“金钱、时间、距离、重量、质量”等的名词做主语时，谓语动词用单数形式。例如：

- *Three years is* enough for us to finish this program.

7）“the + 形容词（或由分词演变而来的形容词）”做主语，指某类人时，谓语动词用复数形式，指具体某人时，谓语动词用单数形式。

8）the rest、part、half、most、all of 加复数名词做主语时，谓语动词用复数形式，加不可数名词做主语时，谓语动词用单数形式。

9）all 做主语时，如果指代人或可数名词含有复数意义，谓语动词用复数形式，如果指代抽象性事物，谓语动词用单数形式。

10）“none of + 不可数名词”做主语时，谓语动词用单数形式；“none of + 代词/复数名词”做主语时，谓语动词用单数形式或复数形式。例如：

- *None of us knows/know* how to work out the problem.

11）trousers、shoes、boots、glasses（眼镜）、socks、gloves 等做名词主语时，谓语动词用复数形式；若这些名词被 this、a pair of 修饰，则谓语动词用单数形式。

12）“a number of + 复数名词”做主语时，谓语动词用复数形式；“the number of + 复数名词”做主语时，谓语动词用单数形式。例如：

- There are *a large number of workers* in the factory, who *are* from America and Germany.
- *The number of the researchers* in that university *is* 500.

13）and 连接两个并列名词做主语，指不同的人、事物或概念时谓语动词用复数形式，指同一人、事物或概念时谓语动词用单数形式。它连接几个被 no、each、every、many a 修饰的单数可数名词做主语时，谓语动词用单数形式。例如：

- Many *boys and girls were* waiting for the arrival of their teacher.
- *Boy and girl is* a good combination.
- *Every boy and girl was* waiting for the arrival of their teacher.

14）Chinese、Japanese、deer、sheep、means 等单、复数同形的名词做主语时，它们自身在句中的内容决定其谓语动词的单、复数形式。

3.9.3 就近一致

就近一致是谓语动词与其最靠近的那个主语的单、复数形式一致，即谓语动词是用单数还是复数形式取决于与该谓语动词最靠近的主语。

1）当主语由 or，either…or…，neither…nor…，not…but…，not only…but（also）…连接时，谓语动词通常和最临近的那个主语一致。例如：

- *Either* he *or you are* to do the work.
- *Neither* the boy *nor his parents know* the calculation method of this problem.
- *Not only* I *but also Mr. Zhang wants* to do that.

2）当 there be、here be 结构后面有并列主语时，谓语和最临近的那个一致。例如：

- *There is a computer* and many journals on the table in the house.
- *There are many journals* and a computer on the table in the house.
- *Here is a computer*，a few pens and some paper for you.

3）做主语的名词或代词后接 with、together with、along with、as well as 等短语时，谓语动词一般和这种名词或代词一致，这些短语前后可用或可不用逗号。例如：

- *The book with* two CDs *is* going to be published next month.
- *The girl*，*as well as* the boys and other girls，*has* learned to drive a car.

第 4 章 英语科技论文词汇

任何一门语言都离不开词汇，写文章就是从词汇库中选择词语来组合、造句，因此了解、掌握词汇的意义和用法对写成、写好文章非常重要。本章试图讲述英语科技论文词汇的意义及用法，但因为英语词汇库实在太大，词汇成员太多，因此开始想用“常用词汇”作为标题的中心词，试图用“常用”加以限定，以将论文中常出现的词语作为讲述对象，缩小词汇范围。然而即便如此，本章写作仍然困难重重，因为常用词汇是一个模糊概念，其成员有多少、是什么实在没法说清楚，如果不对词汇加以归类，恐怕就更难下笔。为此，笔者很是纠结，试图对自认为的部分常用词汇做某种归类，按类别展开叙述。但归类又可以从不同的角度来进行，如果这个角度把握不好，也难以奏效。本章从三个方面对词汇进行归类：英语表达一般词语；英语论文写作常用词汇（分为论文结构功能词汇和论文特别语义词汇）；常见易出错混淆词汇。其实，这种分类只是为了写作方便罢了。

4.1 英语表达一般词汇

4.1.1 说明术语

说明术语是对术语进行修饰、限制，或给予某种解释、说明，或给出某种定义、描述，所用相关词语就是说明术语类词语。这种词语常见的有以下五类。

1. 定义

定义是对事物给予概念性解释，或对事物给出某种定义，这种定义通常是准确、严格或较为准确、严格的，所用相关词语就是定义类词语，常见的有 definition 和 define。

（1）definition

使用 definition 的常见表达有以下几种：

- *Definition*：Psychology is the scientific study of the mind.
- Psychology is by *definition* the scientific study of the mind.
- As mass can vary in several circumstances this *definition* is unsound and misleading.

definition 按顺序（序号）出现时，正确的形式应该是“Definition 1...”“Definition 2 ...”，而不是“One/First definition ...”“Two/Second definition...”。

（2）define

使用 define 的常见表达有以下几种：

- A is *defined* as B.
- A is *defined* as follows：
- A is *defined* to be B.
- A is *defined* by B.
- A is *defined* in terms of B.
- We *define* A as B.

2. 称为

称为是将一事物（A）称作另一事物（B），所用相关词语就是称为类词语。这类表达常见的有以下几种：

- A is *called/termed* B.
- A is *known as* B.
- A is *referred to as* B.
- A is *spoken of as* B.
- The former is *referred to as*..., and the latter *as*...（前者称为……，后者称为……）
- ...is often *shortened to*...（通常简称为……）

称为类词语有时可带修饰语，例如：

- … are *collectively called* …（统称为……）
- … is *variously called* …（有各种叫法……）

3. 意指

意指是指出概念或事物的意思或所指，所用相关词语就是意指类词语。这类表达常见的有以下几种：

- By A one *refers to*…
- By A we *mean/imply that*…
- A is *defined to mean*…

4. 无明确定义

无明确定义是对事物做模糊或不太明确的定义，所用相关词语就是无明确定义类词语。这类表达常见的有以下几种：

- There is *no universally accepted definition to* term A.（对 A 没有普遍公认的定义。）
- A clear definition of A is *difficult to give*.（目前很难给 A 下一个清晰的定义。）

5. 其他定义

除了以上几种说明术语类词语表达外，还有以下几种较为常见的表达：

- So called A is…（所谓 A 是……）
- A, as the name implies/suggests, is…（A，顾名思义就是……）
- A is called …, or alternatively, …（A 称为……，也可称为……）
- A is derived from B, meaning…（A 是由 B 派生而来的，意思是……）

4.1.2 表达时间

表达时间是表达有起点和终点的一段时间，或时间里的某一点，所用相关词语即为表达时间类词语。这种词语常见的有以下五类。

1. 公元

B. C. —Before Christ（公元前），A. D. —Anno Domini（公元后，简称公元）。

- from 500 B. C. to A. D. 600（从公元前 500 年到公元 600 年）（A. D. 写在年份前，B. C. 写在年份后）
- around 1000 B. C.（公元前 1000 年左右）
- as long as 800 B. C.（早在公元前 800 年）
- as long as A. D. 800（早在公元 800 年）

2. 世纪

- in the late eighteenth century（18 世纪后期）
- within the first 20 years of the twentieth century（20 世纪的头 20 年里）
- early in the present century（本世纪初期）
- before the end of this century（在本世纪末之前）
- the second half of the twentieth century（20 世纪后半叶）

3. 年份

- nearly five decades ago（约 50 年前）
- in the past few decades（在过去的几十年里）
- during the past 100 years（在过去的 100 年里）
- for ten years after…（在……之后的 10 年里）
- beginning in the 1990s（从 20 世纪 90 年代开始）
- up to 2020（直到 2020 年）
- dating back to…/go back to…（追溯到）

4. 抽象时间

- for the next few years（今后若干年）
- in the near future（近期内）
- shortly after（在……之后不久）
- at the time of writing this paper（在写本论文时）
- for the time being（目前）

- in today's（当今的）
- at the moment（在……瞬间）
- for the moment（暂时）
- by now（此时）
- at one time（一度）
- the era of（时代）

5. 具体时间

- September 29，2019 或 29 September 2019（2019 年 9 月 29 日）
- Aug. 8，2008 或 8 Aug. 2008（2008 年 8 月 8 日）
- in 2018；in May，2018（有年或月，但无具体日期，用 in）
- on May 10，2020（有具体日期，用 on）
- at 8：30 a. m.；at 16：12 p. m.（有几点几分，用 at）（a. m. 是 ante meridiem 的缩写，p. m. 是 post meridiem 的缩写）

4.1.3　ly 式副词

英语中的副词有很多，其中 ly 式副词较为常见，见表 4-1。

表 4-1　常用的 ly 式副词

副词	中文	副词	中文	副词	中文
actually	实际上	equivalently	等效地	physically	从物理概念上
admittedly	诚然，公认地	essentially	从本质上讲	presumably	可能的情况是
alternatively	换言之	experimentally	从实验上讲	probably	或许
analytically	分析地、解析地	finally	最后	rarely	很少
architecturally	从结构上看	functionally	从功能上讲	roughly	粗略地讲
certainly	当然，直观地讲	graphically	从图形上讲	simultaneously	与此同时
characteristically	从特性上讲	ideally	理想情况下	specifically	具体地讲
chronologically	从年代上看	incidentally	顺便指出	strictly	严格地说
classically	按经典理论	inductively	归纳起来	symbolically	用符号表示的话
clearly	很清楚	logically	从逻辑上讲	technically	从技术上讲
conceptually	从概念上讲	mainly	从主要方面讲	theoretically	从理论上讲
consequently	结果、因此	mathematically	从数学上讲	ultimately	归根结底
conventionally	按照惯例	operationally	从功能上看	unfortunately	遗憾的是
correspondingly	与之对应的是	ordinarily	在正常情况下	usually	通常
curiously	奇怪的是	personally	就个人而言	visually	从视觉效果上看

这类副词通常可以与副词 most/more 组合，如 most notably（最值得注意的是）、more/most importantly（更/最重要的是）、more generally（更普遍的情况）、more precisely/accurately（更确切地说）、more briefly（更简练地讲），等等。

4.1.4　表达分类

表达分类是根据事物的特点分别归类而进行表达，所用相关词语即为表达分类类词语。这种词语常见的有以下几类。

1. 分类常用词

- sort，type，class，variety，category

2. 说明分类及分类的依据

- There are… categories of…（有……类）
- … can be divided into…（能被分解为）
- … can be categorized into…（能被划

分为）

• … can fit into…（能归入）

• … classify…as…types（将……分为）

• … can be categorized according to…（可按……分类）

• … is categorized with respect to…（就……分类）

3. 分类短语

• … fall into/within…（属于）

• … there are many varieties…, from…to…（有许多种，从……到……）

• … be excluded from…（不包含在……中）

4.1.5　表达程度

表达程度是根据事物变化达到的状况而进行表达，所用的相关词语就是表达程度类词语。这种词语常见的有以下九类。

1. 频度程度

• rarely（罕有）、occasionally（偶尔有）、sometimes（有时）、often（常常）、usually（通常）、always（总是）。（程度递增）

2. 不确定程度

• might（可能）、may（可以）、could（能）、can（能）、should（能、该）、ought to（应该）、would（愿意）、will（愿意）、must（必须）。（程度递增）

• maybe（也许）、probably（可能）、certainly（一定）、definitely（肯定）。（程度递增）

3. 定性程度

• bad（劣）、poor（差）、fair 和 ordinary（中）、good（良）、excellent（优）。（好坏程度递增）

• nil（无、零）、slight（轻微的）、moderate（中等的）、severe（严重的）、extreme（极度的）。（程度递增）

4. 定量程度

• negligible（微不足道的）、considerable（相当多的）、substantial（大量的）、appreciable（可观的）、material（重大的）。（程度递增）

• few、not many、a few、some（少量的），a moderate number of、a certain number of（适量的），a large number of、a great many、a lot of、plenty of（大量的）。

5. 某种程度

• to a small extent（在很小程度上）、to what extent（在多大程度上）、to different extent（在不同程度上）、to some extent（在某种程度上）、to a great extent（在很大程度上）、to the extent practicable（在可行程度上）。

6. 强调重要性

• particularly important（特别重要的）、be essential to（对……至关重要）、by far the most important（显然最重要的）、come to play（开始起作用）、play a major role（起主要作用）、play a dominant role（起主导作用）、deserve a special mention（值得专门提到、注意）。

7. 难易程度

• somewhat easier（…）than（多少比……容易一些）、even easier（更容易一些）、be readily to（容易地）、relatively simple（相对简单）、far simpler（简单得多）、rather sophisticated（相当微妙的）、be self-evident（不言自明的）。

8. 数量程度

• no more than（不多于、仅仅）、a few more（比……更多一点）、in part/partially（部分地）、throughout（整个）、be not sufficient to determine（不足以确定）、more or less（或多或少）。

9. 知名程度

• well-known（众所周知）、celebrated

（著名的）、best known（著名的）、most outstanding（最突出的）、remarkable achievement（显赫的成就）、most significant contributions（最重要的贡献）、epoch-making（划时代的）。

4.1.6　of 短语

of 短语分为基本类和复合类，基本类是指含 of 的固定短语，复合类是指在前者的基础上通过在 of 一词的前置词之前加限定语所形成的短语，限定语通常为形容词或形容词性短语。这种短语常见的有以下八类。

1. a number of

a number of 为基本类，复合类是在 number 之前加上某种限定词，如 certain、considerable、great、limited、large、moderate、small、tremendous 等。此类短语常跟复数名词或复数名词短语，做主语时，谓语用复数形式。例如：

- First, we identified *a number of* energy-driven protein/nucleic acid *chaperones* or remodeling *complexes* as stress granule components (Figure 3C).
- For example, only *a limited number of* single-molecule *approaches* have been applied so far to the study of condensins.
- While many men present with localized and curable disease, *a large number of deaths* are driven by the development of metastatic prostate cancer and low curative options.
- We and several other laboratories have also developed SRM assays for human proteins, typically for *a small number of proteins* in the context of a specific biological study.
- The prevalence of alternative splicing and post-translational modifications increase the complexity to *an as-yet-unknown number of* different *proteoforms*, and the annotation of protein-coding regions and experimental evidence for their validity are still being refined.

2. a amount of

a amount of 为基本类，使用不多，复合类多一些。复合类就是在 amount 之前加上某种限定词，如 certain、considerable、fixed、great、large、limited、low、moderate、small、tremendous 等。此类短语后常跟名词或名词短语，名词常为不可数的，有时也可为可数的，做主语时，谓语用单数还是复数，取决于所跟名词或名词短语的表意是单数还是复数。例如：

- They reflected in political entity, political values, political norms, political ability and added political identity, *a amount of ideology*, *security*, political *aegis* of the country.（句中 ideology、security、aegis 为不可数名词）
- In recent years, PSO algorithm has spawned *a considerable amount of researches* on force rules, population topologies and algorithm applications, many of which have made significant improvements on the original PSO algorithm.（句中 research 是可数名词）

In recent years, PSO algorithm has spawned *a considerable amount of research* on force rules, population topologies and algorithm applications, many of which has made significant improvements on the original PSO algorithm.（句中 research 是不可数名词）

- In addition, the worn surfaces of the disks are covered with *a large amount of* loose wear *debris* and magnetorheological (MR) *particles* along the sliding direction, except in the ridges of plastic deformation.（句中 particles 为不可数名词，意思是微粒、粒子）
- Figure 4B shows the ORR performances of the N-GNS powder catalysts, in which the currents are divided by the geometric electrode

surface area (0.283 cm^2), with *a loading amount of* 0.02 mg.（句中 mg 是质量的单位，形式上是单数，相当于一个不可数名词，它与可数名词 milligram 的意思相同）

• When expressed in HEK293 cells both fusions constitutively internalize resulting in *a relatively low amount of* β_2V_2R-βarr1/2 *fusions* being present at the cell membrane.（句中 fusions 是可数名词）

3. a group of

a group of 为基本类，复合类是在 group 之前加上某种限定词，如 certain、interesting、large、small 等。此类短语常跟复数名词或复数名词短语，做主语时，谓语根据主语含义取单数或复数形式。例如：

• The outdoor running trial was conducted with *a group of* 12 *subjects* in which 6 were instructed to drink 150 ml water every 5 min and 6 did not drink water throughout the trial.

• The two-dimensional space-filling of the growing crowns of the tallest individuals relegates *a group of* losing, slow-growing *individuals* to the understory.

• N-glycosylated proteins are secreted or located on the cell surface and constitute *a clinically interesting group of proteins* that is investigated for biomarkers and drug targets.

对于 a subgroup of 结构，subgroup 是一个在 group 前加前缀 sub 而构成的词，表示小群、亚组或隶属的小组织，因此该结构不属于以上所说的复合类。例如：

• *A subgroup of* γ-proteobacteria, which includes Escherichia coli, has a distantly related complex, MukBEF.

4. a lot of

a lot of 常跟名词或名词短语，名词为可数时用复数，不可数时用单数，当其做主语时，谓语与主语保持一致。其同义短语是 lots of、a great deal of、a number of。例如：

• Many scholars have carried out *a lot of research* (*es*) on the mechanism of rolling mill vibration.

• *A lot of research* (*es*) has (have) been carried out on the mechanism of rolling mill vibration by many scholars.

• The study displayed that *a lot of* tumor suppressor *genes* and *protooncogenes* participated in the modulation of cell apoptosis.

• At this time the exhaust gas is very cloudy and contains *a lot of CO and HC*, which will affect the accuracy of the test results.

• Afterwards, *a lot of* new *versions*, which could describe information exchange mechanism among particles, were presented, e.g., double-layer sub-population topology, scale-free network topology, WATTA & STROGATZ small world network topology, which could prevent the particles from getting trapped in local optimum and improve the global search ability.

5. a (the) majority of

a (the) majority of 为基本类，复合类就是在 majority 之前加上限定词，如 vast。此类短语常跟名词或名词短语，名词为单数或复数，做主语时，谓语与主语保持一致。例如：

• A similar bioinformatics analysis of previously known yeast P body proteins revealed that, in contrast to stress granules, *a majority of* P body *proteins* have mRNA-binding activity (73%), which suggests that mRNA-protein interactions may contribute more to P body structure than stress granule structure (Table S1).

• As a consequence of these factors, *the majority of* protein *research* is still focused on the same relatively small subset of proteins for which assays are readily available.

• A similar set of proteins was identified

from both cross-linked and non-cross-linked cells (*p* value of overlap between the two lists $<10^{-99}$), with *the vast majority of proteins* seen in the cross-linked sample also detected in the non-cross-linked experiment but with insufficient peptides to make our statistical cut-offs.

6. a variety of

a variety of 常跟名词或名词短语，名词为单数或复数，做主语时，谓语与主语保持一致。例如：

- Once an assay is developed, it can be applied perpetually in *a variety of studies*.
- Scientists and scholars have endless opportunities to study *a variety of life* in the habitat.
- This implies that stress granules will not be uniform assemblies but that the action of different remodeling complexes will create *a variety of states*, with potentially different biological outcomes.
- Accordingly, here we set out to test this hypothesis by using *a variety of* cellular, biochemical, and biophysical *approaches*.

7. a set of

a set of 为基本类，复合类是在 set 之前加上某种限定词或前缀，如 comprehensive、defined、redundant 等。此类短语常跟名词或名词短语，名词常为可数、复数名词，做主语时，谓语取单数形式。例如：

- However, SRM requires defining a priori *a set of* target *proteins*, optimal *peptides*, and assay *parameters*. (SRM—selected reaction monitoring)
- The pair of mass to charge (*m/z*) values that is isolated in Q1 and Q3 is referred to as a transition, and *a set of transitions* that determine a peptide signature is, in combination with the peptide's elution time, termed SRM assay.
- *A comprehensive set of data* available from these approaches should provide the firm base for mathematical modeling and simulation of chromosome architecture and dynamics.
- Together with the recent finding that mitotic chromatids can be reconstituted in vitro with *a defined set of* purified *components*, it would be fair to conclude that condensins are essentially autonomous, being able to fulfill their basic actions without specialized loading factors.
- First, the dense set of protein-protein interactions could form the basis of *a redundant set of interactions* that drive stress granule or core assembly.

对于 a subset of 结构，因为 subset 是一个在 set 前加前缀 sub 而构成的词，表示子集，因此该结构不属于以上所说的复合类。例如：

- Similar to some yeast granule components, *a subset of* the remaining four *proteins* might localize to stress granules but might not be sufficiently enriched over the cytoplasm to show foci by immune fluorescence (IF) in vivo.

8. either of, any of

either of 和 any of 跟复数名词，谓语取单数形式，但当 any of 表达“其中某些”的意思时，谓语取复数形式。例如：

- *Either of* the *houses* is big enough for all the experimental facilities to be placed in.
- *Any of algorithms* developed recently is an important outcome of this new project in 2020.
- This application could not have been realized by *either of* the *technologies* (flexible sensors and silicon integrated circuits) alone, owing to their respective inherent limitations.
- Notably, no new nuclease-induced off-

target sites were induced by SpCas9-HF1 with *either of* the two *sgRNAs*.

- In this setup, no BRET response was observed following agonist stimulation of *any of* the *receptors* indicating that the BRET detected between barr and both Gαs and Gγ2 reflects molecular proximity consistent with the formation of megaplexes (Figures 4B-4E).
- Genomic sites were excluded from analysis on the basis of overlap with background genomic breakpoint regions detected in *any of* four oligo-only control *samples*, overlap with previously identified Cas9-sgRNA independent breakpoints in human U2OS cells, or as neighbouring genomic window consolidation artefacts likely due to extensive end-resection around breakpoints (Supplementary Table 4).

4.1.7 表达因果

表达因果是根据事物的原因和结果之间的关系进行表达，所用相关词语就是表达因果类词语。这种词语常见的如下：

- as（与 since 基本相同，但没有 since 正式）
- because（原因构成句子的主要部分，因果关系强烈）
- for（原因作补充说明）
- since（原因已知，因果关系较弱）
- so that（以致）
- hence（因而）
- as a result of（由于……的结果）
- partly/mainly because of（部分/主要原因是）
- result from，arise from（因……而……）
- due to（由于）
- owing to，in view of（鉴于）
- on account of，seeing that（由于，因为）

- to be (even) more specific（为了具体起见）
- to be precise（为确切起见）
- to be effective（为有效起见）
- to be complete（为完整起见）
- to be on the safe side（为了保险起见）
- for simplicity（为简单起见）
- for brevity（为简洁起见）
- for clarity（为明确起见）
- for convenience in the present discussion（为目前讨论方便起见）
- for the reason of generality（为一般化起见）
- to be valid for（为对……适用）
- to justify（为证明……是正确的）
- lead to（导致、引出）
- result in（造成、得到）

4.1.8 表达有无

表达有无是根据事物的存在或不存在、拥有什么或未拥有什么而进行表达，所用相关词语就是表达有无类词语。这种词语常见的有以下三类。

1. “有”的基本用语

- have（有，表达“有”最常见的词）
- there is/are/was/were（有，不指出“谁”有）
- there must be（必须有）
- there can be（可以有）
- there may be（可能有）
- there will be（将会有）
- there have been（已经有）

2. “有”的其他用语

- of（在……中有）
- in the presence of（在有……的情况下）
- among ... be（在……中有）
- with［有了……（条件/了解）］
- available（可得到的，可利用的，现

有的）

3. “无”的表达用语

● without（无、没有、缺少）

● in the absence of（无、没有、缺少）

4.1.9 表达关系

表达关系是根据不同事物之间或相同事物的不同方面之间的关系而进行表达，所用相关词语就是表达关系类词语。这种词语常见的有以下五类。

1. 关联

● be related to，be associated with（与……有关的）

● be dependent upon each other（彼此有关的）

● be independent of（与……无关的）

2. 比较

● be compared with，in comparison with（与……相比）

● comparison of ... and ...（与……比较）

● comparison of ... between ... and ...（在……和……之间的比较）

● there is no comparison between ... and ...（在……和……之间不做比较）

3. 对比

● as against（与……相对照）

● in contrast to（与……形成对比、对照、相反）

● versus（对、对于、与……相对或相比）

● with respect to（关于、至于）

4. 类似

● be similar to（与……相似）

● be same as，be identical in（与……相同）

● in an analogous manner（类似的）

● in an analogy to（与……类比）

5. 差别

● be different/distinguished from ... in ...；unlike（不同于）

● contrast with（与……不同之处在于）

● apparent inconsistency（明显的不一致）

● essential difference between ... and ...（在……和……之间的本质区别）

4.1.10 表达手段

表达手段是对为达到某种目的所采取的具体方法而进行表达，所用相关词语就是表达手段类词语。这种词语常见的有以下两类。

1. 基本词汇

● by，with，by means of，with the help of（介词或介词短语）

● approach，choice，method，use，usage，way（名词）

● use，employ［（在实践中）运用］，utilize［利用（有益的方面）］（动词）

2. 其他词语

● from（根据、按照）

● in terms of（根据、按照）

● in accordance with，subject to（根据、按照，使遭受、使服从于、使隶属于）

● according to（根据、按照）

● be based on，on the basis of（基于）

● unless otherwise（除非）

● besides，in addition to（除了……还有……）

● except，apart from（除了……没有……）

4.2 论文结构功能词汇

4.2.1 阐述意义、价值

阐述课题研究的目的、意义、价值（重要性、必要性）等，多用于论文的Introduction和Discussion部分中。

associate with

v. 联合；与……联系在一起。

• Moreover, there may be significant uncertainties *associated with* flood peaks of smaller annual exceedance probabilities due to rating curve extrapolation errors.

• The genes *associated with* these GO terms predominantly encoded expansins, extensins, and xyloglucan endotransglucosylase/hydrolase (XTH) cell-wall-modifying proteins (Table S3).

• The initial increase of the magnetostriction should be *associated with* the partial solution of Tb in the matrix, the maximum value at $x = 0.1$ should be attributed to the < 100 > preferred orientation, and the decrease of the magneto striction is correlated with the appearance of the second phase along the grain boundary.

attention

n. 注意力；关心。

• Recently, with the development of artifical intelligence, researchers have a great deal of *attention* to the solution of nonlinear problems in mechanical properties of alloys.

• Therefore, domestic and abroad researchers pay *attention* to improve the microstructures and mechanical properties of the materials with high pressure treatment.

• Quantum plasmas have drawn reasonable *attention* of many researchers in the last few decades, due to their importance in astrophysical environments, in high intense laser solid density experiments, in ultra cold plasmas, in micro-plasmas and in micro electronics devices.

characterize/characterise

vt. 描绘……的特性；具有……的特征。

过去式 characterized/characterised；过去分词 characterized/characterised；现在分词 characterizing/characterising。

• As a result, *characterizing* mechanisms of acquired resistance in the recurrent disease setting is of great clinical relevance.

• A neural network is *characterised* by its architecture that represents the pattern of connection weights, and the activation function (Fausett, 1994).

• It is seen from Fig. 1a that the surface of NiTi is *characterized* by many intersecting grooves formed during the mechanical polishing.

(be) composed of

由……组成（构成）。

• The experimental material *is composed of* oxygen-free copper (99.97% Cu) and chromium powder (99.9% Cr).

• After preliminary investigations, it was concluded this month that the cause is not seasonal influenza but rather an entirely new hybrid strain *composed of* pig, bird and human viruses.

consist of

由……构成（组成）。

• As shown in Fig. 2, the fully annealed pure copper mainly *consists of* regular and equiaxed grains with a mean size of ~30 μm.

• Immunotherapy *consists of* various forms of vaccination strategies to elicit robust immune response to tumor antigens, and blockade of immune checkpoints to reinstitute host antitumor immunity.

• Develop the generic application *consisting of* the views (user interfaces) and the links between the views, and define the application flow.

evidence

n. 证据，证明；迹象。*vt.* 证明。

过去式 evidenced；过去分词 evidenced；

现在分词 evidencing。

• *Evidence* of obesity as a risk factor for colon, breast, endometrial, and renal cell cancer is increasing, but different results are obtained related to gastric cancer (GC).

• Thus even though residents are confident in their abilities (Table 1), there is little *evidence* that this perception is correct.

• Their authenticity is their strength and without them we could not recall and *evidence* the past.

interest

n. 兴趣，爱好；趣味。*vt.* 使……感兴趣；引起……的关心；使……参与。

• Our further research regarding this phenomenon did not reveal any published data, but indeed represents an area of *interest*.

• Recently, wearable sweat sensors have been developed, with which a variety of biosensors have been used to measure analytes of *interest* (Supplementary Table 1).

• The authors declare competing financial *interests*: details are available in the online version of the paper. Readers are welcome to comment on the online version of the paper.

• What *interests* them most is how China have made such significant advances in just a few decades.

involve

vt. 包含；牵涉；潜心于。

过去式 involved；过去分词 involved；现在分词 involving。

• The reaction is likely to *involve* a composite array of DNA-protein interactions.

• All of these processes *involve* forces separating chromosomes or chromatids, implicating that condensing II might utilize a similar if not identical molecular mechanism in both interphase and mitosis.

• While some mechanisms of therapy resistance have been elucidated, identification of pathways *involved* in platinum-resistant recurrent OC (ovarian cancer) remains in its infancy and existing studies have investigated a relatively small number of relapsed disease samples.

• We developed assays to target N-glycosylated proteins by specifically selecting peptides that span the N-glycosylated sequence motif for use in studies *involving* N-glycosylation affinity approaches.

participate

vi. 参与，参加；分享。

过去式 participated；过去分词 participated；现在分词 participating。

• All patients agreed to *participate* in the present study and signed the informed consent forms.

• The study displayed that a lot of tumor suppressor genes and protooncogenes *participated* in the modulation of cell apoptosis.

• By taking advantage of their distinctive architecture, condensins actively fold, tether, and manipulate DNA strands, *participating* not only in chromosome assembly and segregation during cell divisions, but also in many aspects of large-scale chromosome organization and regulation.

refer to

参考；涉及；指的是。

• Immune checkpoint molecules *refer to* a group of immune receptors that upon engaged with their ligands transmit an inhibitory signal to suppress effector function.

• We *refer to* these sites of high concentration of RNA or of proteins in stress granules as "cores" and the surrounding less concentrated material as the "hell".

relate

vt. 使……有联系。*vi.* 涉及；符合；与……有某种联系。

过去式 related；过去分词 related；现在分词 relating。

- Describe your major professional accomplishments, and *relate* them to why you think you would be an ideal candidate for the listed position.
- Sweat glucose is reported to be metabolically *related* to blood glucose.
- In the regions with large W_2C phase, the removal of the coating materials *is related to* the crack in the binder phase (which becomes brittle due to the dissolution of W and C atoms or the amorphous state), the fracture of carbides under the action of the abrasive, and the disruption of the interfaces between the carbides and the binder phase.

- AP treatment markedly reduced glomerulosclerosis independent of sex and genetic background (Fig. 4a, b), which correlated with attenuated age-*related* increases in blood urea nitrogen (Fig. 4c), indicating preserved kidney function. (AP—AP20187 (B/B Homodimerizer)：小分子靶向基因合成药物。笔者注)
- Hence a secondary horizontal axis is introduced in Figure 5 *relating* the water body element to day 5 of the survey.

role

n. 角色；任务。

- The Asian monsoon is one of the most active components of the global climate system, and plays a significant *role* in global climate modulation.
- The prospective *role* of circulating miRNA (microRNAs) is less evolved and less clear than that of cfDNA (circulating free DNA).
- The *roles* these therapies play in the cycle of cancer and immunity are displayed in Figure 2.

4.2.2 交代现状、计划

交代领域研究的现状、需要解决的问题或作者的研究计划等，常用于论文的 Introduction 和 Discussion 部分中。

address

vt. 从事；忙于；演说；向……致辞。

- Typically, systems biology *addresses* such questions, but the "big data" approaches, excepting the genetic studies, have not yet yielded spectacular new insights into AD pathogenesis.
- To validate the hypothesis of the local relative density mapping (LRDM) method as well as to *address* the effect of the constant k and the distance r_{ij} on the weight function in Eq. (5), a simply supported beam is considered.
- According to my experience of attending all previous strategic dialogues, I feel that the most important issue to be *addressed* by the SED is to how to look at each other. (SED—strategic economic dialogue. 笔者注)
- The key to *addressing* the problem of imbalance is to sample the data set in such a way to form a balanced training set.
- Let us turn to some of the problems you will be *addressing* during this session.

aim

n. 目的；目标。*vt.* 目的在于；引导；把……对准。*vi.* 打算；对准目标。

- The *aim* of this paper is to analyse the

chaotic behaviour and predictability of a streamflow series employing various techniques.

• The present study was *aimed* at analysing the chaotic nature of streamflow series using different techniques.

• In this context, we *aim* to review and discuss obesity and bariatric surgery and its risk factors related to GC (gastric cancer).

• This study *aims* to investigate two important issues: what are the determinants of public goods investment and what is the government's investment behavior in mountainous areas.

• *Aiming* at the problems of product developing and reusing of case model, structure of case-base on case matching, information and standardization is brought forward.

although

conj. 虽然，尽管。

• *Although* no engine is designed to operate at the resonance frequency of any of its components, it is inescapable that the operation frequency is equal to the resonance frequency of the blade in the process of the engine speed changing or the engine design matter.

• Municipality-level incidence data indicate that reports of suspected microcephaly in Brazil best correlate with ZIKV incidence around week 17 of pregnancy, *although* this correlation does not demonstrate causation.

approach

n. 方法；途径。*vt.* 接近。*vi.* 靠近。

过去式 approached；过去分词 approached；现在分词 approaching。

• Historically, there have been several *approaches* for treating heart failure, but loop diuretics has been at the forefront to alleviate the symptoms and has its own side effects as with any medication use; a lesser known and monitored is metabolic alkalosis.

• We used a phylogenetic molecular clock *approach* to further explore the molecular epidemiology of ZIKV in the Americas.

• The door opens by itself as you *approach* it.

• According to *The New York Times*, the worldwide population of hedge funds is *approaching* 6,000.

• That is until Neptune station reports of unidentified spacecraft *approaching*.

demonstrate

vt. 证明；展示；论证。

过去式 demonstrated；过去分词 demonstrated；现在分词 demonstrating。

• To *demonstrate* the applicability of local relative density mapping (LRDM) method for complex 3D structure, a micro Jet engine bracket [32] is designed and optimized by a continuum topology optimization (SIMP).

• It has also been *demonstrated* that hypofractionated radiotherapy is superior to solitary doses in inducing immune responses.

• *Demonstrating* subgroup-specific actionable molecular biology and drug sensitivities in the research setting will be a crucial step in *demonstrating* the need for such stratification in the clinic.

describe

vt. 描述，形容；描绘。

过去式 described；过去分词 described；现在分词 describing。

• In Chinese philosophy, Yin and Yang *describe* how opposite or contrary forces are actually complementary, interconnected, and interdependent in the natural world, and how they give rise to each other as they interrelate to one another.

• The task of science is only to *describe* how things behave.

• Enhanced GFP (EGFP) disruption experiments, in which cleavage and induction of indels by non-homologous end-joining (NHEJ)-mediated repair within a single integrated EGFP reporter gene leads to loss of cell fluorescence, were performed as previously *described*.

• The new methodology is based on the assumptions that salt inflow is the function of space only (during the survey) and the background conductivity can be *described* by the temporal variations observed at a fixed location.

• The conventional definition of spin current is incomplete and unphysical in *describing* spin transport in systems with spin-orbit coupling.

despite

prep. 尽管，不管。

• As compared to the topology optimization, it can be estimated that local relative density mapping (LRDM) saves 9.8% material *despite* its longer generation time.

• *Despite* intense study of the CRISPR-Cas9 system, one additional piece of the puzzle was missing—a small RNA that would come to be called trans-activating CRISPR RNA (tracrRNA).

discover/discovery

discover—*vt.* 发现；发觉。*vi.* 发现。discovery—*n.* 发现；发觉；被发现的事物。

• Given a large number of peptides *discovered* since the peptide selection for the SRMAtlas was performed, we retrospectively investigated the success of selecting suitable peptides that were observed in the recent Peptide Atlas and were not available in the initial selection database.

• The *discovered* markers should be applied in clinic after being verified and confirmed through a series of analysis, which process requires clear and complete preconditions.

• Although these perovskite structures offer high PCEs, reaching >20% PCE with band gaps of around 1.55 eV, fundamental issues have been *discovered* when attempting to tune their band gaps to the optimum 1.7-to 1.8-eV range. (PCEs—power conversion efficiencies. 笔者注)

• The first step in *discovering* disease markers is to screen the candidate molecules and after the quantitative analysis, they are diagnosed to be real biomarkers according to corresponding changes in certain environmental changes, after agent administration and attack of some diseases.

• This perspective aims to fill in this backstory—the history of ideas and the stories of pioneers—and draw lessons about the remarkable ecosystem underlying scientific *discovery*.

• Once a fact is firmly established, the circuitous path that led to its *discovery* is seen as a distraction.

elucidate/elucidation

elucidate—*vt.* 阐明；说明。elucidation—*n.* 说明；阐明。

• The US government's irresponsiveness didn't help *elucidate* whether this situation was a terrorist-targeting project, an isolated event, or a systemic breakdown.

• We used generalized linear models (GLM) to *elucidate* the relationship between the species richness and elevation.

• Although the full potential of checkpoint inhibitors as single therapy has not yet been fully *elucidated*, with many early and late phase clinical trials still ongoing, what is apparent is that not all patients will benefit from this novel class of drugs.

● The main focus of leukemia research has been the *elucidation* of genetic and epigenetic features of these blasts, and how these changes affect cell proliferation, differentiation, and survival.

emphasize/emphasise

vt. 强调，着重。

过去式 emphasized/emphasised；过去分词 emphasized/emphasised；现在分词 emphasizing/emphasising。

● In contrast to traditional economic models, this model would *emphasize* incremental changes and marginal value over a flexible time-horizon.

● Many of these studies *emphasized* the migration of permafrost "boundaries" based on the relationship between climate warming and permafrost "boundaries", which are naturally continuous, inexact representations of permafrost distribution and permafrost degradation (Yang et al., 2010).

● In this admittedly oversimplified model, the division of labor of the two condensin complexes is *emphasized* although their functions are partially overlapped with each other.

● He did repeatedly *emphasise* the "great revival of the Chinese nation" and the need for China to "stand more firmly and powerfully among all nations".

● We coined the term National Competition Policy both to *emphasise* that we needed policies about where and how competition could be better used, and to *emphasise* that those policies should be national.

● *Emphasizing* balance is the general rule of performance appraisal for enterprises.

establish

vt. 建立；创办；安置。

● Non-transgenic mice from C57BL/6J ATTAC transgenic crosses were used to *establish* the control cohort for confirmation that AP had no effect on health span in the absence of the ATTAC transgene.（C57BL/6J—小鼠类别；ATTAC—INK-ATTAC：转基因小鼠。笔者注）

● To process the large number of peptides, we *established* an assay development pipeline including a robotics platform and multiple commonly used MS instruments duplicated at two geographical sites.（MS—mass spectrometry. 笔者注）

● Although the physiological significance of this activity remains to be fully *established*, it is worth speculating how the equivalent action might induce conformational changes in nucleosomal arrays.

● Much has now been *established* about the immune response to tumors and the associated mechanisms of immune escape.

● SRM has been applied for decades in the pharmaceutical industry to quantify small molecules and evolved recently into an *established* technique in the field of proteomics, due to advanced technology and reproducibility across instrument platforms and laboratories.（SRM—selected reaction monitoring, also named multiple reaction monitoring (MRM). 笔者注）

● *Establishing* a joint lab with the nearby Institute of Genetics and Microbiology at Université Paris-Sud, he setout to use tandem-repeat polymorphisms—which were the workhorse of forensic DNA finger printing in humans—to characterize strains of the bacteria responsible for anthrax and plague.

evaluate

vt. 评价；估价。

过去式 evaluated；过去分词 evaluated；现在分词 evaluating。

• It is most important to *evaluate* the crack initiation vibration stress for blade design, aero engine design, and failure analysis.

• The PABST algorithm *evaluated* sequence constraints and ranked observed and predicted peptides, the highest scoring peptides for each protein were selected for SRM assay development. (SRM—selected reaction monitoring. 笔者注)

• First, the type variation of permafrost stability changes over the past 30 years was quantitatively *evaluated* using the transition matrix method, as shown in Tables 4–6.

• After *evaluating* all your expenses you might find you can add a bit more to your slice of savings.

examine

v. 检查；调查；检测。

过去式 examined；过去分词 examined；现在分词 examining。

• This article *examines* the effect of socioeconomic and natural condition elements on public goods investment using recent village survey data and calculated data based on GIS in two different methods, including statistic analysis and regression analysis.

• Hence, using the SRMAtlas it was possible to efficiently profile the changes in protein abundance along a whole pathway, to confirm the co-regulation of novel putative SREBP target genes and to *examine* the relationship of protein regulation in different pathways.

• The alloy microstructure after end quenching test and the primary and secondary phase precipitation locations of supersaturation solid solution were *examined*.

• This benchmark demonstrates the performance of decision-support systems by *examining* large volumes of data, executing complex queries, and providing answers to business questions.

explore

v. 探索；探测；探险。

过去式 explored；过去分词 explored；现在分词 exploring。

• The objective of this paper is to *explore* the expression of p16 in human colorectal cancer and its clinical significance.

• We used a phylogenetic molecular clock approach to further *explore* the molecular epidemiology of ZIKV in the Americas.

• Zhang spent the next year optimizing the system. He *explored* ways to increase the proportion of Cas9 that went to the nucleus.

• These complexities are *explored* in two papers from the early years of the buyback program.

• Scientists have returned from an expedition *exploring* the peaks off the west coast of Scotland.

• Designing and manufacturing such a machine is really worth *exploring* for our scholars.

find

vt. 发现；认为。

过去式 found；过去分词 found；现在分词 finding。

• None of these planets has an environment like that of the earth, so scientisis do not think they will *find* life on them.

• It is *found* that the prepared sample crystallizes in the orthorhombic FeB-type crystal structure (space group *Pnma*).

• Thus, if a causal link between Asian lineage ZIKV and microcephaly is confirmed, it is possible that putative viral genetic determinants of disease will be *found* among the amino acid changes that occur on the ZIKV phylogeny branches ancestral to the French Polynesian and American ZIKV lineages.

• The *finding* that most eukaryotic species

have two different condensing complexes has raised a number of fundamental questions in chromosome biology.

however

conj. 然而；可是。*adv.* 无论如何；不管怎样。

● In western countries, the morbidity of esophageal carcinoma increases annually, and has become the primary pathological pattern of esophageal carcinoma. *However*, squamous carcinoma is the primary pattern in about 90% of Asian countries.

● Figure 2h illustrates that with the increase of temperature, the uncompensated sensor readouts can lead to substantial overestimation of the actual concentration of the given glucose and lactate solutions; *however*, the temperature compensation allows for accurate and consistent readings.

● The disappointing export figures, *however*, were partly offset by strong investment data released the same day.

identify

vt. 确定；鉴定；识别，辨认出；把……看成一样。*vi.* 确定；认同；一致。

● The literature review was to *identify* factors to be included in the resident survey and key stakeholder interviews.

● In this article, we modeled the service specification of each of the *identified* services in detail.

● From these inputs, losses have been consistently *identified* as one of the key downfalls in Australian flood hydrology.

● Human gait recognition is the process of *identifying* individuals by their walking manners.

● After all, people accept influence from those they *identify* with those who are similar to them or people who they aspire to be like.

investigate/investigation

investigate—*v.* 调查；研究。investigation—*n.* 调查；调查研究。

过去式 investigated；过去分词 investigated；现在分词 investigating。

● To further *investigate* the impact of ZIKV infection during neurogenesis, human iPS-derived brain organoids were exposed to ZIKV and observed for 11 DIV (Fig. 4). (ZIKV—Zika virus, iPS—induced pluripotent stem. 笔者注)

● We further *investigated* the intermediates of the ORR by ex situ post-ORR XPS measurements of the HOPG model, which reflects the steady state surface of the N-HOPG model catalyst under ORR and provides mechanistic information about the active sites.

● For this reason, in this paper, the mechanism of quenching sensitivity was *investigated* through the simulated end quenching test method.

● These discoveries have led to an increasing number of studies *investigating* the utility of CFNA (cell free nucleic acids) for the characterization, monitoring and therapeutic targeting of both hematologic and solid malignancies.

● Exploring mechanistic parallels (and conflicts) between transcription-driven and condensin-mediated conformational changes of chromatin would be an exciting target of future *investigation*.

in contrast

与此相反；比较起来。

● While not considering springback, the maximum and minimum error ratios of the formed orthodontic archwire parameters were 22.46% and

10.23%, respectively. *In contrast*, when considering springback, the maximum and minimum error ratios for the height of the dental arch were 11.35% and 6.13%, respectively, demonstrating clear improvement in accuracy.

- *In contrast*, contour curves with different Orthogonal degree based Local Transmission Index (OLTI) values never intersect, and the offset of each curve is quite clear.

monitor

vt. 监控。*n.* 监视（听、控）器。

- Effective policies for public health are required to prevent the smoking initiation so as to reduce the mortality of lung cancer, so American *Food and Drug Administration* (FDA) has introduced a series of measures to *monitor* tobacco products.

- The blood routine examination and hepatic and renal function before and after chemotherapy were *monitored*.

- In total, the recovery yielded 158,015 peptides with verified fragment ion spectra and selected reaction *monitoring* (SRM) assay coordinates corresponding to 95.1% of all selected peptides (Figure 3A).

- That big machine will fail to run if it is started directly by the trigger *monitor*.

present

v. 提出；展（表、呈）现，显示；提交，把……交给；颁发；赠送，授予。

- To this end, several innovating study designs are being developed, which we *present* in the following sections.

- To *present* problems correctly is the first step towards the innovation.

- The methods *presented* here are of particular importance because they enable the quantitative measurement of structures smaller than 10 nm in size with sensitivity to sub-nanometer variations.

- Patients usually *present* with these signs when they have difficulty breathing after contracting pneumonia.

- Through contrast experiment, the researchers studied the effect of different mode of information *presenting* on learning-effect in the e-learning environment.

prompt

adj. 敏捷的，迅速的；立刻的。*vt.* 提示；促进；激起。

- The key to containing any outbreak of plague lies in *prompt* treatment with common antibiotics such as tetracycline and streptomycin, which can reduce death rates from 60% to 15%.

- This review could *prompt* the research on IBD and CRC to deepen the knowledge about the exact action mechanisms of intestinal microbiota, consequently discovering potential biological weapons useful to switch off chronic inflammation and actively preventing carcinogenesis. (IBD—inflammatory bowel disease, CRC—colorectal cancer. 笔者注）

- This *prompted* us to consider cryptic genetic influences that might generate unconventional genetic signals.

- They are rapidly digested and absorbed, raising blood levels of glucose and *prompting* the secretion of insulin to process it.

purpose/purposed

purpose—*n.* 目的；用途；意志。vt. 决心；企图；打算。purposed—*adj.* 打算的；计划的。

过去式 purposed；过去分词 purposed；现在分词 purposing。

- The *purpose* of this study was to assess the effect of long-term cultivation and water erosion on the soil organic carbon (OC) in particle-size fractions.

• Furthermore, while Deep Blue relied on a handcrafted evaluation function, the neural networks of AlphaGo are trained directly from game play purely through general-*purpose* supervised and reinforcement learning methods.

• While SRM Atlas assays are robust and powerful to assess the protein abundance in biological samples for research *purposes*, assays intended to support clinical decisions would require further clinical validation for robustness in large patient cohorts.

• I *purpose* writing a book of how to write and publish a SCI paper.

• This paper is *purposed* on analyzing the difference between two movies, also trying to compare and analyze the two directors' adapting narrative arid intentions which based on the fiction.

remain

vi. 保持；依然；留下；剩余。

• The similarity of the individual curves indicates that the background EC *remains* relatively consistent throughout the five days of the survey. (EC—electrical conductivity. 笔者注)

• However, a real consensus in subgrouping *remains* to be established.

• Despite exciting progress during the past several years, as discussed in the current Review, many outstanding questions *remain* ahead of us.

• In higher eukaryotes, including mammals, a small fraction of condensin II *remained* at kinetochores, but the bulk of condensin II spread along whole arms, acquiring an additional job of axial shortening.

• The purpose of this step is to remove any release agent *remaining* on the specimen surface.

report

v. 报告；报道；报到；举报。*n.* 报告；报道。

• Several authors have *reported* different and innovative techniques to access the excluded stomach.

• This technique can access the excluded stomach even with long limbs and shows high success rate (87.5%), as *reported* by Safatle-Ribeiro et al.

• *Reported* cases of gastric cancer (GC) after bariatric surgery are anecdotal, with low level of evidence, arising between 1 and 10 years postoperatively.

• Retrograde endoscopy has long been *reported*, but failures are frequent due to the angulation of the anastomosis.

• In recent years, some researches on the microstructure and mechanical properties of Cu alloy after high pressure treatment have been *reported*, but the *reports* on the thermal physical properties of Cu alloy after high pressure treatment were seldom involved.

represent

vt. 代表；表现。

• The authors speculate that acquisition from compromised phage might also *represent* the dominant mode of acquisition in wild populations, allowing for a small subset of the population to acquire resistance and escape without needing to outpace a rapidly reproducing phage.

• As cross-linked stress granule preps are likely to more fully *represent* the stress granule proteome, this list was used for all subsequent analyses.

• Another assumption was that all design

inputs, except for loss values, can be *represented* by a single fixed value.

• A recent meta-analysis showed 18 cases of GC after different bariatric procedures, with a mean diagnosis at 8.6 years and adenocarcinoma *representing* 83% of the cases (Table 2).

reveal

vt. 显示；透露；揭露。

• In the past twenty years, a considerable amount of research and development has been conducted by researchers from all over the world on the topic of single-type valveless piezoelectric pumps with cone-shaped tubes, which *reveals* in depth the structural characteristics of the pumps (e.g. piezoelectric vibrator, diffuser/nozzles, pump chamber).

• Our current understanding of the biology behind these rarer subtypes has already *revealed* avenues for potential implemention of targeted therapies.

• More observations of the system are sure to *reveal* more secrets

• Finally the influence law of membrane stress, deployable boom thickness, fabric layers angle of the deployable boom and wall thickness of hub on the fundamental frequency of the whole antenna structure is *revealed*.

study

n. 研究；课题；学习。*vt.* 研究；细察，探究；学习。

• The initial conceptual development of this *study* design dates back to 1970s, when the concept of adaptive randomizations was first introduced.

• We suggest that ecosystem *studies* and management should be viewed in the context of a comprehensive ecosystem classification (e.g., Haeussler 2011).

• Metabonomics is defined as a new branch of science that can provide qualitative and quantitative analysis of all small molecular metabolites of the biological system (cells, tissues and biological organisms) under given time and conditions so as to *study* the entirety of biological endogenous metabolites and describe the response rules of entirety to external and internal stimulation.

• However, the fatigue properties in important severe corrosive environment (like NaCl) were rarely *studied*.

• Finally, the research field of condensins offers an excellent example in which *studying* a diverse array of model organisms greatly enriches our understanding of a biological phenomenon from an evolutionary point of view.

unfortunately

adv. 不幸地。

• *Unfortunately*, the effects of current interventions on growth, immune function, and neuro developmental outcomes have been modest (Dewey and Adu-Afarwuah, 2008).

• *Unfortunately*, the dataset encompasses only 49 brains at the moment, and therefore, statistical power is limited.

4.2.3　记述实验、方法

记述有关实验的原理、方法、技术、设备、步骤或过程等内容，重在表述方法，常用于论文的 Experiment 和 Results 部分中。

according to

根据，按照；取决于；据……所说。

• *According to* the reports in the literatures, high pressure treatment can refine the micro-structures and improve the compactness of metal materials.

● The resistivity was calculated *according to*

$$\rho = RA/L_0,$$

where ρ is resistivity, R is resistance, A is the section area, and L_0 is the effective length of the specimen.

achieve

vt. 达到；完成。*vi.* 达到目的；如愿以偿。

过去式 achieved；过去分词 achieved；现在分词 achieving。

● In a nutshell, to *achieve* balanced regional development, there are still a lot of efforts to be done in improving the level of public goods investment in rural communities, especially in backward rural communities of mountainous area of Sichuan in China.

● As prior studies in mice and humans indicate that DYSTROPHIN levels as low as 3% to 15% of wild type are sufficient to ameliorate pathologic symptoms in the heart and skeletal muscle and that levels as low as 30% can suppress the dystrophic phenotype altogether, the restoration of DYSTROPHIN *achieved* here by one-time administration of AAV-Dmd CRISPR clearly encourages further evaluation and optimization of this system as a new candidate modality for the treatment of DMD (see supplementary text).

● As shown, this could be *achieved* by increasing the harvested area or time of harvesting.

● The deadline for *achieving* the goals is 2020.

● The communique was a watered-down version of what the British had hoped to *achieve*.

acquire

vt. 获得；取得；学到；捕获。

过去式 acquired；过去分词 acquired；现在分词 acquiring。

● The primary goal of this Review is to discuss how the unique architecture of condensins has *acquired* a diverse array of chromosomal functions during evolution.

● For 98.9% of the predicted human proteome, we were able to *acquire* high-quality SRM traces with at least one peptide per protein, while 90.3% of the proteome is represented by three SRM assays.

● Spectra were *acquired* in a data-directed approach using an exclusive precursor selection Auto MS/MS mode.

● The real challenge for these investigations will be *acquiring* sufficiently large cohorts to make meaningful conclusions that can pave the way for stratification of therapy, which will undoubtedly require international collaborative efforts.

afford

vt. 买得起；给予，提供。

● Few people can *afford* these consumer goods only on salary.

● These observations may not be a coincidence: the settings may have *afforded* greater freedom to pursue less trendy topics but less support about how to overcome skepticism by journals and reviewers.

● Where fuel competition is *afforded* and the monopoly gasoline currently enjoying in the United States is broken, the costs of powering the transportation sector are dramatically reduced.

● Visible light is part of the electromagnetic spectrum and 10,000 times bigger than the radio spectrum, *affording* potentially unlimited capacity.

allow

vt. 允许；给予；认可。

● Genome barcoding *allows* for tracking lineages originating from individual cells, facilitating studies of population evolution, cancer, development, and infection.

• These in turn *allow* better targeting of the high salt inflow zones for salt interception.

• For a small number of peptides, we *allowed* less strict criteria with regard to length, hydrophobicity, and charge state, to be able to select several peptides per protein and ensure as many proteins as possible are considered for SRM assay development.

• This shows that the effective time interval of earthquake events is only about 1,000 years, and the maximum cycle of seismic activity that we are *allowed* to infer does not exceed 500 years.

• John van der Oost had just received a major grant from the Dutch National Science Foundation. In addition to working on the problem described in his proposal, he decided to use some of the funding to study CRISPR. (In his report to the agency 5 years later, he underscored the value of the agency's policy of *allowing* researchers the freedom to shift their scientific plans.).

assess

vt. 评定；估价。

• We used the β_2V_2R as our model class B GPCR to be consistent with the biophysical experiments in this study where we *assessed* the ability of a purified β_2V_2R-barr1 complex to interact with, and activate, Gs (see below).

• The development of a new methodology to interpret run of river salinity data to *assess* salt inflow to the River Murray. (论文题目)

• With the advent of powerful computing facilities over the last couple of decades, computer models have been extremely useful in devising and *assessing* the various engineering and operational strategies for managing water quality and quantity.

• The skill of the prediction is *assessed* with reference to the climatological (observed) values as the control or reference forecasts.

carry out

执行，实行；实现；完成。

• The experiment is *carried out* by a homemade DBD load as shown in Fig. 2.

• Qualitative and quantitative analysis was *carried out* to answer the above-cited questions.

collect

vt. 收集。

• Some modules and submodules *collect* data from run time entities, while the others aggregate the data over the run time entities.

• It offers a good way to *collect* the energy from the sunlight and change it into electricity.

• The majority of the field samples were of glacial origin but were *collected* from a range of landforms.

• The accuracy of on-body measurements was verified through the comparison of on-body sensor readings from the forehead with ex situ (off-body) measurements from *collected* sweat samples (Fig. 3b).

• By *collecting* the data from thousands of experiments, they can gain valuable insights in an entirely new way.

conduct

v. 组织，实施，进行；引导。

• We *conducted* face-face structured interviews from July to August 2012 and investigated 79 households, including 30 households in Village No. 27, 30 households in Qipanshan Village, and 19 households in Shizhuozi Village.

• A series of laboratory tests were *conducted* with 55 different sets of data.

do/doing/undo

do—*v.* 做；进行；完成。doing—*n.* 活动；所作所为。undo—*vt.* 取消；解开。

• Therefore, many countries start to *do* research on projectile core of the composites in recent years.

• Peng et al. *did* numerous researches through experiments and finite element method and discussed the influences of processing conditions such as pressing orientation and groove width on constrained groove pressing (CGP) in detail.

• Even if they wanted to select an environmentally friendly product, it is easier said than *done*.

• Until now, few researches have *done* to discuss how the mechanical parameters measured on coating surface and cross section affect the wear resistance of the coating.

• Modelling of many weather phenomena has been *done* so far employing the concept of stochastic systems.

• Not only would this represent the first instance of a new epoch having been witnessed firsthand by advanced human societies, it would be one stemming from the consequences of their own *doing*.

• The threads in COF-505 have many degrees of freedom for enormous deviations to take place between them, throughout the material, without *undoing* the weaving of the overall structure.

expose

vt. 显示；揭露；使曝光。

过去式 exposed；过去分词 exposed；现在分词 exposing。

• To investigate SNPs resulting in non-synonymous sequence variants, we developed assays for the most frequent mutations, which typically result in protein malfunction and may *expose* an altered phenotype. (SNPs—single nucleotide polymorphisms：单核苷酸多态位点。笔者注)

• Furthermore, we *exposed* a similar $FA_{0.83}Cs_{0.17}Pb(I_{0.6}Br_{0.4})_3$ film to monochromatic irradiance of much higher irradiance of $5Wcm^{-2}$ and observed no red shift after 240 s of illumination (fig. S9).

• Figure 1F shows the oxygen reduction reaction (ORR) curves obtained under oxygen-saturated conditions with the subtraction of data under nitrogen-saturated conditions as the background, in which the currents are divided by the geometric surface areas of electrodes (the same as the *exposed* catalyst surface areas) as described in the experimental method of the supplementary materials.

• Some receptors are only indirectly affected by Aβ, inferred by physiological alterations of cells *exposed* to Aβ.

• Cas10 cleaves DNA *exposed* by a transcription bubble using a single catalytic site in its palm polymerase domain.

• But the two teams of scientists, have found that some genes are left *exposed* in sperm, in an "open conformation", allowing them to play an important role in the development of the embryo.

• Due to the *exposing* nature of the terrain, air power is critical in controlling this major artery.

focus

n. 焦点；中心；焦距。*vt.* 使集中；使聚焦。*vi.* 集中；聚焦。复数为 focuses 或 foci。

过去式 focused/focussed；过去分词 focused/focussed；现在分词 focusing/focussing。

• This motivated the *focus* in the current work on cryptic genetic influences in this region that could cause unconventional association signals that do not resemble the LD patterns of individual variants.

• Do one thing, with all your *focus* and energy, and do it right.

• However, there is one very funny thing that happens whenever the researchers *focus* their study on one of these molecular motions.

• This review *focuses* on the currently available literature *focusing* on CFNA (cell free nucleic acids) in hematologic malignancies.

• Next, we turn from the disk subsystem to *focus* on the processors.

• For residential energy consumption, our surveys *focused* on (a) the main energy sources; (b) the magnitude of different energy consumption per year or per month; (c) the end use of different energy resources (heating, cooking and home appliances); (d) the duration of use of different energy sources; and (e) the utilization efficiency of different energy resources (the ratio of supplied energy to the total energy content of the consumed fuel).

• However, the current research on quenching sensitivity of aluminum alloys is mainly *focused* on the effects of quenching velocity on the residual stress of thick plates and the properties of alloys as well as the relationship between quenching sensitivity and the addition of trace alloying element (such as Zr, Cr, etc.).

• In this article we are *focusing* on the non-technical theoretical knowledge needed; the technical aspects of these systems have been covered in previous articles within this series.

from

prep. 来自，从。

• Moreover, the elevated levels of heat shock protein (Hsp) 60, Hsp10, Hsp70 and Hsp90 have been found in serum and colonic biopsies derived *from* IBD patients. (IBD—inflammatory bowel disease. 笔者注)

• *From* the cross-sectional evaluation results, it can be seen that both the hardness and elastic modulus of the ultrafine-structured coating are obviously higher than those of the conventional coating.

• Our analysis is more consistent with a beginning in the mid-20th century, and a number of options have already been suggested within that interval, ranging *from* 1945 to 1964 CE.

• During pregnancy, the primary etiology of microcephaly varies *from* genetic mutations to external insults.

furnish

vt. 提供；供应；装备。

• The General Dynamics-led team will provide training for hardware and software, and also *furnish* independent verification and validation of sensors and systems.

• Visitors can spend their time here by not only observing the game and comparing their acquired experiences, but they can also take a rest in the functionally *furnished* rooms.

• Like other nations, the United States uses tax revenues to maintain a government, to *furnish* common services and to implement social policy and economic objectives.

generate

vt. 使形成；发生。

过去式 generated；过去分词 generated；现在分词 generating。

● It is clear that qualitative aspects also *generate strong constraints to water availability* and should be accounted for in water resource planning and management, especially in semiarid environments, where vulnerability is higher.

● To *generate* SRM assays, the peptides selected above were chemically synthesized and used to *generate* fragment ion spectra that were processed into consensus spectra and ultimately SRM assays.（SRM—selected reaction monitoring. 笔者注）

● To investigate whether GPCR-barr complexes can interact with, and activate, G protein in a cellular environment we *generated* fusion proteins of GPCR-barr and investigated their ability to activate Gs in HEK293 cells.

● In space, however, these rockets are inefficient—they burn through huge quantities of fuel while *generating* more thrust than necessary.

keep

vt. 保持；遵守。*vi.* 保持；继续不断。

过去式 kept；过去分词 kept；现在分词 keeping。

● Additionally, the application *keeps* track of the duration of exercise as well as the distance travelled.

● Test and calibration of laboratory equipment and instruments were performed in a regular basis. Samples were properly preserved to *keep* the availability.

● The volumes of water droplets were all *kept* at 2 μL, and the spreading time was set at 1,000 ms.

● When positive super helical torsion is applied with magnetic tweezers, nucleosomal arrays are subjected to plectonemic supercoiling while *keeping* their left-handed state of DNA wrapping.

● Every one of us must love trees, protect forests, and *keep* on planting trees each year.

obtain

vt. 获得。

● For most alloy steel, even the material of heat-treatment steel with 900 MPa of tensile strength, aluminum alloy with low elongation and etc., it is expected to *obtain* high accuracy internal thread by adopting the cold-extrusion processing method.

● To determine the surface topography of the samples,（SEM）images were *obtained* using JSM-5800（JEOL, Japan）.

● The exact reasons why plants have evolved multiple ways to regulate the same growth process will remain unclear until additional data have been *obtained*.

● Our findings demonstrate that specific bacterial members of the gut microbiota *obtained* from an undernourished Malawian infant are able to metabolize S-BMO-derived sialyllactose to its constituent monosaccharides.

● These results demonstrate the feasibility of *obtaining* a combined tandem solar cell efficiency ranging from 19.8%, if we combine with the stabilized power output of the semi-transparent cell, to 25.2% if we combine with the highest current density-voltage（*J-V*）measured efficiency of the $FA_{0.83}Cs_{0.17}Pb(I_{0.6}Br_{0.4})_3$ cell.

perform

vt. 执行；完成；演奏。*vi.* 执行，机器运转；表演。

● Self-efficacy is the confidence that can *perform* a specific behaviour.

● To determine whether the GEF（guanine

exchange factor) plays a role in apoptosis, experiments were *performed* with an alky lating drug bendamustine.

• CRISPR was soon adapted for a vast range of applications—creating complex animal models of human-inherited diseases and cancers; *performing* genome-wide screens in human cells to pinpoint the genes underlying biological processes; turning specific genes on or off; and genetically modifying plants—and is being used in thousands of labs worldwide.

• This unit determines which of these operations to *perform*.

• They brought him to court. The judge sentenced him to *perform* in the park for the children of the city.

prepare

vt. 准备；使适合。*vi.* 预备；做好思想准备。

过去式 prepared；过去分词 prepared；现在分词 preparing。

• What evolutionists want and have always wanted, is to *prepare* our generations for a future Atheistic society, a world without God in it.

• To compare the HOPG model catalysts with powder catalysts, we *prepared* nitrogen-doped graphene nanosheets (N-GNS) and measured their ORR activities by the rotating disc method in 0.1 M H_2SO_4.

• Choose some food that can be *prepared* in advance so you're not trapped in the kitchen.

• The thermal diffusion coefficient, thermal conductivity, and thermal expansion coefficient of CuCr alloy *prepared* by infiltration were measured by thermal constant tester and dilatometer before and after high pressure heat treatment.

• Basement drainage is best addressed at the time of *preparing* a stormwater plan for a development but retrofit solutions may also be required.

• Each member of the research group should *prepare* well for the project and research.

produce

v. 生产，产生；引起，导致。

过去式 produced；过去分词 produced；现在分词 producing。

• Step-up direct current/direct current converters were used to *produce* a fixed, regulated output of +5 V for the microcontroller and +3.3 V for the Blue tooth modules.

• These results suggest that C4 is *produced* by, or deposited on, neurons and synapses.

• When such crowded helices are exposed to a high-salt buffer, their hydrophobic surfaces would mediate and stabilize artificial protein-protein interactions, there by *producing* large protein-DNA aggregates.

• The generalization performance of a trained network is measured on the error it *produces* with the new data set.

proceed

vi. 开始；继续进行；发生；行进。

• The APV-MCTS algorithm *proceeds* in the four stages outlined in Fig. 3. (APV-MCTS—asynchronous policy and value MCTS algorithm. 笔者注)

• The path you have entered is too long. To *proceed*, enter a shorter path.

• This kind of network analysis does not *proceed* by formal rules

• Opinion on how regulators should *proceed* on individual cases, though, is varied and heated.

• We'll take a quick look at the code you just imported before *proceeding*.

provide

vt. 提供；准备。

过去式 provided；过去分词 provided；现在分词 providing。

● The radome is a cover made of natural or manmade dielectric material, or a special shaped electromagnetic window made of a truss-supported dielectric housing. It *provides* a suitable interface for maintaining the structure, temperature and aerodynamic characteristics.

● In the next sections, we will *provide* an overview of such innovative study designs, once we have *provided* a summary of the molecular landscapes of breast cancer.

● The new methodology, in this example, has *provided* a much-improved resolution and narrowed the zone of high salt inflow from a 4 km long to a 1 km long river reach.

● The reflector surface and torus are deployed gradually by shape recovery and thrust *provided* by the inflatable struts.

● Translational medicine devotes itself to establishing a bridge for laboratory research and clinical practice, which can rapidly translate the basic research results of medical biology into the theory, techniques, methods and agents that can be applied in clinic, thus *providing* references for the connection of disease markers in research. Therefore, this study mainly summarized the exploration of disease markers under translational medicine models so as to *provide* a basis for the translation of basic research results into clinical application.

purchase

n. 购买。*vt.* 购买。

过去式 purchased；过去分词 purchased；现在分词 purchasing。

● Consider the case of a program sending a *purchase* request to another program over a network.

● The term "take-out" describes both a style of eating and a growing list of prepared foods that consumers *purchase* from a restaurant or food stand and eat in another location.

● The customer continues in this fashion for all items he wants to *purchase*.

● The human-derived, HT-29 cell line was *purchased* from ATCC (American Type Culture Collection).

● In Greece and elsewhere in the EU, the banks support the government by *purchasing* its bonds, and the government guarantees the banks.

record

n. 纪录；档案。*vt.* 记录，记载；标明。

● There is no *record* of any similar medieval map with such scale or accuracy.

● We review anthropogenic markers of functional changes in the Earth system through the stratigraphic *record*.

● Climatic *records* derived from ice cores in the TP (Tibetan Plateau) provide valuable information on the atmospheric circulation, such as the strength and variability of the westerlies and the intensity of the summer monsoon system (Kang *et al.* 2002, 2003, 2010).

● Randomly 10 visual fields were selected and positive cells were *recorded* according to cell percentage.

● The warming took place over about 1600 years and is *recorded* by a variety of stratigraphic signals that are not all globally synchronous.

● The posterior distribution for the age of clade B encompasses the *recorded* duration of the ZIKV outbreak in three of five island groups of French Polynesia (4) (Fig. 3C).

• These alterations are currently often studied by *recording* synaptic function in neuronal cell and hippocampal slice models treated with Aβ peptides, either synthetically generated, isolated from " natural " brain material, or expressed in transgenic mouse models.

subject to

使服从；使遭受。

• The evaluating result of vibration stress of in-service blade *subjected to* centrifugal force and bending vibration stress agrees with aero engine test result.

• Both of these impacts of small reservoirs are *subject to* climate change.

• During the wear test, the coating was *subjected to* the continous extrusion stress and scratching of abrasives at a certain load.

treat/treatment

treat—*vt.* 治疗；对待。treatment—*n.* 处理；对待；治疗；疗法。

• As a result, existing bovine milk-based infant formulas and complementary/therapeutic foods used to *treat* under nutrition are deficient in these important human milk components.

• Above all else, we believe in the importance of integrity—not only integrity to our customers but also integrity among our designers and integrity in how we *treat* our employees.

• We observed that, when yeast cells were *treated* with sodiumazide (NaN_3, which induces stress granules; Buchan et al., 2011) prior to lysis, GFP-positive granules were observed in the lysate.

• The overall on-target mutagenesis frequencies of GUIDE-seq tag-*treated* samples was determined by T7 endonuclease I assay as described above. (GUIDE-seq—the genome-wide unbiased identification of double-stranded breaks enabled by sequencing. 笔者注)

• Then transesterification takes place by heating the crude biodiesel to about 100 degrees Celsius to remove any water, and *treating* it with methanol and a catalyst.

• For these patients, multidisciplinary palliative *treatment* must be provided by surgeons, oncologists, radiotherapists, and nutritionists.

undertake

vt. 承担；从事。

过去式 undertook；过去分词 undertaken；现在分词 undertaking。

• The bulk miner (BM) is a large track mounted cutting machine that will *undertake* the bulk of cutting operations.

• With the help of several noted researchers, I founded the Allen Institute for Brain Science in 2003 to *undertake* this project.

• To further confirm the ability of the β_2V_2R-βarr1/2 fusions to activate Gs, real-time kinetic studies of ISO-stimulated cAMP production were *undertaken*.

• So on behalf of the United States and our people I give each of you, and your nations, my heartiest welcome and my heart felt thanks for being here and *undertaking* this great mission with us.

yield

v. 出产；屈服；投降。*n.* 产量；收益。

• As with wild-type SpCas9, the eighth sgRNA (FANCF site 4) did not *yield* any detectable off-target cleavage events when tested with SpCas9-HF1 (Fig. 2a).

• There are many reasons why systems biology approaches are currently not *yielding* novel insights in our field.

• Schizophrenia risk and *C4A* expression levels *yielded* the same ordering of the *C4* allelic series（Fig. 5a，b）.

● To strive, to seek, to find, and not to *yield*.

● We use a plastic PC-ABS material with a *yield* strength of 4.1E+7 N/m^2.

● Prolonged exposure to shade is costly. Shaded plants have accelerated reproductive development, leading to lower biomass and seed *yield*.

4.2.4 展现实验、结果

展现或记述由实验所得到的定量或定性结果，包括数字数据、现象记述、文本记录和图形图像等，常用于论文的 Results 和 Discussion 部分中。

affect

vt. 影响。

● So, any interruption to one of the access path does not *affect* storage access from the host.

● As the size of semiconductor device and interconnection shrinks, more and more critical design performance will be seriously *affected* by the interconnections.

● Some factors *affecting* the azimuthal uniformity of feed currents are simulated and discussed.

alter

vt. 改变，更改，切除。

● Due to their mechanism of action they can *alter* the electrolyte balance in the body leading to various electrolyte losses and side effects per black box warning.

● On the morning of day 6 of her admission, she was found to be *altered*, somnolent, and drowsy by the morning staff.

analyze/analyse（analysis/analyses）

analyze/analyse—*vt.* 对……进行分析，分析；分解。analysis/analyses—*n.* 分析；解析；分解。analyses 是 analysis 的复数形式。

过去式 analyzed/analysed；过去分词 analyzed/analysed；现在分词 analyzing/analysing。

● Apoptotic nuclei, a hallmark of cell death, were observed in all ZIKV-infected neurospheres that we *analyzed*（Fig. 3B）.

● We now switch our attention to numerically *analyze* the mB equation.

● The ray-transfer matrices of the model were used to *analyze* the expected focal position inside the eye using several commercially available objectives (and their combinations with appropriate ETL and concave offset lenses, together referred to as ‘combined offset and ETL’ in Figure 1c).

● The N1s spectra were *analyzed* by least-squares fitting analysis, which includes the components of pyridinic N（398.5 eV）, graphitic N（401.1 eV）, pyrrolic N（400.1 eV）, and oxidic N（403.2 eV）.

● In this case, the architect adds value by *analyzing* the impact and complexity of the requirements.

● The researchers say the new method is faster and cheaper than current methods because it can *analyse* multiple sequences in a short time.

● As cross-linked stress granule preps are likely to more fully represent the stress granule proteome, this list was used for all subsequent *analyses*。

confirm

vt. 证实；确认；确定。

● Failure mode and fracture characteristics should be analyzed finely before applying this method to evaluate the vibration stress so as to *confirm* that the blade fractures are due to bending resonance.

• This can be *confirmed* by Fig. 4e, f, in which the oxygen content in wear debris of NiTi is higher than that of Nb-NiTi.

consider

vt. 考虑，考虑到；认为。

• Cutting of all available reed areas for summer or winter harvest seems to be unrealistic, *considering* the effects on wildlife and reed vitality.

• The microscopic holes of material can be *considered* as a zero-expansion phase.

detect/detection

detect—*vt.* 发现；探测；察觉。detection—*n.* 侦查，探测；发觉，发现；察觉。

• During the development and progression, diseases are often accompanied with a series of molecular substance changes, so *detecting* specific biomarkers can be used to diagnose the disease condition and evaluate therapeutic efficacy or assist with the development of new agents in clinic.

• In order to avoid false positive results, we additionally *detected* the p16 expression of HT-29 cell line, which was derived from human colorectal cancer tissues by western-blotting assay.

• In this study, p16 expression in 30 colorectal cancer samples was *detected* for exploring its clinical significance so as to provide the experimental basis for the early diagnosis, treatment and prognosis of colorectal cancer.

• Abrosia was conducted at least 6 h before examination, and the *detected* blood glucose level was <200 mg/dL.

• It is attributed in part to the difference in ethnicity of the cohorts, but it could also be explained by the diversity of PD-L1 (programmed cell death ligand-1) antibody clones and plethora of *detection* systems (Table 4)

determine

vt. 决定；确定。

过去式 determined；过去分词 determined；现在分词 determining。

• The deformation homogeneity largely *determines* the performance of constrained groove pressing (CGP) processed materials.

• Thus, we conclude that the processing rate is *determined* not only by the pass number, but also by the die structure, such as groove width and constraint.

• The grain size distribution for the four available samples was only *determined* up to sieve No. 200 (75 μm) (Figure 3) and, therefore, the grain size analysis of fines is unknown.

• Modelling these losses is important in *determining* the shape and magnitude of the flood hydrograph.

develop

vt. 开发；进步；使成长。*vi.* 发育；生长；进化。

• The aim of this paper, therefore, is to *develop* a prototypical quasi-economic model for medium-to long-term management of the southern Murray-Darling Basin.

• A new method has therefore been *developed* to assign the salt inflow more closely to the location where it actually occurs and at the correct rate.

• The self-discipline I am speaking of is all about *developing* a mentality where you can fix your mind on something and achieve it.

• These data provide direct evidence that C4 mediates synaptic refinement in the *developing* brain.

• Consistent with this idea is the observation

that interference with senescent cell accumulation in BubR1 progeroid mice delays several of the ageing-associated disorders that these animals *develop*.

distinguish

v. 区别，区分；辨别；表现突出。

● In this study, the 3D Hashin-type and Maximum stress failure criteria are used in failure prediction, both of which *distinguish* various failure modes.

● Human C4 exists as two paralogous genes (isotypes), C4A and C4B; the encoded proteins are *distinguished* at a key site that determines which molecular targets they bind.

● To measure the copy number of compound structural forms of C4 (involving combinations of L/S and A/B), we perform long-range PCR followed by quantitative measurement of the A/B isotype-*distinguishing* sequences in droplets.

elevate

vt. 提升。

过去式 elevated；过去分词 elevated；现在分词 elevating。

● The decreased levels of fatty acid metabolites in the liver and serum of these nonfasted mice could reflect a more normalized anabolic state. Accordingly, we documented that these animals had significantly *elevated* concentrations of serum insulin, leptin, and triglycerides (Figures S2B-S2D).

● Deformation behavior at *elevated* temperature and processing maps of flange blanks on casting-rolling compound forming technology. (论文标题)

● The median expression of *C4A* in brain tissues from schizophrenia patients was 1.4-fold greater ($P = 2 \times 10^{-5}$ by Mann–Whitney *U*-test; Fig. 5d) and was *elevated* in each of the five brain regions assayed (Extended Data Fig. 8).

● Previous studies have hypothesized that noise exposure could cause hypertension by persistently *elevating* the body's level of stress hormones.

enhance

vt. 提高；加强；增加。

过去式 enhanced；过去分词 enhanced；现在分词 enhancing。

● The formed strengthened layer enhances the material's yield strength and fatigue strength, and the reduction of surface roughness can decrease the surface defect, namely decrease stress concentration, which is beneficial to *enhance* the material's fatigue strength.

● In this study, a smaller groove width of 2 mm and full constraint from the dies can *enhance* the effect of interface regions on the deformation of adjacent shear regions.

● The salt accumulation process in the basin has been naturally *enhanced* during times of high groundwater levels and arid periods.

● This article uses example code to demonstrate keystroke dynamics for *enhancing* the security of your applications in authentication and continuous data entry contexts.

exert

vt. 发挥，运用；施以影响。

● Post-deformation annealing process *exerts* significant influence on the crystallization temperature which climbs up with the increase of annealing temperature.

● Many questions have emerged in relation to the possible pathologic effects *exerted* by dys regulated circulating miRNA (microRNA).

● The method of Google is to *exert* a

subtle influence on youth's minds.

• As altitude increases, atmospheric pressure decreases, thus *exerting* less "push" on the water entering the pump suction.

exhibit

vt. 展览，展出；展现，显示。

• Besides Blue-and-White Chinese porcelain that was transported by caravan over the Silk Road, a representative selection of imperial treasure is also *exhibited* here.

• After a couple dismal years of sales due to a global recession, manufacturers are *exhibiting* a renewed sense of creativity.

• These compounds *exhibit* complex magnetic structures and possess different magnetic-phase transition, which will induce interesting physical properties.

• A multi-objective optimization strategy is applied to the catenary system, which *exhibits* the improvement of all three design objectives.

• As for the remaining participants, about 15 percent *exhibited* grief symptoms that were moderately high at 6 months but almost completely gone by 18 months.

illustrate

vt. 阐明，举例说明；图解。

过去式 illustrated；过去分词 illustrated；现在分词 illustrating。

• Fig. 19 *illustrates* the attitude tracking error, from which we can observe that the attitude error temporarily increases in the presence of interference moment and force but soon converges to approximately zero; it also shows that the attitude tracking error is relatively small.

• The code above is, of course, nonsense, but its structure is common, so it shall serve to *illustrate* the point.

• This is futher *illustrated* in Figure 3 (c), assuming a simple background EC (electrical conductivety) pattern (linearly decreasing EC in time and increasing downstream).

• The remainder of this article focuses on three examples, each *illustrating* some problems I have encountered in my work.

interact/interaction

interact—*vi.* 互相影响；互相作用。

interaction—*n.* 互相影响；互相作用。

• These super-complexes or "megaplexes" more readily form at receptors that *interact* strongly with β-arrestins via a C-terminal tail containing clusters of serine/threonine phosphorylation sites.

• Currently, many research groups are focusing on improving the quadrotor UAV's ability to aggressively maneuver and *interact* with the environment. (UAV—unmanned aerial vehicles. 笔者注)

• Because these visitors are already on your site, you have an opportunity to *interact* with them in a more meaningful way.

• The microbial associated molecular patterns (MAMPs) are responsible for the activation of immune system through *interacting* with pattern recognition receptors (PRRs) and subsequent triggering of inflammatory processes.

• This indicates that these proteins can *interact* in planta and that the *interaction* occurs in the nucleus.

• From a systems perspective, biological processes constitutenet works of *interacting* molecules, and changes in network state are informative about the biochemical state of the cell.

• Indeed, evidence is available that an SMC dimmer has multiple DNA-*interacting* regions (Hirano and Hirano, 2006; Kim and

Loparo, 2016), and a recent study has identified a DNA-binding activity in the HEAT subunits of condensin I as well (Piazza et al., 2014). (SMC—structural maintenance of chromosomes. 笔者注)

lead to

导致;通向。

• In the case studied, artificial neural network (ANN) models were suitable considering mixing of different solids in the sewer system which *leads to* different viscosity fluid flow conditions.

• Advances in gene therapy have *led to* its application in pancreatic cancer, malignant gliomas, and lung cancer.

• Water erosion is the main factor *leading to* land degradation and poor fertility in the long-term eroded cultivated lands of the Loess Plateau and constrains the sustainable agricultural development in this region.

manipulate

vt. 操纵;操作;巧妙处理。

过去式 manipulated;过去分词 manipulated;现在分词 manipulating。

• By taking advantage of their distinctive architecture, condensins actively fold, tether, and *manipulate* DNA strands, participating not only in chromosome assembly and segregation during cell divisions, but also in many aspects of large-scale chromosome organization and regulation.

• The eukaryotic HEAT subunits are drastically different from the bacterial kite subunits in their sizes and structures, implicating that acquisition of the HEAT subunits during evolution might have provided condensins with elaborate abilities to *manipulate* and handle increasing lengths of eukaryotic chromosomes.

• Expression levels of inflammation markers in *unmanipulated* 18-month-old C57BL/6 females suggests that repeated vehicle injections were not a source of tissue inflammation.

• Information architecture is a set of related data and content that's accessed, *manipulated*, and stored to provide state and meaning to the enterprise operations.

• There, scientists have been *manipulating* the levels of bone-morphogenetic protein or BMP in the brains of laboratory mice.

mediate

v. 调停;调解;斡旋。*adj.* 间接的;居间的。

过去式 mediated;过去分词 mediated;现在分词 mediating。

• When such crowded helices are exposed to a high-salt buffer, their hydrophobic surfaces would *mediate* and stabilize artificial protein-protein interactions, thereby producing large protein-DNA aggregates.

• We concluded that (1) colonization of the gut per se is not sufficient to produce the S-BMO-enhancement of growth, and (2) other members of the community and/or higher-order interactions between B. fragilis, E. coli and these other members are required to *mediate* this growth promotion.

• In mice, C4 *mediated* synapse elimination during postnatal development.

• Cas9-*mediated* DNA cleavage is dependent on DNA strand separation.

• Stress granule assembly is *mediated* by prion-like aggregation of TIA-1.

• After he left office, he was involved in *mediating* conflicts from Africa to the Middle East.

• Conference host Egyptian President Hosni Mubarak said it is his "priority to reach a truce between Israel and the Palestinians," despite the multiple setbacks in negotiations Egypt has been *mediating*.

• Two talks are to cooperate to perhaps negotiate, and tripartite talk is *mediate*.

modify

v. 修改，修饰；更改。

过去式 modified；过去分词 modified；现在分词 modifying。

• The prospect that CRISPR might be used to *modify* the human germline has stimulated international debate.

• The common reference electrode for the Na^+ and K^+ ISEs was *modified* by casting 10 μl of reference solution onto the Ag/AgCl electrode.

• Carbon, nitrogen, and phosphorus cycles have been substantially *modified* over the past century.

• As with many other Cas proteins, the Cas1-Cas2 integrase complex shows promise for use in *modifying* genomes.

modulate/modulation

modulate—*v.* 调节；调制；调整。modulation—*n.* 调制；调整。

过去式 modulated；过去分词 modulated；现在分词 modulating。

• Moreover, probiotics *modulate* the expression of genes encoding junction proteins in colocytes and stimulate the mucosal immune system in the patient's intestinal tract to secrete protective immunoglobulins, such as secretory IgA and protective defensins in the colonic lumen.

• As soon as the earthquake struck, the reactors scrammed: The control rods, used to *modulate* the speed of the nuclear reaction, were inserted into the reactor cores, shutting off the nuclear reactions.

• Several observations indicate that stress granules are *modulated* by a variety of ATP-driven remodeling complexes.

• Hegazy group investigated the effect of *Lactobacillus delbruekii* and *Lactobacillus fermentum* administration on 30 patients with mild to moderate UC, evaluating their potential immune-*modulating* effects.

• Probiotics moved to improve the clinical symptoms of IBD through GALT immune *modulation* based approaches using *Lactobacilli* and *Bifidobacteria*. （IBD—inflammatory bowel disease. 笔者注）

modulator

n. 调制器，调幅器。

• MCPH1 regulates chromosome condensation and shaping as a composite *modulator* of condensin II.

• The cooling followed an orbitally related insolation decline, with small fluctuations representing changes in solar intensity, which are controlled by *modulators* such as the 208-year Suess cycle.

observe

v. 观察；注意到。

过去式 observed；过去分词 observed；现在分词 observing。

• To further investigate the effect of $p16^{Ink4a+}$ cells on physiological functions that change with age, we focused on kidney and heart, two vital organs in which we *observed* ATTAC-mediated clearance.

• Figure 1 shows the surface topographies of NiTi and Nb-NiTi samples *observed* by SEM.

• From these observations, it is *observed* that the best network is a layer recurrent neural network having two hidden layers consisting of 7 neurons in each, and the training algorithm is trainrp, and the transfer function is Log-Sigmoid.

• On the other hand, a substantial increase in sweat [Na^+] and a smaller increase in sweat [K^+] (no clear increase in [K^+] was *observed* in two out of six subjects) were *observed* in dehydration trials (without water intake) after 80 min when subjects had lost a large amount of water (2.5% of body weight) (Fig. 4d and e).

• Astronomers got information about the planet by *observing* its effect on its star's light when it moved in front of the star from our point of view.

• For a real researcher, first he should *observe* for himself, do his own research, and then see if his findings can stand scientific verification.

result in

导致，结果是。

• Cluster analysis of species abundances *resulted in* the identification of five major vegetation groups based on both the most constant and dominant species.

• In our electrical current measurements, the direction of the current was from the shared Ag/AgCl reference/counter electrode towards the working electrode of each of the glucose and lactate sensors, which would *result in* a negative transimpedance output voltage.

retain

vt. 保持。

• Here, the "solid structure" refers to a three-dimensional structure that the cone-shaped tube, as a flow-resistance-type non-movable valve, *retains*, such as cone, square cone and so on.

• We envisioned that disruption of one or more of these contacts might alter the energetics of the SpCas9-sgRNA complex so that it might *retain* enough for robust on-target activity but have a diminished ability to cleave mismatched off-target sites.

• We expanded this assay to assess whether SpCas9 (K855A), eSpCas9 (1.0), and eSpCas9 (1.1) broadly *retained* efficient nuclease activity, measuring on-target indel generation at 24 target sites spanning 10 genomic loci (Fig. 2A).

• The search tree is reused at subsequent time steps: the child node corresponding to the played action becomes the new root node; the subtree below this child is *retained* along with all its statistics, while the remainder of the tree is discarded.

• For example, Listing 32 shows how to get the first two elements from a list while *retaining* the remainder.

show

vt. 显示；说明；展出。

过去式 showed；过去分词 showed/shown；现在分词 showing。

• The results *show* that at higher inflation pressure, the surface shape error is greater, and asymmetric inflation pressure on the inflatable torus causes larger reflector surface deformations than symmetrical pressure.

• First, granules in cells treated with $NaAsO_2$ for 60 min were observed to move and fuse in the cytoplasm (Movie S2), while granules in ATP-depleted cells were static and *showed* no fusion events (Movie S3).

• Bootstrap scores are *shown* next to well-supported nodes, and the phylogeny was midpoint rooted.

• One smart cutting tool (namely Smart Tool 1) as *shown* in Fig. 2 uses the force shunt and the indirect force measurement to detect the cutting force respectively.

• Our results, together with recent studies *showing* brain calcification in micro cephalic fetuses and newborns infected with ZIKV, reinforce the growing body of evidence connecting the ZIKV outbreak to the increased reports of congenital brain malformations in Brazil.

4.2.5 进行讨论、分析

讨论、分析和总结实验或仿真的意义及创新之处，常用于论文的 Discussion 部分中。

accompany

vt. 伴随，陪伴。

过去式 accompanied；过去分词 accompanied；现在分词 accompanying。

• Economic globalism involves long-distance flows of goods, services and capital, and the information and perceptions that *accompany* market exchange.

• The explosion of debt that is certain to *accompany* the enactment of this national health care bill can only add to that nervousness.

• Most patients with esophageal carcinoma are *accompanied* by lymph node metastases or distant metastases when being diagnosed, so the prognosis is poor.

• "This is a huge milestone in this type of laser research," said Stephen Durbin, a scientist at Purdue University, in an *accompanying* article in the journal *Nature Physics* about the diamond mirror.

account

n. 账户；解释；理由。*vi.* 解释。

• A simple example of an activity service would be a bank transaction in which a customer transfers money from one *account* to another.

• Taking the utilization efficiency into *account*, energy fuels cannot provide all the energy stored in them.

• Taking *account* of the fact that areas further than 11 m and less than 1.0 m from the fence received almost no snow accumulation, we treated the site 11 m from the snow fence as soil with shallow snow cover (n. b., the site less than 1.0 m from the fence was greatly influenced by the snow fence in its soil temperature and moisture) and the site 3 m from the snow fence as soil with deep snow cover.

• There are two possible reasons that *account* for this difference.

anticipate

vt. 预期，期望。

过去式 anticipated；过去分词 anticipated；现在分词 anticipating。

• Researchers at the Mill Hill laboratory *anticipate* that the Turkish virus will also have this characteristic.

• It is *anticipated* that a combinational approach would be superior by preventing relapse due to the development of microenvironment-mediated survival of leukemia cells.

• Some are still waiting to buy, *anticipating* even better bargains in 2012 as the European debt crisis continues to unfold.

argue

v. 辩论，争论；证明；提出理由。

过去式 argued；过去分词 argued；现在分词 arguing。

• More modern interpretations *argue* that the cancer cell uses a modification of metabolic pathways adaptively, shifting in favor of the production of macro molecules needed to meet the biomass demands of rapidly dividing cells.

• This perspective *argues* for a hydrological model of a river-centric and seasonal nature.

• I believe, as John Gray has *argued*, that we humans, like most creatures, are preoccupied with the needs of the moment.

• The groups then fell to *arguing* amongst themselves each insisting its definition was correct and all the others were wrong.

attribute

n. 属性；特质。*vt.* 归属；把……归于。

过去式 attributed；过去分词 attributed；现在分词 attributing。

• These distinctive *attributes* of the recent geological record support the formalization of the Anthropocene as a stratigraphic entity equivalent to other formally defined geological epochs.

• These phenomena were partly *attributed* to variations in vegetation coverage and soil properties.

• Dalton, the chemist, repudiated the notion of his being " a genius ", *attributing* everything which he had accomplished to simple industry and accumulation.

aspect

n. 方面；方向；外貌。

• But how to ensure the electrical performance of the antenna from the *aspect* of electromechanical coupling is the content that we want to present in the following.

• In 2015, Ullmann, et al. [51], systematically discussed the valveless piezoelectric pumps with cone-shaped tubes in a series structure and in a parallel structure from the *aspects* of fluid mechanics and circuit theory, and got a series of related conclusions (see Fig. 36 and Fig. 37).

• Their disadvantages include the following: (1) small measuring range; (2) affected by the tested material and diffraction limit; (3) cannot measure the features with high *aspect* ratio, such as deep holes, lateral-walls and micro grooves.

become

v. 成为；变得；变成。

过去式 became；过去分词 become；现在分词 becoming。

• As more simulations are executed, the search tree grows larger and the relevant values *become* more accurate.

• If so, when did this stratigraphic signal (not necessarily the first detectable anthropogenic change) *become* recognizable world wide?

• In humans, adolescence and early adulthood bring extensive elimination of synapses in distributed association regions of cerebral cortex, such as the prefrontal cortex, that have greatly expanded in recent human evolution and appear to *become* impaired in schizophrenia.

• Many scientists recognize that environmental issues are *becoming* more and more important for the deep sea mining industry and a variety of tailing placements and pollution treatment programs were put forward for discussion.

• The study nicely illustrates how work in preclinical animal models can translate into novel insights relevant to human pathogenesis: epileptic seizures and neuronal hyperexcitability in AD patients *became* a recognized part of the condition. (AD—Alzheimer's disease)

conceive

v. 构思；设想；考虑；怀孕。

过去式 conceived；过去分词 conceived；现在分词 conceiving。

• **Author Contributions** W. G., S. E.

and A. J. *conceived* the idea and designed the experiments. W. G. , S. E. , H. Y. Y. N. and S. C. led the experiments (with assistance from K. C. , A. P. , H. M. F. , H. O. , H. S. , H. O. , D. K. , D. -H. L.). W. G. , S. E. , A. P. , G. A. B. , R. W. D. and A. J. contributed to data analysis and interpretation. W. G. , S. E. , H. Y. Y. N. , G. A. B. and A. J. wrote the paper and all authors provided feedback.

● It really is a miracle that is part of our toolkit, that we can use to manifest anything we can *conceive* at any time.

● The finding is in line with other recent work suggesting that stress relief might increase the success rate for women who have trouble *conceiving*.

● So by pre-planning the time you want to *conceive*, you should be able to decide the gender of your baby, or in theory anyway.

exploit/exploitation

exploit—*vt.* 开发，开拓；开采。exploitation—*n.* 开发；开采；开垦。

● Development and application of various omics as well as molecular biological database have been used to screen various biomarkers, evaluate the diagnosis, classification, therapeutic response and prognosis of disease and *exploit* new therapeutic methods and agents.

● Since 2006, the technical preparation for commercial *exploitation* of seafloor massive sulfide (SMS) deposits was significantly strengthened by Nautilus Minerals in the southwest Pacific area.

focus（复数 focuses/foci）

n. 焦点；中心。*v.* （使）集中；（使）聚焦。

过去式 focused/focussed；过去分词 focused/focussed；现在分词 focusing/focussing。

● Ecosystem response to climate change in high-altitude regions is a *focus* on global change research.

● Peter (2011) *focused* his investigation on the relationship between the volume of stored water and the largest simulated cluster of connected reservoirs.

● The hotspot of the MCE (magnetocaloric effect) research is *focused* on materials which undergo a first-order magnetic-phase transition.

● Trained as a microbiologist in Germany, Vogel had begun *focusing* on finding RNAs in pathogens during his postdoctoral work in Uppsala and Jerusalem and had continued this work when he started his own group in 2004 at the Max Planck Institute for Infection Biology in Berlin.

given

prep. 考虑到。*v.* 给予（give 的过去分词）。

● *Given* the lack of surveillance program after bariatric surgery, a delay in diagnosis caused by unspecific symptoms is common.

● *Given* that reed summer harvesting is recommended to restore the lake, questions must be raised about the effect on the local economy.

● Label-free quantification can also be performed by SRM but is only accurate for relative quantification *given* there is no standardization anchor point. (SRM—selected reaction monitoring. 笔者注)

● A schematic description of the layers used in the present study is *given* in Fig. 1.

● For time-lapse images, cells were fixed at time points *given* in Figure 5, as described in Buchan et al. (2013) (these images were taken using the Deltavision).

hypothesize/hypothesis

hypothesize—*v.* 假设，假定。hypothesis—*n.* 假设。

过去式 hypothesized；过去分词 hypothesized；现在分词 hypothesizing。

• We initially *hypothesized* that off-target effects of Streptococcus pyogenes Cas9 (SpCas9) might be minimized by decreasing non-specific interactions with its target DNA site.

• Other questions in the section, for example, ask users *hypothesize* about a modern day nuclear strike and tell of good deeds gone wrong.

• To facilitate below discussion, in this paper, the four pieces of *hypothesis* are defined premise *hypothesis*.

imply

vt. 意味；暗示；隐含。

过去式 implied；过去分词 implied；现在分词 implying。

• This finding *implies* that the earliest fatigue crack must be initiated from the epoxy-coating film.

• This *implied* a highly charged concept of home—although home did not necessarily mean birthplace.

• In India reports of consumption drops of 25 percent have caused domestic prices to fall 12–13 percent, *implying* lower production prospects.

in the presence of/in the absence of

有……时，在……存在时；有……在场/缺乏，不存在；无……时，缺少……时。

• Indeed, TLR2 and TLR4 also play a crucial role in tumor formation, especially *in the presence of* specific human genetic polymorphisms, such as TLR4 299Gly.

• We found that in mice deficient in C4 (ref. 36), C3 immunostaining in the dLGN was greatly reduced compared to wild-type littermates (Fig. 7a, b), with fewer synaptic inputs being C3-positive *in the absence of* C4 (Fig. 7c).

indicate

vt. 表明；指出；指示；预示。

过去式 indicated；过去分词 indicated；现在分词 indicating。

• The results *indicated* there was a remarkable main peak around 60 ~ 700 μm in the profile, which reflected the relatively stable sedimentary environment and single material source (Figures 3a–f).

• The exponential divergence of the nearby trajectories and hence an unstable orbit (chaos) is *indicated* by a positive λ.

• Downstream of 417 km, the daily curves remain essentially parallel; hence averaging over the five days provides robust results, as *indicated* in Figure 7.

• The outcomes of these studies affirm the existence of low-dimensional chaos, thus *indicating* the possibility of only short-term predictions.

insight

n. 洞察力；洞悉。

• These findings also provide *insight* into the mechanism of Cas9 targeting and nuclease activity.

• Measurements of human sweat could enable such *insight*, because it contains physiologically and metabolically rich information that can be retrieved non-invasively.

• Further experimentation and validation of hypotheses at the protein and in vivo level will still use classical reductionist approaches and yield individual *insights*, but being able to fit such pieces into the time, cell, and network dependent atlas that describes the evolution of Alzheimer ' s disease will be transformative for the field.

interpret

vt. 解释；说明。

- As a part of agent development, non-invasive imaging biomarkers can be used to *interpret* the developmental methods of new agents based in biological distribution, target participation, pharmacological/functional efficacy, efficacy/patient stratification or analysis on adverse responses.
- Because attitude values represented by a quaternion are not easily *interpreted*, the attitude tracking error is observed via the following method.

notion

n. 概念；见解。

- Despite their differences, common to these theories is the *notion* that the predicted size structure is a property of steady-state forests far removed from the influence of disturbance.
- These findings suggested that B. fragilis is a primary consumer of S-BMO, involved in a food web where E. coli acts as a secondary consumer. To test this *notion*, we incubated B. fragilis or E. coli in PBS containing S-BMO and assayed the supernatants for sialyllactose and sialic acid using ultra high performance liquid chromatography-mass spectrometry (UPLC-MS).

predict

vt. 预报，预言；预知。

- The model can successfully *predict* the experimental results with more than 90% accuracy with an average absolute percentage error of around 7%.
- This demonstrates the ability of the model to *predict* sewer solid capture efficiencies of the device under real-world conditions.
- They all stand on the shoulders of giants: British economist Thomas Malthus *predicted* in the 19th century that the rise in population would widespread famine and catastrophe.
- Another attempt is that the training algorithm was changed in the Network 16 (i. e., double layer-layer recurrent neural network) in order to get the further improvement in *predicting* the output result.

propose

vt. 提出；建议；打算，计划。

过去式 proposed；过去分词 proposed；现在分词 proposing。

- Previous papers *proposed* flow softening and microcracking as softening mechanisms.
- To efficiently reduce computational costs and handle the large number of design variables, a local relative density mapping (LRDM) method is *proposed* for lattice structures to be fabricated by additive manufacturing.
- Floatable control is preferred by most *proposed* and existing environmental regulations.
- The government is *proposing* increasing permanent residence for highly educated and highly skilled foreigners.

prove

vt. 证明。*vi.* 证明是。

过去式 proved；过去分词 proved/proven；现在分词 proving。

- Faraday used no mathematics at all to *prove* his rule.
- I had no choice but to write it down, so I could *prove* to my friends that nothing ever changes.
- Passports are frequently serviceable in

proving the identity of the traveller.

● These signatures have been produced using training datasets of varying numbers and many have *proven* prognostic in independent datasets.

● Last decade, a newly developed severe plastic deformation (SPD) technique named "constrained groove pressing (CGP)" was *proved* to be more suitable for producing ultra-fine grained (UFG) metal sheets than repetitive corrugation and straightening and accumulative roll bonding in ths aspects of processing result and practical application.

● But most of these guesses rested on little or no evidence, and one by one they *proved* to be wrong.

● As shown in Fig. 3, a diverse array of organic frameworks *proved* to be competent coupling partners for the C-H arylation protocol.

reflect

vt. 反映；反射。

● Members of the collection represent the microbial exposures experienced by the host, including inheritance of microbes from the mother, and *reflect* selection placed on the microbiota by regional dietary practices.

● To objectively *reflect* the influence of snow cover on soil water in the freeze-thaw process, a number of indices were selected to analyze soil water dynamics.

● Photosynthetic pigments within the plants absorb most of the visible light (red and blue), whereas far-red (FR) light is poorly absorbed and most is transmitted through or *reflected* by leaves.

● Figure 2g shows that the responses of glucose and lactate sensors increase rapidly upon elevation of the solution temperature from 22 ℃ to 40 ℃, *reflecting* the effect of increased enzyme activities.

shed light on

阐明；使……清楚地显出。

● While most mechanistic work on acquisition has been performed in Type I systems, recent studies in Type II systems have also *shed light on* key aspects of spacer acquisition.

● There is no doubt that further comparison of bacterial and eukaryotic condensins not only will *shed* new *light on* the evolutionary origins of the segregation machineries, but also will help us to understand the fundamental aspects of chromosome segregation through both theoretical and experimental approaches.

suggest

vt. 暗示；提议，建议；启发；使人想起。

● Since strains defective in the CCT complex form more stress granules, while granule clearance during recovery is unaffected (Figures 6A and 6B), we *suggest* that the CCT complex plays a role in inhibiting granule assembly, perhaps by limiting interactions between polyQ domains on stress granule components (Tam et al., 2006). (CCT complex—chaperonin-containing T complex. 笔者注)

● It *suggests* that, during the whole constrained groove pressing (CGP) process, the drop of elongation caused by work hardening and microcracking always exceeds the rise due to grain refinement.

● However, a higher close on Friday would begin to *suggest* a market low is in place.

● In addition, we identify new stress granule components DPYSL3, DCTN1, and TUBA4A, which have been *suggested* to modify ALS susceptibility (Blasco et al.,

2013; Münch et al., 2004; Smith et al., 2014).

• Although the variation at *C4* and in the distal extended MHC region associated with schizophrenia with similar strengths ($P = 3.6 \times 10^{-24}$ and 5.5×10^{-28}, respectively), their correlation with each other was low ($r_2 = 0.18$, Fig. 4b), *suggesting* that they reflect distinct genetic influences.

support

vt. 支持，支撑。*n.* 支持，支撑；支持者，支撑物。

• Although it has been proposed that the N-terminal tail of histone H2A acts as a chromatin "receptor" for condensin I (Tada et al., 2011), recent data from the chromatid reconstitution system do not *support* such a proposal (Shintomi et al., 2015).

• This was further *supported* by the different evaporation values at different landforms.

• These observations, although some what counter intuitive at first sight, implicate that the early-acting machineries have evolved to increase the fidelity of chromosome segregation *supported* by the more general, late-acting machineries.

• Finally, emerging functions of condensins in *supporting* interphase chromosome organization will be discussed.

• This indicates that the realization of the benefits of genomic knowledge for experimental protein research critically depends on the availability of assays *supporting* the quantification of any human protein.

• The genetic *support* for a role of autophagy in AD compared to Parkinson disease, for example, remains surprisingly limited, however.

via

prep. 通过；按照；经由。

• Emerging evidence indicates that one of the mechanisms of actions of chemotherapy is *via* activation of the immune system through multiple pathways.

• The proposed value-based model seeks to extend this plan *via* the principles of prioritization, optimization and unification.

• Endoscopic evaluation of the excluded stomach is described both *via* retrograde approach and double-balloon technique.

4.3　论文特别语义词汇

4.3.1　模糊时间

表达过去和现在的不确指时间，主要是一些同词根的形容词/副词，还有一些表达不确指时间的短语。

current/currently

current—*adj.* 现在的；最近的。currently—*adv.* 当前。

• In view of the *current* situation of Sichuan Provice, public goods investment should ideally be focused on increasing productivity and further concerned about the coordinated development of the region.

• There are *currently* three methods typically employed to measure circulating microRNAs (miRNA).

instant/instantly

instant—*n.* 瞬间；立即；片刻。*adj.* 立即的；紧急的；紧迫的。instantly—*adv.* 立即地；马上地；即刻地

• With the continuous increase of the bending moment, the bending angle and the springback angle also change. When the bending moment increases enough to cause the archwire to experience plastic deformation, the bending moment remains constant. We assumed that the bending moment was M_L at this *instant*.

● In previous centuries, anyone entering the Forbidden City without permission would have faced *instant* death.

● For instantaneous failure, the material is assumed to fail immediately in a mode at damage initiation, and thus the material properties associated with that failure mode are degraded *instantly*.

● In February 2011, Zhang heard a talk about CRISPR from Michael Gilmore, a Harvard microbiologist, and was *instantly* captivated.

late/lately

late—*adj.* 晚的；迟的；最近的。*adv.* 晚；迟；最近；在晚期。lately—*adv.* 近来，不久前。

● Given that the MCM and RVB complexes are suggested to serve as remodelers of protein-nucleic acid complexes, they may limit stress granule assembly by either promoting a *late* step in granule assembly, leading to more stable granules, or by inhibiting granule disassembly.

● First, this good message come *late* this month.

● *Lately*, we have been witnessing success stories of personalized medicine in oncology, through the registration of highly potent molecularly targeted agents for patients with tumors bearing specific molecular aberrations.

present/presently

present—*adj.* 现在的；出席的。*n.* 现在。presently—*adv.*（美）目前；不久。

● In the *present* study 47 sets of experimental data were collected.

● It may be stressed here that the results of the *present* investigation should be useful for understanding the essence of science.

● At *present*, three methods can be used to obtain vibration stress: experimental methods, three-dimensional (3-D) finite element analysis, and quantitative fractography.

● *Presently*, a new lifestyle called low carbon life is spreading every corner of our country.

● Staff at the WHO country office in Niger reported today that no human cases are *presently* under investigation for possible H5N1 infection.

previous/previously

previous—*adj.* 以前的；早先的。*adv.* 在先；在……以前。previously—*adv.* 以前；预先；仓促地。

● Many *previous* researches have confirmed the influence from sub-cloud evaporation on isotopic fractionation in precipitation.

● This list of proteins is significantly enriched for *previously* identified stress granule proteins ($p = 7.303 \times 10^{-61}$).

recent/recently

recent—*adj.* 最近的；近代的。recently—*adv.* 最近；新近。

● *Recent* studies have shown that the novel coronavirus has an average time interval of 7.5 days between human-to-human transmission.

● Many severe and critically-ill COVID-19 patients have been found *recently* with a higher level of IL-6 in their blood, the increasing level of IL-6 may be recommended as a warning sign that the patient's situation could possibly deteriorate.

to date

至今；迄今为止。

• *To date*, two different mechanisms of PD-L1 (programmed cell death ligand-1) expression on tumors have been described: innate immune resistance and adaptive immune resistance.

• *To date*, few genome-wide association study (GWAS) association shave been explained by specific functional alleles.

4.3.2 罗列举例

陈列、列举或提出案例或例子。

like

v. 喜欢；希望；想。*adj.* 同样的；相似的。*prep.* 像；如同。*conj.* 好像。

• Ideally, we would *like* to know the optimal value function under perfect play $v^*(s)$.

• Once the micro crack initiates, these pores are more *like* a "window".

• ANN has already been successfully used in similar kinds of environmental problems *like* water level predictions, flood forecasting and control in combined sewers in the works of Chiang et al. (2010), Bruen& Yang (2006) and Weyand (2002).

• In addition, RNP granules are observed to display liquid-*like* behavior of fission and fusion, and they flow *like* liquids under mechanical pressure

especially

adv. 特别；尤其；格外。

• The need for a comprehensive understanding of the processes and patterns of carbon allocation, *especially* against the background of global climate change has been emphasized.

• The existing methodology is robust for analysing cumulative salt inflows over river reaches but as signs salt inflows up to several kilometres downstream from where they actually occur, *especially* if river velocity is high through a narrow river cross-section.

for example

例如，比如，譬如。

• Thus differences between EC measurements of days 2 and 1 at the same location, *for example*, can be written as $EC(r,t_2)-EC(r,t_1) = BEC(r,t_1)$.

• Different microstructures will form during different SPD (severe plastic deformation), *for example*, amorphous or nanocrystalline structures.

• Its (Genomics) branches include pharmacological genomics, environmental genomics, nutritional genomics, functional genomics and toxicological genomics. *For example*, the terminal target of functional genomics is to destruct or decompose the gene interaction network in order to make the action mechanism of the physiology and disease clear, and a total of 20,000–25,000 coded protein genes and derivatives (*for example*, RNA, splice variant and gene mutation) in human genomes may become the potential biomarkers for some diseases, such as the specific mutation on epidermal growth factor receptor (EGFR) gene, p53 and K-ras have significant connection with various cancers.

for instance

例如，比如，譬如。

• Although this scenario emphasizes division of labor of the two condensin complexes, it is important to note that they have overlapping functions, too, and that condensin I can compensate lack of condensin II at least partially, *for instance*, in Xenopus egg cell-free extracts.

• The subunit organization of cohesins

shares many similarities to that of condensins: a mitotic cohesin complex, *for instance*, is composed of a distinct pair of SMC subunits (SMC1 and SMC3): a kleisin subunit (Scc1/Rad21) and a HEAT subunit (Scc3/SA). (SMC—structural maintenance of chromosomes. 笔者注)

in case/in the case of/in the case that

万一，假使/至于，在……的情况下/在这种情况下。

- *In case* amenable to surgical treatment, extent of resection is determined by tumor location.
- *In case* the reviewers have completed their reviews, you are requested to make your decision online.
- *In the case* of stochastic system, due to the change in the spectral slope on the differentiation, the correlation dimension of the differentiated signal will be much larger than that of the original signal.
- *In the case that* one or more of the reviewers assigned decline to review this paper, you should select one or more alternative reviewers.

include

vt. 包含，包括。

- Some of the common drawbacks in the available commercial devices *include* inadequate screening capacity, external power needs and high cost.
- Gene-therapy refers to the genetic modification of normal or diseased cells to treat diseases, *including* cancers.
- Given that the limted number of studies *included* molecularly annotated cohorts, a meta analysis to evaluate correlation of PD-L1 (programmed cell death ligand-1) espression with molecular alterations has not been conducted.

in particular

尤其，特别。

- This paper presents some innovative design concepts and, *in particular*, the development of four types of smart cutting tools, including a force-based smart cutting tool, a temperature-based internally-cooled cutting tool, a fast tool servo (FTS) and smart collets for ultraprecision and micro manufacturing purposes.
- Therefore, surface acoustic wave (SAW) sensor as an alternative is further researched in the smart tool design and development, which may also lead to the smart cutting tool for ultraprecision and micro machining *in particular*.
- What is worth mentioning *in particular* is that the valveless piezoelectric pump with a single raindrop-shaped tube developed by Huang, Wang and Yang [29, 30].

particularity/particularities

n. 特质；个性。

- The *particularity* and practicality of harmony operations of close-coupling multiple helicopters indicate that research on the relevant issues are urgent and necessary.
- The universality of contradiction resides in the *particularity* of contradiction.
- As a famous philosopher and mathematician in the world, Bertrand Russell's thoughts on education have their *particularities*.

specifically/specially

adv. 特别地；明确地。

- However, some markers *specifically* expressed in tumor cells using by Cell Search system in CTC detection have significant difference between western and eastern populations.
- This study analyzed the distribution characteristics of metastatic nidi esophageal

carcinoma in the aspect of positron emission tomography（PET）/CT, in which the metastatic nidi in rare locations were *specially* analyzed.

such as

比如；诸如。

• More recently, novel therapeutic strategies, *such as* immunotherapy, have been investigated.

• Wherever possible, the selection of the filter parameter value should be informed by field studies using techniques *such as* chemical tracers and reach water balances.

• At present, commercially available wearable sensors are only capable of tracking an individual's physical activities and vital signs（*such as* heart rate）, and fail to provide insight into the user's health state at molecular levels.

4.3.3 描写顺序

描写事物的先后次序，或事物结构、组成的某种关联或顺序。

follow

vt. 遵循；跟随。*vi.* 跟随；接着。

• The majority of participants（70.4%）indicated that they read and *follow* the package instructions on how to calculate the proper amount of fertilizer to apply.

• The current land use that delivers the majority of the nutrients to the estuary is grazing（39%）*following* by residential uses（17%）.

• Patients were *followed* up every month after the end of treatment for recording reoccurrence, metastasis and deaths.

immediate/immediately

immediate—*adj.* 立即的；直接的；最接近的。immediately—*adv.* 立即，立刻；直接地。

• The peak of successively *immediate* urinary flow rate curve during micturition process was traced by urinary flowmeter for calculating the maximum flow rate.

• As known to all, if diseases can be diagnosed early, and treated and controlled by corresponding measures *immediately*, the survival rate and quality of life（QOL）of patients will be greatly improved.

in the course of/during the course of

在……的过程中；在……期间。

• Information from the input layer is then processed *in the course of* one hidden layer; following which output vector（stress）is computed in the final（output）layer.

• *During the course of* the project, updated versions of the UniProtKB/Swiss-Prot database were released.

simultaneous/simultaneously

simultaneous—*adj.* 同时的；同时发生的。simultaneously—*adv.* 同时地。

• There is, however, a trend to develop *simultaneous* multi element measurements through the use of multichannel detection.

• Temperatures of the upper and lower surface of the CFRP boom change *simultaneously*.

• With respect to this, we believe it logical to develop combinational therapies that *simultaneously* target survival pathways in the leukemic cells and the cytoprotective mechanisms afforded by their interaction with their stromal neighbors.

subsequent/subsequently

subsequent—*adj.* 后来的，随后的。subsequently—*adv.* 随后，后来；接着。

• Interestingly, *subsequent* studies that coupled gene expression profiling analysis with that of genome copy number provided evidence for distinct profiles of copy number aberrations among the afore mentioned BC intrinsic subtypes.

• Once the application is deployed and *subsequently* started in an enterprise environment, it can then serve its clients, possibly for an extended period of time.

then

adv. 然后；那么；于是。*conj.* 然后，当时。

• The subsequent corrosion property will *then* be dependent on how fast corrosion enables penetration to the next layer of carbides in terms of different matrix compositions.

• We introduce a new technique that caches all moves from the search tree and *then* plays similar moves during rollouts; a generalization of the 'last good reply' heuristic.

• At each step of the rollout, the pattern context is matched against the hash table; if a match is found *then* the stored move is played with high probability.

4.3.4 对照比较

对不同事物或事物的不同方面作对比，揭示、突出所比较对象间的差异或共同点。

agree

vt. 同意；赞成；承认。*vi.* 同意；意见一致。

• While we *agree* that PIFs likely play a general role in growth under many conditions, we suggest that the upstream pathways acting on PIFs are diverse and condition specific rather than linear. (PIF—phytochrome interacting factors. 笔者注)

• Siksnys submitted his paper to *Cell* on April 6, 2012. Six days later, the journal rejected the paper without external review. (In hindsight, *Cell*'s editor *agrees* the paper turned out to be very important.) Siksnys condensed the manuscript and sent it on May 21 to the *Proceedings of the National Academy of Sciences*, which published it online on September 4.

• Yet market forces alone are *agreed* to be insufficient.

• The GSSP chosen to define the base of the Holocene was *agreed* to lie within the NGRIP2 ice core from central Greenland (NGRIP, North Greenland Ice Core Project). (GSSP—Global Boundary Stratotype Section and Point)

• We thank Fan Hui for *agreeing* to play against AlphaGo; T. Manning for refereeing the match; R. Munos and T. Schaul for helpful discussions and advice; A. Cain and M. Cant for work on the visuals; P. Dayan, G. Wayne, D. Kumaran, D. Purves, H. van Hasselt, A. Barreto and G. Ostrovski for reviewing the paper; and the rest of the DeepMind team for their support, ideas and encouragement.

compare

vt. 比较；对照。*vi.* 比较；相比。

• We *compare* our results *with* other available studies summarized in Table 9.

• To validate the accuracy and agreement between the optical models, we *compared* the model-based predictions both inside and outside of the eye, including the dependence on the ETL focal distance of the axial shift (Figure 2a) and the lateral scan range (Figure 2b, defined as the lateral displacement per scan angle).

• *Compared with* soil temperature, the effect of snow on soil moisture in the active layer of permafrost regions was more straightforward.

consistent with

符合；与……一致。

● *Consistent with* this, kidney transcript levels of a key component of this system, angiotensin receptor 1a (Agtr1a), increased between 12 and 18 months (Fig. 4g). By contrast, no such increase was observed in AP-treated mice.

● This interaction network argues that some proteins will be recruited to stress granules independent of binding RNA, which is *consistent with* the observation that roughly half of the proteins in the yeast or mammalian stress granule proteomes are not known to bind RNA (Castello et al., 2012; Mitchell et al., 2013) (Figure 4C; Tables S1 and S2).

● In this setup, no BRET response was observed following agonist stimulation of any of the receptors indicating that the BRET detected between barr and both Gαs and Gγ2 reflects molecular proximity *consistent with* the formation of megaplexes (Figures 4B-4E). (BRET—bioluminescence resonance energy transfer. 笔者注)

equivalent

adj. 等价的，相等的；同意义的。

● To obtain the desired structures, we adopt the method of adding a kinematic limb that contains an actuated joint to an existing RPR-*equivalent* PM. (PM—parallel mechanisms. 笔者注)

● Therefore, the appropriate relative density may be required to ensure that the elastic modulus of the generated lattice structure is *equivalent* to the corresponding elastic constants of the topology optimization.

fold

vt. 折叠；合拢。*vi.* 折叠起来。*n.* 褶层；折痕；倍。

● Parabolic membrane antennas are *folded* or wrapped when in stowed condition.

● Because of the great weight pressing down on them, these layers tend to *fold* downward at weak spots, and this finally causes an actual break in the crust.

● Five of the 32 single amino acid mutants reduced activity at all three off-target sites by at least 10-*fold* compared to wild-type (WT) SpCas9 while maintaining on-target cleavage efficiency, and 6 others improved specificity 2 to 5-*fold*.

● The copper(I) ions can be reversibly removed and added without loss of the COF structure, for which a 10-*fold* increase in elasticity accompanies its demetalation. (COF—covalent organic framework. 笔者注)

in line with

符合；与……一致。

● Sweat [Na^+] increases and [K^+] decreases in the beginning of perspiration, *in line with* the previous *ex situ* studies from the collected sweat samples.

● And we have announced our intention to cut our emissions in the range of 17 percent below 2005 levels by 2020 and ultimately *in line with* final climate and energy legislation.

liken

vt. 比拟；把……比作。

● One way of thinking about Chinese currency policy is to *liken* it to a form of industrial policy.

● ANN is a mathematical model consisting of a number of interconnected processing elements organized into layers; the geometry and functionality of which were *likened* to that of the human brain.

match

v. 相配；相称；比得上；相似。*n.* 匹配。

过去式 matched；过去分词 matched；现在分词 matching。

- This pump can be used in structural design to realize the special fluid delivery or energy transfer. It can *match* the energy consumption between diffuser and nozzle, which may become a design direction for the valveless piezoelectric pumps with cone-shaped tubes in the future.
- A non-iterative size, *matching*, and scaling (SMS) method was proposed by Graf et al.
- Our observations suggest that off-target mutations might be minimized by using SpCas9-HF1 to target non-repetitive sequences that do not have closely *matched* sites (for example, bearing 1 or 2 *mismatches*) elsewhere in the genome; such sites can be easily identified using existing publicly available software programs.
- The *match* version of AlphaGo continues searching during the opponent's move.

resemble

vt. 类似，像。

过去式 resembled；过去分词 resembled；现在分词 resembling。

- Seedlings grown at a constant fluence rate of red light, but specifically deprived of blue light, *resemble* in many ways plants exposed to low R：FR. （R：FR—red to far-red ratio. 笔者注）
- CRISPR-Cas9 nucleases enable highly efficient genome editing in a wide variety of organisms, but can also cause unwanted mutations at off-target sites that *resemble* the on-target sequence.
- Such an explosion should have *resembled* a basketball in shape.
- A new mineral facies has been found for the first time in mathiasite, resembling priderite in composition.

respective/respectively

respective—*adj.* 分别的，各自的。respectively—*adv.* 分别地；各自地。

- Furthermore, the slow changing patterns of the signals after 2.5 s and 5.5 s in *respective* Fig. 4 (a) and 4 (b) are not related to the cutting force, but most probably due to the pyroelectric effect of the piezoelectric material.
- Daily variation of the temperature is within a narrow range: long-term average values of minimum and maximum daily temperature in the region are 18 and 36 ℃, *respectively*.
- For the uncoated specimens, the fatigue limits (FLs) are 163.89 MPa in air and 67.35 MPa in 3.5 wt% NaCl solution, *respectively*, whereas those for the coated specimens are 175.95 and 171.97 MPa, *respectively*.

similar

adj. 相似的。比较级 more similar；最高级 most similar。

- The concept is *similar* to that of a cumulative volume, and knowing the river geometry the location of each element can be determined at each day.
- As figure 7 indicates, the overall results of the existing and new methodology are *similar* on a 50 + km scale.
- They are *more similar* to normal human languages than assembly or machine languages and are therefore easier to be used for writing complicated program.
- The problem can be described as:

searching the sequence *most similar* to a given time series from a large time series database.

versus

prep. 对；与……相对。

- Fig. 24 shows the relationship curves between cone angle and flow resistance coefficient for diffuser and nozzle, that is, curves of flow resistance *versus* cone angle.
- A Prussian blue mediator layer was deposited onto the Au electrodes by cyclic voltammetry from 0 V to 0.5 V (*versus* Ag/AgCl) for one cycle at a scan rate of 20 mV s^{-1} in a fresh solution containing 2.5 mM $FeCl_3$, 100 mM KCl, 2.5 mM $K_3Fe(CN)_6$, and 100 mM HCl.

4.3.5 猜测想象

对不能确定或准确把握的事物，给出或描绘出可能、猜想的结果或状态。

appear

vi. 出现；显得。

- Indeed, the well recognized albeit rare abscopal response *appears* to be immune-mediated.
- There are few pits *appearing* in ultrafine-structured coating after wear, whereas the binder phase is almost removed by the cutting action of abrasives leading to the exposure of hard phases.
- In this study, the samples underwent a four-pass constrained groove pressing (CGP) with a theoretically total strain of 4.64 until obvious cracks *appeared* on the surface of pressed samples at the last pass.

assume

vt. 假定，假设。

过去式 assumed；过去分词 assumed；现在分词 assuming。

- In fact, in this study we *assumed* there are going to be technological improvements.
- In the upper part (figure 3 (a)), a unit salt inflow is *assumed* between river 30 and 29 km, and no background EC (electrical conductivety) is present.
- Atovaquone is prone to resistance, and it has been *assumed* that this resistance will spread, as it has for other antimalarials.
- It is *assumed* that the last eukaryotic common ancestor (LECA) had ancestral forms of both condensins I and II and that some species, including fungi, have lost condensin II during evolution (Hirano, 2012).
- *Assuming* a single-centred flow model, the best-case sampling interval can be calculated to be less than 1 min, based on the sweat rate (~3–4 mg $min^{-1}cm^{-2}$) and the pad size (1.5 cm × 2 cm × 50 μm).

perhaps

adv. 也许；可能。

- During the match against Fan Hui, AlphaGo evaluated thousands of times fewer positions than Deep Blue did in its chess match against Kasparov; compensating by selecting those positions more intelligently, using the policy network, and evaluating them more precisely, using the value network—an approach that is *perhaps* closer to how humans play.
- *Perhaps* a more fundamental measure of human perturbation of the climate system is the human-driven change to the planetary energy balance at Earth's surface, as measured by changes in radiative forcing.

presume

vt. 假定；推测；意味着。

过去式 presumed；过去分词 presumed；

现在分词 presuming。

● On account to four kinds of knowledge, we *presume* many complexes about nuclear lamina have relationship with nuclear structure, signaling and gene regulation.

● It is wrong to *presume* guilt or innocence without due process.

● It is *presumed* by most test approaches that users are trying to use the program to help them get some work done.

presumably

adv. 大概；推测起来；可假定。

● Kinesins are key components in spindle movements during the cell cycle and can *presumably* meliorate the action of docetaxel. Inhibition of the kinesin complex or key members may interfere with the resistance mechanism to docetaxel and highlight a possible avenue for therapeutic intervention.

● It is worth noting that the SL policy network p_σ performed better in AlphaGo than the stronger RL policy network p_ρ, *presumably* because humans select a diverse beam of promising moves, whereas RL optimizes for the single best move.

seem

vi. 似乎；像是。

● There also *seems* to be a lack of regulation and guidance for basement drainage.

● One might have predicted that the LBL pathway has faster kinetics than low R：FR, a pathway that *seems* to be more complex. (LBL—low blue light; R：FR—red to far-red ratio. 笔者注)

speculative

adj. 投机的；推测的；思索性的。

● These types of *speculative* ventures can now be de-risked, and market proof validated before the project creator starts to "cut wood".

● The negative effects of global warming are *speculative*, and there and then.

● It rejects all dogmatic and *speculative* assertions in philosophy.

suppose

vt. 假设；认为；推想。

过去式 supposed；过去分词 supposed；现在分词 supposing。

● *Suppose* that the enterprise invested RMB 250,000 Yuan in machine tool product configuration design, must the design work be finished within 10 d?

● In this case, the telescopic arm is *supposed* to be a rigid body, and the schematic of the robot mechanism can be simplified.

● It is possible to develop the theory of consumer choice without *supposing* that a utility function exists at all.

● Anyone *supposing* that Afghanistan is ready for instant democracy is deluding themselves, and attempts to institute it are less than likely to succeed.

4.3.6　修饰限制

对名词或名词短语（主语、宾语），动词或动词短语（谓语），形容词或形容词短语（状语）进行修饰限定，使语义表达更加准确、细致和生动。

apparent/apparently

apparent—*adj.* 明显的；表面上的。apparently—*adv.* 明显地；表面上。

● Moreover, if the GFP-tagged protein only occupies part of the core structure, the *apparent* volume occupied by GFP, as measured by microscopy will be smaller than the hydrodynamic volume measured by particle tracking.

• Thus, the two HEAT subunits of condensin Ⅰ have *apparently* antagonistic roles in regulating dynamic assembly of chromosome axes.

approximate/approximately

approximate—*adj.* 近似的；大概的。approximately—*adv.* 大约，近似地；近于。

• The *approximate* invariance of the middle frequencies energy relationship between video adjacent frames under photometric distortion and spatial desynchronization is discovered by analysis.

• The conversion between the two units may be expressed *approximately* by TDS [mg/L] = $K \times EC$ [μS/cm].

clear/clearly

clear—*adj.* 清楚的；明显的。clearly—*adv.* 清晰地；明显地；无疑地。

• Aggregate yields from thermonuclear weapon tests that began in 1952 CE and peaked in 1961–1962 CE left a *clear* and global signature, concentrated in the mid-latitudes and highest in the Northern Hemisphere, where most of the testing occurred (Fig. 4B) (CE 和 AD 同义，公元纪年中公元后的表示方法，CE 是英语 Common Era 的缩写，AD 是拉丁文 Anno Domini 的缩写。笔者注)

• The process of preferential adsorption is not *clearly* described.

• However, the ORR active site (or sites) is *unclear*, which retards further developments of high-performance catalysts. (ORR—oxygen reduction reaction. 笔者注)

• Figure 2 shows *clearly* that the passive corrosion behavior of the S45C steel substrate does not resemble the corrosion behavior of the coatings, and the differences of the electrochemical response of the coatings arouse from the different multi-phase microstructures of the corroding coatings.

compelling/compellingly

compelling—*adj.* 强制的；引人注目的。compellingly—*adv.* 咄咄逼人地。

• One of the more *compelling* arguments to increase focus on oligodendrocytes and possible demyelination is the strongly accentuated age-related breakdown of myelin in AD and in APOE4 patients. (APOE4—Apolipoprotein E4. 笔者注)

• The Sound Bar produces *compellingly* rich audio with 2.1 channels, 310 W speakers and LG's advanced 3D Surround Processor.

complete/completely

complete—*adj.* 完全的；彻底的；完整的。completely—*adv.* 完全地，彻底地；完整地。

• We now have the *complete* quaternion-based nonlinear controller for the quadrotor UAV. (UAV—unmanned aerial vehicles. 笔者注)

• In the cold ring rolling, the bonding strength of interface is low due to *completely* mechanical connection, which can not meet the manufacturing requirements for high-performance duplex-metal composite rings.

considerable/considerably

considerable—*adj.* 相当大的；重要的，值得考虑的。considerably—*adv.* 相当地；非常。

• In addition, the process is complex with characteristics such as difficult to preform, huge cost and *considerable* wasting material and energy.

• The area of lattice structures has received *considerable* attention owing to their excellent properties: they can be designed and used for multiple purposes, such as weight reduction, heat transfer, energy absorption, thermal protection.

● The following must be done to increase farmers' incomes *considerably* as fast as possible.

critical/critically

critical—*adj.* 关键的；决定性的。critically—*adv.* 很大程度上；极为重要地。

● Therefore, the investigation of the stability of supersaturated α solid solution and the process of precipitation is *critical* to understand the quenching sensitivity of aluminum alloys.

● Continuous efforts should *critically* compare and contrast the actions of condensins and cohesins and further clarify their similarities and differences (see also Box 1).

crucial/crucially

crucial—*adj.* 重要的；决定性的。crucially—*adv.* 关键地；至关重要地。

● The vibration stress in resonance condition of a blade is *crucial* to determine the rationality of the engine design and is the basic work of blade failure analysis.

● *Crucially*, however, the same learning goals are set for all children, and the quality of their learning is assessed against the same criteria.

distinguished

adj. 著名的；卓著的。

● Professor Liu is one of the most *distinguished* scholars in international mechanical engineering field.

● An ancient Greek city of Cyrenaica, founded in 630 B. C., it was noted as an intellectual center with *distinguished* schools of medicine and philosophy.

dramatic/dramatically

dramatic—*adj.* 戏剧的；引人注目的。dramatically—*adv.* 戏剧地；引人注目地。

● By virtue of the *dramatic* development of high-throughput biological technique, biologists have established protein interaction network.

● If two machines are operated in coordination, as in Fig. 5 (b), the efficiency can be *dramatically* improved.

essential/essentially

essential—*adj.* 基本的；必要的；本质的。essentially—*adv.* 本质上；本来。

● The fast tool servo (FTS) can be considered as one type of smart tool systems to position the tool precisely and dynamically, which is *essential* for machining micro-structured surfaces and special-featured components.

● Secondly, the key principles of relative density mapping (RDM) and local relative density mapping (LRDM) are *essentially* different.

exact/exactly

exact—*adj.* 准确的，精密的；精确的。exactly—*adv.* 恰好，正是；精确地；正确地。

● To eliminate the non-ideal effects such as voltage offset and to obtain precise signal readings, the *exact* numerical linear relationship between output and input was obtained to map the original input signal to the analogue circuit readouts, which in turn allowed for subsequent signal calibration and processing at the software level.

● It is believed that microstructures of *XY*-plane describe the basic features of constrained groove pressing (CGP) deformation more *exactly* than those of others.

exclusive/exclusively

exclusive—*adj.* 独有的；排外的；专一的。exclusively—*adv.* 唯一地；专有地；排外地。

• The Nautilus Minerals was approved by the government of Papua New Guinea (PNG) and given exploration licenses for *exclusive* economic zones (EEZs) for SMS deposits.

• Although terms such as interweaving (3), poly catenated (2), and interpenetrating (4-6) have been used to describe interlocking of 2D and 3D extended objects (Fig. 1, C and D), most commonly found in MOFs, we reserve the term " weaving " to describe *exclusively* the interlacing of 1D units to make 2D and 3D structures (Fig. 1, A and B).

extensive/extensively

extensive—*adj.* 广泛的；大量的；广阔的。extensively—*adv.* 广阔地；广大地。

• *Extensive* research and testing have since demonstrated that well-managed indoor residual spraying programmes using DDT pose no harm to wildlife or to humans.

• Nanocomposite magnets, comprising a mixture of hard and soft magnetic phases with nano-scale grains, were *extensively* studied for their noticeable merits like large saturation magnetization, large maximum energy product, and low content of rare earth.

extreme/extremely

extreme—*adj.* 极端的；极度的；偏激的。extremely—*adv.* 非常，极其；极端地。

• Sometimes researchers should take *extreme* measures to solve some difficult problems.

• This decrease is mainly due to the degradation of the lower zone of permafrost, especially the degradation of the *extremely* unstable type, about 44.2% of which has changed into seasonal frozen ground.

high/highly

high—*adj.* 高的；高级的。adv. 高。highly—*adv.* 高度地；非常。

• After the cold-extrusion forming of the internal thread, the metal in surface layer appears cold hardening, the surface structure fiber is refined, reasonable streamline distribution forms along the thread form, and residual stress field exists in certain depth on the surface layer, and the tensile strength and anti-fatigue performance of cold-extruded internal thread is made far *higher* than the tensile strength and anti-fatigue performance of cold-extruded internal thread processed by other methods.

• In order to achieve *high*-precision metrology of micro/nano CMMs, the probing head needs a *highly* precise ball-ended stylus tip with accuracy in the sub-micro order. (CMM—Coordinate Measuring Machining. 笔者注)

increasingly

adv. 越来越多地；渐增地。

• High-altitude ecosystem response to climate change is becoming *increasingly* important in the study of ecological changes on different time scales.

• The demand for large-aperture (hundreds of square meters or more) antennas in future space missions has *increasingly* stimulated the development of deployable membrane antenna structures due to their light weight and small stowage volume.

intense/intensely

intense—*adj.* 强烈的；紧张的；非常的。intensely—*adv.* 强烈地；紧张地。

• The rapid development of technology derived from CRISPR-Cas systems, most notably Cas9 but also Cas6f/Csy4, Cascade, and Cpf1, has fueled *intense* interest in the field.

• Flow structure alters *intensely* in ignition process, which results in remarkable differences between flow structures of reaction and cold flow.

interesting/interestingly

interesting—*adj.* 有趣的。interestingly—*adv.* 有趣地。

● An *interesting* question will be to determine whether SpCas9-HF1 induces off-target mutations at frequencies below the detection limit of existing unbiased genome-wide methods (Supplementary Discussion).

● *Interestingly*, a subpopulation of Th17 (supTh17) cells exhibits immune suppressive properties because it expresses high levels of both CD39 and FOXP3 and consequently produces extracellular adenosine.

like/likely

like—*adj.* 像，相似的；和……一样，同样的。likely—*adj.* 很可能的。*adv.* 很可能地；或许。

● We provide evidence that stress granules contain a dynamic shell-*like* structure surrounding stable cores.

● People are more *likely* to take action if they believe their behaviour is a significant part of the problem and that by changing their behaviour they could make a difference.

● At 25 to 30 years after disturbance, the power function extends through the full range of diameters present, and *unlike* in younger patches, a power law is a *likely* model of the data.

● At the propagation region, typical beach marks are observed visually, which is *likely* due to low-frequency engine start-up cycles.

● In sharp contrast, modern rates of atmospheric C emission (~9 Pg $year^{-1}$) are probably the highest of the Cenozoic era (the past 65 Ma), *likely* surpassing even those of the Paleocene-Eocene Thermal Maximum.

● Quantification of exon23 excision revealed variable efficiencies (Fig. S10, G and J), which *likely* reflected targeting of only a subset of endogenous satellite cells that may be variably represented among the isolated and cultured cells.

near/nearly

near—*adj.* 近的；近似的。*adv.* 近；接近。nearly—*adv.* 差不多，几乎。

● The most strongly associated markers in several large case/control cohorts were *near* a complex, multi-allelic, and only partially characterized form of genome variation that affects the *C4* gene encoding complement component 4 (Extended Data Fig. 1).

● Among them, when driving frequency of 3 000 Hz, the pump with a minimum side length of 143 μm had the maximum flow rate of *near* 400 ml/min.

● Previous analyses of outbreaks of related flaviviruses suggest that, to be informative, molecular epidemio logical studies of the current Zika virus (ZIKV) epidemic should use full or *near*-complete coding region sequences.

● *Nearly* all probing styluses with a ruby ball tip in the market are made by this way.

● Both thrust forces remain negative at the early stage but increase gradually and turn into positive, reaching *nearly* 0.5 N and 3.5 N, respectively, at a later stage.

nice/nicely

nice—*adj.* 令人愉快的；美好的；友好的；细微的。nicely—*adv.* 漂亮地；恰好地；精细地。

● All the same, it would really be *nice* of the leader to let employees know the goals of company development.

● From a psychological point of view,

animals, especially birds and mammals, have evolved to be *nice* to their kin, especially to their children.

• Recent biochemical studies have *nicely* recapitulated the entrapment reaction using purified proteins in vitro (Murayama and Uhlmann, 2014, 2015), although the question of exactly how cohesin holds two sister chromatids together remains under debate (e. g., Eng et al., 2015).

• The study *nicely* illustrates how work in preclinical animal models can translate into novel insights relevant to human pathogenesis: epileptic seizures and neuronal hyper excitability in AD patients became a recognized part of the condition (Bakker et al., 2012).

permanent/permanently

permanent—*adj.* 永久的，永恒的；不变的。permanently—*adv.* 永久地，长期不变地。

• China will launch a spacecraft this month to conduct its first manned space docking, state media reported on Saturday, the latest step in the country's plan to build a *permanent* space station by 2020.

• Who ever thought that China would send people into space and the USA would *permanently* end their space program?

potential/potentially

potential—*adj.* 潜在的；可能的。potentially—*adv.* 可能地，潜在地。

• Smart tooling and smart machining have tremendous *potential* development and are drawing attention as one of next generation precision machining technologies particularly in the Industry 4.0 context.

• Exchange rates constitute a *potentially* valuable but controversial tool for hydrological managment.

precise/precisely

precise—*adj.* 精确的；明确的。precisely—*adv.* 精确地；恰恰。

• For the results that follow below, we used the *precise* composition $FA_{0.83}Cs_{0.17}Pb(I_{0.6}Br_{0.4})_3$, which has an optical band gap of 1.74 eV as determined by a Tauc plot (Fig. S7).

• However, most of them cannot be used to measure the real micro geometrical features high *precisely* because the parameters of the ball tips are not appropriate.

• By controlling *precisely* the volume fraction in solid phase, the low-density defects due to insufficient feeding can be avoided.

predominant/predominantly

predominant—*adj.* 主要的；支配的；有影响的。predominantly—*adv.* 主要地；显著地。

• The prominent yellowish color to these stones is consistent with their *predominant* cholesterol component.

• An average global temperature increase of 0.6 ℃ to 0.9 ℃ from 1900 to the present, occurring *predominantly* in the past 50 years, is now rising beyond the Holocene variation of the past 1400 years, accompanied by a modest enrichment of $\delta^{18}O$ in Greenland ice starting at ~1900.

probable/probably

probable—*adj.* 很可能的；可信的。probably—*adv.* 大概；或许；很可能。

• At every step of the tree traversal, the most *probable* action is inserted into a hash table, along with the 3 × 3 pattern context (colour, liberty and stone counts) around both the previous move and the current move.

• It is a remarkable fact that the distribution of synergy angle β_s at a blending angle of 45° and 50° is different from other. It is also *probable* that the spiraling flow around rod changed.

● *Probably* influenced a lot by the research from Stemme [1] and Gerlach [19], early valveless piezoelectric pumps with cone-shaped tubes were mostly in the form of solid.

● These trends are *probably* caused by increased blood serum [Na^+] and [K^+] with dehydration and increased neural stimulation, a conclusion in agreement with previous *ex situ* sweat analyses.

● The paradoxes in the different studies will *probably* only be resolved when the progressive and divergent cellular responses in Alzheimer's disease (AD) are systematically and comprehensively delineated.

profound/profoundly

profound—*adj.* 深厚的；深远的；渊博的。profoundly—*adv.* 深刻地；深深地；极度地。

● As a patient-oriented science, translational medicine has established a bidirectional channel between clinical treatment and basic research, in which problems in clinical practice are discovered and proposed, whose corresponding measures are found and established in basic research and then applied in clinic to resolve the practical problems in clinic, so it has favorable application prospect and *profound* social significance.

● Reflecting back on this era is of course a reminder of the many ways in which China has changed *profoundly* in a relatively short period of time.

radical/radically

radical—*adj.* 根本的；彻底的；激进的。radically—*adv.* 根本上；彻底地；激进地。

● The XRD (X-Ray Diffraction) patterns are obtained from the *radical* direction of the ring.

● An updated family tree of the animal kingdom could *radically* change the way we think about the evolution of species.

ready/readily

ready—*adj.* 准备好的；现成的；快要……的。readily—*adv.* 容易地；乐意地。

● However, the new Ares-Orion vehicle is not expected to be *ready* until at least 2015.

● These low-temperature natural spherules, which are *readily* distinguishable from high-temperature industrial SCPs, demonstrate the likely persistence of SCPs as a stratigraphic marker. (SCPs—spherical carbonaceous particles. 笔者注)

● Human C4A and C4B proteins, whose functional specialization appears to be evolutionarily recent (Extended Data Fig. 10a), show striking biochemical differences: C4A more *readily* forms amide bonds with proteins, while C4B favours binding to carbohydrate surfaces, differences with an established basis in C4 protein sequence and structure.

reasonable/reasonably

reasonable—*adj.* 合理的，公道的。reasonably—*adv.* 合理地；适度地。

● On the other hand, the arrangement of cables should be more difficult in view of the recover patient's safety. Therefore, the number of cables can be down to a *reasonable* number if possible.

● The notion of *reasonably* widespread entrepreneurial opportunity underpins basic faith in the free-market system.

remarkable/remarkably

remarkable—*adj.* 卓越的；非凡的；值

得注意的。remarkably—*adv.* 显著地；非常地；引人注目地。

● The *remarkable* thing about online media is that conversations can bring in so many people.

● Tropical tree size distributions are *remarkably* consistent despite differences in the environments that support them.

rough/roughly

rough—*adj.* 粗糙的；粗略的；未经加工的。roughly—*adv.* 粗糙地；概略地。

● You can paint them with two-dimensional images, or add *rough* textures to their surfaces.

● In the first DNA fragment he examined, Mojica found a curious structure—multiple copies of a near-perfect, *roughly* palindromic, repeated sequence of 30 bases, separated by spacers of *roughly* 36 bases—that did not resemble any family of repeats known in microbes (Mojica et al., 1993).

spontaneous/spontaneously

spontaneous—*adj.* 自发的；自然的；无意识的。spontaneously—*adv.* 自发地；自然地；不由自主地。

● Economic globalization is the *spontaneous* result of the development of productivity.

● One hypothesis is that RNP granules are liquid-liquid phase separations, in which high concentrations of assembly components reach a critical threshold and then *spontaneously* assemble into RNP granules through weak multivalent interactions (Brangwynne et al., 2011; Weber and Brangwynne, 2012).

strong/strongly

strong—*adj.* 坚强的；强壮的。*adv.* 强劲地；猛烈地。strongly—*adv.* 强有力地；坚强地；激烈地。

● The disappointing export figures, however, were partly offset by *strong* investment data released the same day

● The quadrotor UAV is nevertheless an under-actuated, nonlinear, and *strongly* coupled system with 6 degrees of freedom (DOF). (UAV—unmanned aerial vehicles. 笔者注)

● The performance and reliability of the ring products have a close relationship with the final microstructure, which depends *strongly* on the microstructure evolution history and processing parameters in the accumulative and multi-pass processing.

substantial/substantially

substantial—*adj.* 实质的；充实的；大量的。substantially—*adv.* 实质上；大体；充分地。

● Immunotherapy is a rapidly evolving and complex field that offers great potential to deliver *substantial* benefits to patients with a range of different cancers.

● By electronic guidance of the radar beam this technology enables the sensor to fulfill several tasks at the same time while increasing the detection capability *substantially*.

surprising/surprisingly

surprising—*adj.* 令人惊讶的；意外的。surprisingly—*adv.* 惊人地；出人意外地。

● Experimenters have teased out this *surprising* conclusion by inviting subjects to gamble on the throw of dice.

● *Surprisingly*, while the 3′ hairpins of the tracrRNA have been shown to provide nearly all of the binding energy and specificity for Cas9, the repeat-anti-repeat region of the sgRNA as well as the seed sequence were required to induce the conformational rearrangement.

● Not *surprisingly*, condensing II's interphase functions are tightly regulated.

temporary/temporally（temporarily）

temporary—*adj.* 暂时的，临时的。temporarlly（temporaily）—*adv.* 现世地；暂时地；临时地，临时。

• There is plenty of evidence to support the fact that such *temporary* measures can make a permanent difference.

• Although there is no doubt that marked environmental perturbations occurred at both 8200 and 4200 yr B. P., most proxies indicate subsequent recovery in a matter of centuries, implying that these were *temporally* discrete paleoclimatic events as opposed to truly novel states within the Earth system. （yr B. P.—years before the present. 笔者注）

• Fig. 19 illustrates the attitude tracking error, from which we can observe that the attitude error *temporarily* increases in the presence of interference moment and force but soon converges to approximately zero; it also shows that the attitude tracking error is relatively small.

• The dewatered ores are stored *temporarily* in the hull of PSV and then discharged to a transportation vessel moored alongside the PSV every 5–7 days. （PSV—production support vessel. 笔者注）

thorough/thoroughly

thorough—*adj.* 彻底的；周密的。thoroughly—*adv.* 彻底地，完全地。

• Due to limitations of space, not all aspects of condensin biology, e. g., a *thorough* treatment of meiotic roles in different organisms, will be covered here.

• To prepare the glucose sensors, the chitosan/carbon nanotube solution was mixed *thoroughly* with glucose oxidase solution（10 mg ml^{-1} in PBS of pH 7.2）in the ratio 2∶1 (volume by volume).

virtual/virtually

virtual—*adj.* 虚拟的；事实上的；实质上的。virtually—*adv.* 事实上；实质上；几乎。

• The torque ***M*** and force ***F*** are used as *virtual* controls in Eqs.（5）and（7）.

• Plastics spread rapidly via rivers into lakes, and they are now also widespread in both shallow-and deep-water marine sediments as macroscopic fragments and as *virtually* ubiquitous microplastic particles（microbeads, "nurdles," and fibers）（Fig. 2A）, which are dispersed by both physical and biological processes.

4.3.7　增加减少

表述增加、减少或相关的语义（如增大、增强、加强、促进，降低、减弱、抑制、阻止）。

decrease

n. 减少，减小；减少量。*v.* 减少，减小。过去式 decreased；过去分词 decreased；现在分词 decreasing。

• Meanwhile, Table 2 also shows that the elongation *decreases* continuously as strain increase, and the highest rate of *decrease* is observed at Pass 1.

• The effects of distance from quenched end hardness（Fig. 2b）indicate that the hardness *decreases* with the increase of distance from quenched end.

• Microcephaly is associated with *decreased* neuronal production as a consequence of proliferative defects and death of cortical progenitor cells

• We initially hypothesized that off-target effects of SpCas9 might be minimized by *decreasing* non-specific interactions with its target DNA site.

hamper

vt. 妨碍；阻止；束缚。

- A recent research finds that listening to loud music while driving can seriously *hamper* reaction times and cause accidents.
- Further US dollar strength will *hamper* China's efforts to stabilise the renminbi and limit capital flight, with the risk that the country raises short-term borrowing costs.
- Cool and gas outburst is one of the most serious natural disasters to *hamper* the security status during mining.
- With databases growing at an unrelenting pace, the ability to collect statistics by accessing all of the data may be *hampered* by fixed batch windows or memory and CPU constraints.
- Such bottlenecks have been *hampering* the opening of new mines and the expansion of existing ones.

increase

n. 增加，增长；提高。*v.* 增加，增大，加大。

过去式 increased；过去分词 increased；现在分词 increasing。

- Fossil fuel combustion has disseminated black carbon, inorganic ash spheres, and spherical carbonaceous particles worldwide, with a near-synchronous global *increase* around 1950.
- The *increase* of supersaturated solid will *increase* the distortion of lattice to impede the movement of conduction electrons, reduce the electron mean free path, and decrease the electrical conductivity.
- It has been argued that ~8000 years ago, with a global population estimated at less than 18 million, the initiation of agricultural practices and forest clearances began to gradually *increase* atmospheric CO_2 levels.
- This pattern is consistent with more efficient switching from anabolic storage of fat in the fed state to its oxidation in the fasted state, or *increased* metabolic flexibility (Muoio, 2014).
- Global sea levels *increased* at 3.2 ± 0.4 mm/year from 1993 to 2010 and are now rising above Late Holocene rates.
- Although this population growth is commonly thought to have *increased* exponentially through the 19th and 20th centuries, recent analyses suggest that it can be differentiated into a period of relatively slower growth from 1750 to 1940 CE and one of more rapid growth from 1950 to 2010 CE.
- The reduced sediment flux to major deltas, combined with *increasing* extraction of groundwater, hydrocarbons, and sediments (for aggregates), has caused many large deltas to subside more quickly, a process beginning in the 1930s at rates faster than modern eustatic sea-level rise.

inhibit

vt. 抑制；禁止。

- Therefore, it would *inhibit* grains growth to some extent.
- The results show that the ultrafine-structured coating has much higher density and *inhibited* decarburization than the conventional coating, which thus results in higher hardness and elastic modulus values than the micronsized coating.
- But their response was dampened or *inhibited*, when the researchers subsequently scratched the itchy skin.
- The new gel has a double action, *inhibiting* two enzymes involved in virus replication and infection of other cells.

prevent

v. 预防，防止；阻止。

过去式 prevented；过去分词 prevented；现在分词 preventing。

• The results showed that the small reservoirs *prevent* the formation of large clusters during non-extreme rainfall events, which reduces the system vulnerability to flooding.

• The researchers say thousands of deaths could be *prevented* if people ate less meat.

• Despite a lack of overt difference at 18 months, AP-treatment *prevented* age-dependent reductions in both spontaneous activity and exploratory behaviour measured by open-field testing (Fig. 3c), which was independent of sex and genetic background.

• Thus, investigating and *preventing* the fracture failures of blades as a result of vibration fatigue are of great significance.

• Although treatments exist for the psychotic symptoms of schizophrenia, there is no mechanistic understanding of, nor effective therapies to *prevent* or treat, the cognitive impairments and deficit symptoms of schizophrenia, which are the earliest and most constant features of the disorder.

raise

n. 加薪；上升。*v.* 提高；提升；养育；升起；上升。

过去式 raised；过去分词 raised；现在分词 raising。

• A few months later, members of Congress voted a pay *raise* for themselves and the executive branch of government.

• During wet weather conditions, sewer overflows to receiving water bodies *raise* serious environmental, aesthetic and public health problems.

• After the power is *raised*, the sweat rate visibly increases, followed by a sharp increase in skin temperature and sweat [Na^+] as well as a slight increase in [K^+] (in three of the seven subjects, [K^+] remained stable)

• The finding that most eukaryotic species have two different condensing complexes has *raised* a number of fundamental questions in chromosome biology.

• A previous study of germ-free and conventionally *raised* mice indicated that the microbiota affects bone mass.

• Furthermore, recent studies showing that senescent cells have beneficial effects in injury repair and tissue remodeling have called into question the simplistic view of senescence as only a driver of age-dependent pathologies, *raising* the specter that senescent cell clearance might remove useful cells in addition to detrimental ones.

reduce

v. 减少；降低；缩小。

过去式 reduced；过去分词 reduced；现在分词 reducing。

• This means that comparison of baseflow indices between studies and over time are difficult and this *reduces* confidence in the approach.

• As a result of ZIKV infection, the average growth area of ZIKV-exposed organoids was *reduced* by 40% compared with brain organoids under mock conditions [$0.624 \pm 0.064\ mm^2$ for ZIKV exposed organoids versus $1.051 \pm 0.1084\ mm^2$ for mock-infected organoids (normalized); Fig. 4E].

• Studies on murine models reveal that dysbiosis "alone" is able to *induce* CRC formation in presence of polymorphisms responsible for *reduced* activity of NOD2 (CRC—colorectal cancer. 笔者注)

• It is possible to form a lower-band-gap

triiodide perovskite material and current-match the top and bottom junctions in a monolithic architecture by simply *reducing* the thickness of the top cell.

stimulate

v. 刺激；激励。

过去式 stimulated；过去分词 stimulated；现在分词 stimulating。

- These results demonstrate that both the β_2V_2R and V_2R *stimulate* Gs signaling from internalized compartments, whereas the β_2AR does not seem to exhibit such behavior using this method.
- Inshort, agonists were allowed to *stimulate* receptors for varying time intervals (0.5 to 14 min), followed by a washout of the agonists.
- The demand for large-aperture (hundreds of square meters or more) antennasin future space missions has increasingly *stimulated* the development of deployable membrane antenna structures due to their light weight and small stowage volume.
- Disintegration activity was *stimulated* when using the correct leader-repeat border sequences, highlighting intrinsic sequence-specific recognition by Cas1.
- Aware of Zhang's efforts and *stimulated* by Charpentier and Doudna's paper, Church set out to test crRNA-tracrRNA fusions in mammalian cells.
- The hormone plays a key role in *stimulating* release of the hormones which control the menstrual cycle.

suppress

vt. 抑制；镇压。

- Homologous chromosomes are paired in somatic diploid cells in this organism, and condensin II subunits were identified as factors that *suppress* such somatic homolog pairing (Joyce et al., 2012) (Figure 4A).
- Attempts to *suppress* an idea will lead to the idea appearing in your dreams.
- From the figures, it can be observed that the velocity, attitude and angular velocity initially exhibit large fluctuations in response to torque and force disturbances, but these fluctuations are quickly *suppressed*.
- For example, recent work suggests that reactive astroglia in these mice secrete GABA, which paradoxically could contribute to seizure-like activity by *suppressing* inhibitory input on excitatory circuitry (Jo et al., 2014). (GABA—Gammer Amino Butyric Acid：氨基丁酸。笔者注)

4.4　常见易出错混淆词汇

4.4.1　a 为首字母

英语中有的词在构型上相近，意义却不同，有的词构型不同，意义却相同或相近，使用时若不加区分，就容易出错。以下给出一些常见的易出错和易混淆的词语，并简单辨析。

above

above 常指前面提到过，如 as discussed above、the above method 等，类似的词语有 such as the former、as stated in the latter 等。使用这类词语时必须以能够确指为前提，所指事物在上文中明确写出过，或由语境自然明确确定，若不能确指，虽使用起来方便，却易造成表达不清楚，进而造成理解困难。例如，以下句子中的 above 就可能存在指代不明的问题：

- As discussed *above*, project and technical parameters are random variables whose variance decreases over the project lifecycle.

• Compared with other stratigraphic changes described *above*, the climate and sea-level signals of the Anthropocene are not yet as strongly expressed, in part because they reflect the combined effects of fast and slow climate feedback mechanisms.

• So what should we do to promote peace, inaddition to the proposals mentioned in *above* paragraph?

absorption/adsorption

这两个词表面上看只有一个字母的差别，均表示“吸纳”，但准确地说，前者侧重吸收，后者侧重吸附，区别在于吸纳的程度，前者到了里面，后者却只到表面。例如：

• This controls the *absorption* of liquids.（液体的吸收）

• Transmission electron microscopy creates images by shooting trillions of electrons through an object and measuring their *absorption*, deflection and energy loss.（电子的吸收）

• *Adsorption* is to the surface, not inside. So there is a clear difference between the two words: *adsorption* and *absorption*.（表面吸附，这种吸附未到里面）

• The various methods of purification through the use of selective *adsorption* have several features in common.（液体的吸附）

access/assess

access 做动词时指“接近、进入、使用、获取”，assess 只能做动词，指“评估、估价”。例如：

• Computed tomography (CT) scan can help to locally *access* tumor extension and distant metastasis.

• From there you feed electricity into the grid and you can *access* any power outlet and feed our battery.

• These patterns represent a few “abstract gauges” that help the steering process to *assess* scope management, process control, progress, and quality control.

• Once again *assess* your life mirrors and write down a list of all who are healthy in the physical or who have recovered from an ailment of their own.

adapt/adept/adopt

adapt 是动词，指“适应”；adept 是形容词，指“熟练的、擅长的”；adopt 是动词，指“采用、采取”。例如：

• That's a real critical issue—what is the range of temperature or climate conditions to which we can *adapt*, and when do we exceed those?

• As people age, however, the body becomes more *adept* at regulating temperature, so brown fat stores shrink and white fat starts to emerge.

• We must *adopt* modern tools at every stage of the product development lifecycle, from requirements to design, development, and testing.

• A linear relaxation method is *adopted* to simplify the computational complexity.

affect/effect/impact/influence

affect 是动词，指“影响”（influence）；effect 是名词，指“结果的影响”，有时可做动词，指“招致”，论文中很少用；impact 是动词或名词，指“冲击、碰撞”，自然科学中不同参数之间的相互作用一般不用 impact，而多用 affect 或 influence；influence 是名词或动词，指“影响”或“改变”。例如：

• But what Einstein tells us is that the path you take through space and time can dramatically *affect* the time that you feel elapsing.

● The scientists doing the experiment have pounced on the *effect* of the particular moment.

● So changes to this strategy need not be introduced into the analysis model, but rather in the design model where they would have the most *impact*.

● Seasonal snow is one of the most important *influences* on the development and distribution of permafrost and the hydrothermal regime in surface soil.

● How do these factors *influence* government's investment behavior?

agree to/agree with

agree to 指"答应、同意、赞成、接受(某事)", agree with 指"(对意见、看法等)表示赞同、赞成、同意"。例如:

● Under the treaty, countries with nuclear weapons *agree to* move toward disarmament, while countries without nuclear weapons *agree not to* acquire them, and all have the right to peaceful nuclear energy.

● I don't *agree with* him on that although he might be right.

alternate/alternative

alternate 指"交替的、轮流的", alternative 指"另外的、选择性的"。例如:

● I present an *alternate* classification system of language style for science and technology in my new monograph.

● Photo synthetic rate of *alternate* irrigation was the highest, fixed irrigation was the lowest in all treatment.

● It might be reasonable to simply pick an *alternative* port number so you can have a dedicated server for your application.

● Magnetic refrigeration based on the MCE (magnetocaloric effect) is expected to be a promising *alternative* technology to the conventional gas compression refrigeration due to its higher energy efficiency and friendly environment.

and

and 使用广泛,连接类似的两个或多个词、短语或句子,被连接的各部分相互关联但又各自独立。连接三个及以上的部分时,最后一个部分的前面应该加 and,其他部分的后面加逗号。and 前是否加逗号,要按表达的需要而定,通常加逗号是美式英语写法,不加逗号是英式英语写法。

and 连接句子时,若两个句子都很简单,其间可以不加逗号,加上也可以;当两个句子都比较复杂时,通常需要加逗号;但连接一个不完整的句子时,不宜加逗号。用 and 来开始一个句子也是可以的,这时它起着对上文中相关或并列描述进行连接的作用。例如:

● We review evidence supporting a long, complex cellular phase consisting of feedback *and* feed forward responses of astrocytes, microglia, *and* vasculature.

● Now when we look at the horizon of the Industry 4.0 manufacturing scenarios *and* the emerging/evolving technologies, what are the challenges facing to our digital design *and* manufacturing research community *and* the industry?

● Retrofit solutions are likely to be complex *and* costly. Sealing the existing floor drain *and* installing a sump *and* pump may be feasible but is likely to require indoor excavation works.

● Examples of these technologies are novel manufacturing technology such as ① additive manufacturing, ② ICTs such as Cyber-Physics Systems (CPS), Big Data, the Internet of Things

(IoTs), Artificial Intelligence (AI), Digital Twin, *and* SMAC (Social, Mobile, Analytics, Cloud), *and*③product *and* design technology such as Smart Products, User Experience (UX) *and* Human Centered Design (HCD). (ICT—information and computing technologies)

● In high precision machining, however, to position the cutting tool with such high accuracy *and* repeatability is the key *and* this is normally undertaken in a "passive" manner, i. e. the tool's position relying on the slideways' positioning accuracy but without measuring the tool cutting behaviour *and* process conditions.

● The tremendous increase in knowledge on the molecular biology, pathophysiology, *and* diagnosis of Alzheimer's disease (AD) is exciting *and* holds promise for future prevention *and* therapies but also starts to erode the assumptions of its main theoretical foundation.

● *And*, their seminal papers were often rejected by leading journals—appearing only after considerable delay *and* in less prominent venues.

apparent/apparently

apparent 有"明显的"(obvious、clear)和"貌似的"(seeming)两种意思，使用时应以语义明确为前提，否则宜直接用 obvious、clear 或 seeming。apparently 是 apparent 的副词形式，相应的词有 obviously、clearly 或 seemingly。例如：

● With regard to the persuasion achieved by proof or *apparent* proof: just as in dialectic there is induction on the one hand and syllogism or *apparent* syllogism on the other, so it is in rhetoric.

● Not only might it have applications in fields such as microscopy, but it *apparently* also has the ability to optically store data forever.

appear

appear 有"出现"(to come into view)和"好像、似乎"(seem)两种意思，后一种更为常见。例如：

● The first signs of the dawn *appear* on the horizon.

● It *appears* that the location is determined via cell-tower triangulation, but the timing of these recordings varies.

as

as 可做连词表示因果关系，或表示"在……的时候"，如同 when。(表示因果关系时，其同义词有 because、because of、for、since 和 as，其中 as 表示的因果关系最弱，for 和 since 次之，because 和 because of 最强。because 后面跟句子，because of 后面跟名词或名词短语。since 常强调当时的情况、时间、地点，相当于"既然"。)as 还可做副词，表示"同样地；像……一样"，如同汉语比喻中的喻词。还可做介词，表"如同；当作；以……的身份"。例如：

● None of the initial effects of protopathy should be considered overwhelming or irreversible, *as* brain cells can apparently survive this stress for many years. (连词，表示因果关系)

● Furthermore, *as* iodide is substituted with bromide, a crystal phase transition occurs from a trigonal to a cubic structure; in compositions near the transition, the material is unable to crystallize, resulting in an apparently "amorphous" phase with high levels of energetic disorder and unexpectedly low absorption. (连词，表示"在……的时候"，如同 when)

● *As* illustrated in Fig. 1a, the flexible integrated sensing array (FISA) allows simultaneous and selective measurement of a panel of metabolites and electrolytes in human perspiration as well as skin temperature during prolonged indoor and outdoor physical activities.（副词，表示“同样地；像……一样”）

• GUIDE-seq experiments were performed and analysed essentially *as* previously described.（GUIDE-seq—the genome-wide unbiased identification of double-stranded breaks enabled by sequencing. 笔者注）（副词，表示“同样地；像……一样”）

• Once a fact is firmly established, the circuitous path that led to its discovery is seen *as* a distraction.（介词，表示“如同；当作”）

• The coordinate frame mark {S} is attached to the base coordinate frame of the robot while the coordinate frame mark {M} is set *as* the measuring coordinate frame of the FARO Arm.（介词，表示“如同；当作”）

augment

augment 的意思是“增加、增大，继续增长，进一步扩大”。例如：

• Caffeine can *augment* this effect—increasing our amount of available fuel.

• To *augment* the assay development beyond a representative product of the 20,277 UniProtKB/Swiss-Prot proteins, we selected peptides identifying splice and sequence variants and N-glycosylation sites (Figure 1, step 3).

average/mean/median

average 和 mean 都可表示“平均的”之意，前者还可表示“普通的”，后者还是一个数学用语，意为“平均值”；median 是中值、中位数之意，指某一系列值中的中间的那个值，也可做形容词，表示“中值的；中位数的”。例如：

• But, this trend does not seem to threaten the *average* American.（普通的）

• *Mean* daily temperature in the semiarid region of Brazil is no lower than 20 ℃, and usually of the order of 24 to 27 ℃.（平均的）

• The *median* age in Afghanistan is under 18.（中值的）（阿富汗人口的年龄中值低于 18 岁。）

4.4.2　b 为首字母

below

below 常用来提示下文将“阐述；提到；给出……”，常见短语有 as below、see below、as shown below、indicated below、are shown below、as described below，表示“如下所述、如下所示”之类的意思，与 as follows 同义。例如：

• Some combinations are *below*.

• The created models can be analyzed and used as a basis for code generation (see *below*).

• The inflection point at ~1950 CE coincides with the Great Acceleration, a prominent rise in economic activity and resource consumption that accounts for the marked mid-20th century upturns in or inceptions of the anthropogenic signals detailed *below*.

但是，当 below 不能确指时，虽然使用起来方便，但阅读、理解上可能费劲，因此少用为好。

beside/besides

beside 指“在……旁边”，besides 指“除了……之外还”。例如：

• We both looked down at the newspaper *beside* him.

• *Besides* which major or which university to choose, students and parents face another dilemma: the location of the university.

between/among

between 指“在两事物之间”，among 指“在三个或三个以上的事物之间”。例如：

• 14 locations were measured uniformly with the 2 mm gap *between* two consecutive locations.

• The Surgeon General reports that both voluntary and involuntary smoking show close association *between* smoke exposure and life of the smokers.

• This type of study designs corresponds to trials that allow modifications in the study during its conduct, related *among* other parameters to the study population, or the statistical framework.

• *Among* the many distinct geochemical signatures that human activities have introduced into the sedimentary record are elevated concentrations of polyaromatic hydrocarbons, polychlorinated biphenyls, and diverse pesticide residues, each beginning at ~1945 to 1950 CE.

but

but 同 and 一样，也是一个广泛使用的连接词，不过其连接的句子有对照和相反之意。连接较为简单的两个句子且不影响句子流畅时，but 前面可不加逗号；相反，连接较长、复杂的两个句子时，but 前面应加逗号，以便把两个句子分开。but 有时也可放在句子的开头。例如：

• The Fourier spectra shown in Fig. 11 indicate that the high frequency 5X component does exists *but* is relatively weak which leads to the small inner loop.

• The reason may lie in that in Ref. [26] the oil film force generated by the journal bearings affected the rotor orbits, *but* in this research the high precision ball bearings are used which can help to reduce the additional force.

• *But* we need to do something more significant than we have done so far.

4.4.3　c 为首字母

can not/cannot

can not 和 cannot 意义相同，均为情态动词 can 的否定形式，缩写为 can't，意思是“不能、不可以、无法或不可能”（表示“不可能”时，其反义词是 must）。美式英语常分开写，英式英语除强调 not 时分开写，一般连写。例如：

• We *can not* (*cannot*) compel you to do it, but we think you should.（不能、无法）

• They *can not* (*cannot*) have gone out because the light's on.（不可能）

• They *must* have gone out because the light's not on.（上一句的反义句）

compare with/compare to

compare with 是“比较”，而 compare to 是“比作、比喻为”。例如：

• *Compared with* other stratigraphic changes described above, the climate and sea-level signals of the Anthropocene are not yet as strongly expressed, in part because they reflect the combined effects of fast and slow climate feedback mechanisms.

• *Compared to* ground investigations and visual interpretation of aerial photographs or satellite data, digital interpretation of satellite imagery is a less expensive means of mapping and monitoring land cover changes over large areas.

• Figure 1F shows that the pyri-HOPG model catalyst displays high activity at high voltages, *compared to* the very low ORR activities of the N-free model catalysts.（ORR—the oxygen reduction reaction；HOPG—highly oriented pyrolytic graphite，pyri-HOPG—pyridinic N-dominated HOPG. 笔者注）

compose/consist/comprise

compose 指“组成、构成”时，是及物动词，句式一般为“...is composed of...”；consist 指“由……组成”时，是不及物动词，句式一般为“... consists of ...”；comprise 指“包括、包含”时，也有“由……组成”之意，但用法不易弄明白，尽量少用。例如：

• Grid computing is a form of distributed computing where the centralized computer *is composed of* a cluster of networks.

• Grid computing is a form of distributed computing where the centralized computer *consists of* a cluster of networks.

• These rules *comprise* the grammar of a language.

conserved/conservative

conserved 指“保存的、保持的”，conservative 指“保守的、守旧的”。例如：

• The method was used to design proteins that bind a *conserved* surface patch on the stem of the influenza hemagglutinin (HA) from the 1918 H1N1 pandemic virus.

• *Conservative* treatment of a PDPH in the ambulatory patient includes traditional analgesics, fluids, and bed rest. (PDPH—Post Dural Puncture Headache. 笔者注)

continual/continuous

continual 指“频繁的、经常发生的、持续不断的”，continuous 指“连续的、不间断的”。例如：

• Once waves have been generated, gravity is the force that drives them in a *continual* attempt to restore the ocean surface to a flat plain.

• Unexpectedly, we did not simply shift the region of structural instability to higher energy, but we observed a *continuous* series of dark films throughout this entire compositional range.

correlate with/correlate to

correlated with 的意思是“使……与……发生关系；把……与……关联起来；与……相符（相近）”。correlate to 的意思是“使相互关联”。例如：

• Low serum leptin levels were a major predictor of childhood mortality in a cohort of undernourished Ugandan children, and leptin levels *correlate with* bone mineral density in women.

• Moreover, the insertion of multiple spacers *correlated with* increased resistance.

• The result of this experiment does not *correlate with* that of earlier one.

• The number of cores is linearly *correlated with* stress granule volume (average of 4.7 cores/mm^3 granule volume, r^2 = 0.90), demonstrating the consistent nature of these substructures.

• Form (format) and meaning (content) *correlate to* each other.

• The double asterisk (**) indicates total copy number of *C4* is also strongly *correlated to* copy number of the *CYP21A2P* pseudogene, which is present on duplicated copies of the sequence shown in Extended Data Fig. 1g.

4.4.4 d 为首字母

deduct/deduce

deduct 表示“扣除、减去，演绎”，deduce 表示“推理、推论、推断，演绎出”。例如：

• Taxpayers affected by a natural disaster can *deduct* casualty losses on a federal income tax return.

• Math is almost always about formalizing our “common sense” about some domain, so that we can *deduce* and/or prove new things about that domain.

design/designate

design 是“设计、构思”之意，designate 是“指定、指出”之意。例如：

- The authors had *designed* two versions of the CRISPR array—one in the anti-sense direction (complementary to both the mRNA and coding strand of the DNA locus) and one in the sense direction (complementary only to the other DNA strand).
- The effective linkage optimization methods are needed for *designing* high performance drawing servo presses.
- If two terms are not synonyms, but are related in some other way that is important, you can *designate* them as related terms.

different from/different than

different from 是英式英语，different than 是美式英语，均表示“不同于；与……不同”，一般没有差别。如果比较的双方都是人或物，则用 different from 较好；如果其中一方不是人或物而是从句，则用 different than 较好。例如：

- The eukaryotic HEAT subunits are drastically *different from* the bacterial kite subunits in their sizes and structures, implicating that acquisition of the HEAT subunits during evolution might have provided condensins with elaborate abilities to manipulate and handle increasing lengths of eukaryotic chromosomes.
- Sweat analyte levels on the wrist follow similar trends but with concentrations *different from* those obtained at the forehead (Extended Data Fig. 9).
- This leads to the second reason why China is fundamentally *different than* America: economic geography.
- WMA is a *different* format of digital music *than* MP3.
- The research result is totally *different than* people expected.

diminish/decrease

diminish 意为“削弱、减为很少”；decrease 意为“减少”，但并不一定减为很少。例如：

- During the measuring process, a long small piece of the sample was picked to *diminish* the demagnetization field.
- The water flow into the Wuliangsuhai Lake showed a *decreasing* trend, falling from a peak of 733 million m^3 in 1995 to around 448 million m^3 in 2008.

dramatically/drastically

dramatically 表示“显著地、戏剧地、引人注目地”；drastically 表示“激烈地、彻底地”。例如：

- Because we can make all three of these changes, and if we do so, we can *dramatically* reduce how often we have a crisis and how severe those crises are.
- During the Ordovician, most life was in the sea, so it was sea creatures such as trilobites, brachiopods and graptolites that were *drastically* reduced in number.

due to

due to 表示“应归于……”或“由……引起”(caused by)，传统语法上主要引导表语，后面跟名词，其语义有因果关系，但具体语义不能确定时，不宜用 due to，而应该用 because of 或 caused by。例如：

- The maximum displacement takes place at the gimbal area *due to* the long feeder line without the bellow support.
- The increasing range of application of titanium alloys is *due to* their very good combinations of high strength to weight ratio,

low density, exceptional corrosion-resisting properties.

• Among the studies, the core-shell catalyst has attracted more and more attention *because of* its special characteristics of large surface-to-volume ratios and the aggregation resistance performance. （这里用 because of 比用 due to 的效果好，因为其具体原因可能还未最终确定）

4.4.5　e 为首字母

equipment

equipment 表示“设备、装备和器材”，是不可数名词，也是集合名词，equipments 是错误的写法。例如：

• These controllers and associated *equipment* should all be housed together in a control room.

• Therefore, the meaning of the research in fouling monitoring technique and *equipment* is great.

4.4.6　f 为首字母

few/a few（little/a little）

few 指“很少”，强调没有多少，有否定义；a few 也指“很少”，但强调有一些，虽然不多，有肯定义。表示“量”时，little、a little 的用法同 few、a few，但 few 修饰可数名词，little 修饰不可数名词。little 还可做形容词，意思是“小的”。例如：

• 18 months, an age at which relatively *few* mice in each of the cohorts had died.

• Alternatively, absolute label-free quantification based on *few* anchor points can be pursued.

• The complexity of how neurons are progressively affected by Aβ stress has been addressed by *a few* groups.

• A framework could be generated by measuring all changes in the different cell types of the brain over the different Braak stages in *a few* relevant brain areas, which would constitute a valuable resource for the field and propel AD research into the complex biology of the 21st century.

• The answer to the problem has caused *a little* trouble for some researchers recently.

• To avoid contaminating the *little* critters, scientists have to shower and scrub before entering mouse quarters.

flammable/inflammable/nonflammable

flammable 和 inflammable 做形容词时都表示“易燃的、可燃的”意思，不要误将后者作为前者的反义词来用。nonflammable 是前两个词的反义词，意思是“不易燃的、不可燃的”。例如：

• It is very necessary and important to separate *flammable* gas cylinders from oxygen and other oxidizing gas cylinders during storage.

• Dangerous goods are *inflammable*, explosive, toxic, corrosive and radiating and have other characteristics.

• Gaseous argon is tasteless, colorless, odorless, non corrosive, and *nonflammable*.

follows/following

follows 是名词 follow 的复数，表“跟随、追随、遵照”。following 是形容词，表“接着的、下面的、其次的”，也可为动词 follow 的动名词形式。例如：

• The paper is organized as *follows*. *Following* an introduction of the experiment setup in section 2, the spectral lines measured on the EAST tokamak are presented in section 3, including carbon, oxygen, lithium, nitrogen, argon, iron and molybdenum lines.

• Substituting striation spacing *S* into Eq. (1), the *following* equations can be obtained.

4.4.7　i 为首字母

in contrast/on the contrary

这两个词都有与此相反的意思。但是，in contrast 表示“比较而言；比较起来”，加强比较，以突出差异，既可以对相同点进行比较，也可以对不同点（包括完全相反）进行比较；on the contrary 表示“反之”，意思是相反、正相反，强调完全不同。例如：

● *In contrast*, the hinge of condensins is “closed”, making them a rod-shaped complex.

● A similar bioinformatics analysis of previously known yeast P body proteins revealed that, *in contrast to* stress granules, a majority of P body proteins have mRNA-binding activity (~73%), which suggests that mRNA-protein interactions may contribute more to P body structure than stress granule structure (Table S1).

● *On the contrary*, a man with a dream is like a warrior armed with ambition, foresight and gallantry, daring to step into an unknown domain to make a journey of adventure.

● This is not an economical way to get more water; *on the contrary*, it is very expensive.

注意：不要将 in contrast 误用为 in contrary，因为 in contrary 是一个错误的短语，但 in contrary to 是正确的，表示“与……形成对照”。

induce (provoke)

induce 表示“引发、诱导”，provoke 表示“驱使”。例如：

● By contrast, in adaptive immune resistance, PD-L1 (programmed cell death ligand-1) expression is *induced* on tumor cells secondary to local inflammatory signals.

● Although vaccines are able to *provoke* immune response, translating this observation into clinical benefit has been challenging.

4.4.8　m 为首字母

minimal/trivial

minimal 意思是“最小的、最低的”，trivial 是“轻微的、琐碎的、不重要的”。例如：

● Residual leukemic cells residing in these protective niches after treatment are therefore potentially contributing to *minimal* residual disease (MRD) persistence and to disease relapse in patients after chemotherapy.

● Science is still a very mysterious subject so there are millions of *trivial* facts about it—this will be the first of many scientific fact lists in the future.

4.4.9　p 为首字母

percent/percentage/percentile

percent 跟在数字后面，以代替%；percentage 意思是“百分率、百分比”，不能与数字一起用，但可以说 small（或 large）percentage；percentile 是统计学用语，意思是“百分位、百分位数”，表示在 100 个分组中事物出现的概率。例如：

● In their first year, 33 *percent* of students report that they never talk with professors outside of class, while 42 *percent* do so only sometimes.

● In order to test the working ability of the equipment under rated load, the load test can be carried out according to the *percentage* increasing mode of the design center.

● The factors of human body dimension should be considered during the furniture designing, the selection of *percentile* of the dimension must be reasonable.

preceding/proceeding

preceding 意思是“前面的、前述的”；proceeding 是名词（表示“进行、程序、诉讼”）或动词 proceed（表示“开始、继续做、行进”）的现在分词。例如：

• Inflation rate is the percentage change in the price index from the *preceding* period.

• Second, the modernization is *proceeding* slowly and in a piecemeal manner.

present/represent

present 意思是“呈现、提出、介绍”，表示存在，represent 是“代表、表现”。例如：

• Missing flow data *presents* a number of challenges for estimating base flow.

• A new sewer overflow device is *presented* which consists of a rectangular tank and a sharp crested weir with a series of vertical combs.

• Local farmers and fishers are engaged in the reed harvesting, *representing* an important additional income in winter times.

• And the most difficult problem is that the value of vibration stress of the experimental blade can not *represent* the realistic vibration stress of the failure blade.

protect/preserve

protect 是“保护、防卫”，preserve 是“保存、保护、维持、保持不变”。例如：

• Microenvironment, cell adhesion, and chemotaxis *protect* B-cell leukemias from spontaneous and drug-included apoptosis.

• Care must be taken to *preserve* the vascular supply of the greater curvature, particularly the right gastric and gastroepiploic arteries.

provided/providing

这两个词均可做连词，指“假如、倘若，以……为条件”，但 provided (that) 比 providing (that) 更常用。注意，这两个词在形式上与 provide 的过去时（过去分词）和进行时（现在分词）相同（表示“提供、给予”之意），但连词与动词完全是两回事，不可混淆。例如：

• The base CE, i. e., the CE value calculated from the default CE versus precursor ion mass function, *provided* the highest abundance signal for the majority of fragment ions.（连词）

• A strong correlation between genetic divergence and sampling time within the outbreak lineage (Fig. 2, inset) shows that our approach is appropriate, *provided* that whole genomes are used.（连词）

• I do believe in people being able to do what they want to do, *providing* they're not hurting someone else.（连词）

• By combining tree search with policy and value networks, AlphaGo has finally reached a professional level in Go, *providing* hope that human-level performance can now be achieved in other seemingly intractable artificial intelligence domains.（现在分词）

4.4.10　r 为首字母

remainder/remaining

这两个词均可做形容词，表示“剩余的、剩下的”，前者还可做名词，表示“残余、剩余物、其余的人”。例如：

• The method of claim 1, further comprising paying a *remainder* bandwidth for the prefetched data when the prefetched data is consumed by the UE, wherein the *remainder* bandwidth is a difference between the reduced bandwidth and the network bandwidth. (UE—user equipment. 笔者注）

• The *remainder* of this article talks about the roles that enterprise architects, application architects, and developers need to play on a project.

• This code is essentially the same for all

the document models in this article, so I'll skip over it in the *remaining* examples without any discussion.

4.4.11　s为首字母

since

since 做介词表"自从……；自……以来；自……以后"，表示时间，也可用来表示因果关系，同 because、as 等。例如：

● *Since* the discovery of the condensin complexes almost two decades ago, our understanding of large-scale chromosome structures has been facilitated and deepened.

● *Since* the initial release in 2005, the Protein Atlas evolved into a knowledge base that includes a diverse collection of 25,039 monoclonal and polyclonal antibodies, collectively targeting 17,005 proteins corresponding to 84% of the predicted proteome (v. 15).

● *Since* yeast stress granules are of similar size in cells and in lysates, yeast stress granules may have proportionally smaller dynamic shells than mammalian stress granules.

● These effects on granule dynamics are likely to be direct *since* we confirmed their localization to both yeast and mammalian stress granules.

subsequent/subsequently, consequent/consequently

subsequent/subsequently 表示时间或空间顺序，表示"后来的、随后的；随后、其后、后来"之类的意思；consequent/consequently 表示逻辑推理结果，表示"随之发生的、作为结果的；因此、结果、所以"之类的意思。例如：

● *Subsequent* higher-order assembly leads to linear organization and axial shortening of chromosomes.

● *Subsequently*, the fine particles are widely dispersed in the atmosphere and descend to earth very slowly.

● The conglutination phenomenon often appears in actual processing of insect images, which may cause difficulty for *consequent* process if not be disposed.

● Compared with the large chamfers, the small chamfer leads to a greater tensile stress and, *consequently*, induces more cracks, due to its narrow channel and severe corner.

such as/including

such as 用来举例，表示"比如、诸如"之意，including 用来列出所包含的各个组成部分，但不一定全部列出。这两个词语所修饰的名词或短语即后面所列出的对象应紧跟在这两个词语的后面，而且这些名词或短语所指的事物、概念应属于同一类别。当 such as 后面排列的部分较多或相对独立时，应该用逗号将 such as 与其前面的语句分开。例如：

● According to this model, condensin II initiates loop formation through a mechanism *such as* chiral looping (Hirano, 2012) or loop extrusion (Alipourand Marko, 2012).

● G protein subunits then interact with a variety of effectors, *such as* enzymes and ion channels, to initiate downstream responses.

● To process the large number of peptides, we established an assay development pipeline *including* a robotics platform and multiple commonly used mass spectrometry (MS) instruments duplicated at two geographical sites.

symptom/syndrome

symptom 意思是"症状、征兆"，syndrome 是"综合症状、并发症状"。例如：

● A behavior of one area might be just the *symptom* of another incorrectly configured or misbehaving area.

• Furthermore, regular monitoring of the organisms causing each *syndrome* should be conducted on a regular basis to validate the treatment recommendations.

synergic/combined

synergic 意思是“协作的、合作的、协同的”，combined 是“联合的、结合的、组合的”，前者表示的事物间的关系比后者的要更加紧密、密切一些。例如：

• From the viewpoint of *synergic* control, a microcomputer control system of pulse MIG welding is developed in this paper, and the hardware structure and control principle of the system are introduced.

• These treatment modalities may be *combined* with established treatments such as surgery, radiation, and chemotherapy to effectively treat bladder cancer.

4.4.12　t 为首字母

than

than 用来比较，所比较的两个对象应该具有一致性，对不具一致性、不在一个层面的对象通常是不能比较的。（有些词的意思代表最终状态，不能进行比较，这样的词有 absolute、complete、extinct、full、permanent、unique、universal 等。）例如：

• So our equations are more direct and brief *than* the expressions with higher powers in the earlier literature.

• Generally, it is much more convenient to measure the position of the end-effector *than* to measure the pose.

• Divergence date estimates are robust among different combinations of prior distributions, molecular clock models, and coalescent models (supplementary materials sections 4 and 5) and are more likely to shift into the past *than* toward the present as virus genomes accumulate through time.

that/which

that、which 常用来引导修饰从句。修饰从句时，如果可以省略而不影响句子的完整性，这时的从句就是非限定性从句，用 which，并用逗号把从句与主句分开；如果不可以省略，这时的从句就是限定性从句，用 that，但无须用逗号把从句与主句分开。例如：

• Metabonomics, *which* was initially proposed by Nicholson group, is an important part of the systematic biology.

• Occurrence and progression of laryngeal carcinoma result from mutual action of various pathological factors, *in which* disordered cell apoptosis plays a pivot role.

• Laryngeal carcinoma has been one of the diseases *that* severely affect human health and quality of life.

• It is worth mentioning that better accuracy and stiffness can be achieved when there are fewer single-DOF joints, because the kinematic joint is a weak point *that* is the main cause of deformation and clearance issues.

toward/towards

toward、towards 用法相同，指“朝、向、趋向”，美式英语用 toward，英式英语用 towards。例如：

• According to the above conclusion, under the same magnetic field, GMM whose easy magnetization direction *toward* <111> crystal orientation has a larger magnetostriction ability. (GMM—giant magnetostrictive materials. 笔者注)

• This adjusts the policy *towards* the correct goal of winning games, rather than maximizing predictive accuracy.

• But in fact, defects of GMM bring about a part of the magnetic domain remaining in their initial states. And then it leads to the process that magnetic domain always motions *towards* to the direction of possession the minimum free energy is not continuous.

4.4.13 u 为首字母

underlining/underlying

underlining 是动词 underline（在……下划线、画底线标出、强调）的分词形式，表示“强调的、加强的、突出的”，underlying 是形容词，表示“基本的、根本的、潜在的、在下面的”。例如：

• Once you pinpoint the *underlining* emotions, you can work on better ways for dealing with your feelings.

• The *underlying* theory behind immune surveillance is that many tumors are eliminated by the immune system, while some cancers develop ways and means to escape the immune response.

use (using)/utilize/employ

use（using）和 utilize 均指使用、运用和利用，但 utilize 更强调“利用、有效使用”之意，严格意义上来讲二者不宜相互代替。employ 指“使用、采用，雇用、使从事于……”。例如：

• We *used* data retrieved for 166 IPS (invasive plant species) recorded in Nepal which were introduced to an ecosystem other than their natural home (Tiwari et al. 2005).

• The theorem shows how to find the mean value *using* a definite integral.

• Immunotherapy has been *utilized* as a strategy for treating cancer for over 100 years since the use of Coley's toxins for the treatment of sarcoma.

• Vectors are *employed* to transduce a target cell with foreign DNA.

4.4.14 v 为首字母

vary/change

vary、change 都有“变化”的意思，但 vary 更强调变化的多样性，表示变化的不同，而 change 侧重变化这一动作本身，表示发生了改变、转变。例如：

• The experimental conditions were *varied* with different flow volumes, number and spacing of combs layers.

• As well known, grain refinement is an important process to not only improve mechanical properties, but also *change* physical and chemical performances.

• Another attempt is that the training algorithm was *changed* in the Network 16 (i.e., double layer-layer recurrent neural network) in order to get the further improvement in predicting the output result.

4.4.15 w 为首字母

while

while 可做表示时间的连接词（当……的时候），也可做表示转折或逻辑的连接词（然而、虽然、尽管）。例如：

• *While* I do not present any quantitative data on speed, the chart is arranged in order of speed, from fastest to slowest.（表示时间）

• The circuits are configured to ensure that the final analogue output of each path is finely resolved *while* staying within the input voltage range of the analogue-to-digital converter.（表示时间）

• For these applications, the sample collection and analysis are performed separately, failing to provide areal-time profile of sweat content secretion, *while* requiring extensive laboratory analysis using bulky instrumentation.

（表示转折）

• With continued perspiration, the skin temperature rises at about 400 s because of muscle heat conductance to skin and then remains stable, *while* the concentration of both lactate and glucose in sweat decrease gradually.（表示转折）

为避免出现语义不清，实际写作中提倡少用 while，多用 and、although、but、when 等连接词。

第5章 英语科技论文标点

标点符号通过其在句中的类别及所处位置的不同，对单词、短语、句子进行分隔和组合，与语言文字紧密恰当地组合在一起，赋予书面语以不同的意义和情感，并达到句子结构清晰、行文准确流畅、有助于阅读理解的目的。英语标点符号主要有逗号、分号、冒号、破折号、连字符、括号、引号、斜线号、撇号、省略号、句号、问号、叹号等，其中分号、冒号、破折号、连字符、撇号（表所有格）主要用于连接词或承接句子各部分，成对出现的逗号、破折号、引号、括号主要用于封闭句子各部分，省略号、句号（缩写点）、撇号（表缩写）主要用于表示省略，句号、问号、叹号主要用于表示句子的结束。英语标点符号在类别和形式上基本同中文标点符号，但其间还是有差别的，必须加以区分，避免中、英语标点混用。本章介绍14种常见的英语标点符号的使用场合，并给出它们在英语科技论文中的实例语句。

5.1 逗号

逗号（comma，“,”）用来分隔句子或句子的各种成分，表示较小的停顿。英语中逗号使用很广，规则较多，再加上与中文逗号的形式相同，写作时容易混淆二者用法的差异，而在需要逗号处遗漏逗号，在不需要处使用逗号。逗号主要用在以下场合：

1）在由多个同等成分（如单词、短语、子句以及数字、名称、量值、符号等）组成的句子中，除最后一个成分外，其他成分的后面都要用逗号，以分隔这些成分。例如：

- Water, sodium hydroxide, and ammonia were the solvents.（分隔单词）
- Keywords: thick-thinned contraction, basement structures, salt structures, physical modeling, Kuqa depression.（分隔关键词）
- Parallel mechanisms are suited to applications that require high structural rigidity and accuracy, fast dynamic response, and large load-to-weight ratios.（分隔短语）
- Rolling velocity is 10, 20, 30, 40 m/s, respectively, inlet oil temperature 27, 60, 90, 125 ℃, maximum Hertz pressure 0.8, 1.0, 1.1, 1.2, 1.35, 1.5 GPa.（分隔数字）
- Shaolin Zheng, Lidong Zhang, Wu Zhang, and Yajun Yang.（分隔人名）
- Yixin Yu[1,2,*], Liangjie Tang[1,2], Wenjing Yang[3], Wenzheng Jin[1,2], Gengxin Peng[3], and Ganglin Lei[3].（文章署名中分隔作者姓名）
- The proposed SLA values are close to those expected for observations on infertile soils (*A. elatius* 35–37, *F. rubra* 13–15, *M. caerulea* 21–24; Poorter and de Jong, 1999).（分隔类名称）
- For the SRM, the rated power, rated rotate speed, and rated torque is 26.2 kW, 2 500 r/min, and 100 N·m, respectively.（分隔量名称；分隔量值）
- The wavelet network has six inputs and six outputs corresponding with the link length variables (l_1, l_2, l_3, l_4, l_5, l_6) and the position and orientation variables (x, y, z, ϕ, θ, ψ).（分隔量符号）

此类表达中，连词and前面是否加逗号

所表达的含义有可能是不同的。例如：

• The complex consists of three conformable, well-layered units of gabbro, diorite and granodiorite and granophyre.

此句的 units 是由"①gabbro，②diorite，③granodiorite and granophyre"组成，还是由"① gabbro，② diorite and granodiorite，③granophyre"组成，还得考虑一番。

2）在并列句中使用逗号分隔分句，如有并列连词（如 and、but、for、nor、or、so、yet 等），逗号就用在并列连词的前面，若没有并列连词，就直接用在分句间。例如：

• The theory on nucleation and growth of martensite transformation is the core part of martensite theory, but it has been incomplete until now.

• Field relations indicate divergent geomorphic histories for the two formations, yet over broad areas they are nearly coextensive.

• Water is a compound, it is made up of hydrogen and oxygen.

• Compared with the quenched martensites, the size of fresh martensites is smaller, it is about 0.3–0.5μm.

以上四句中每个句子的两个画线部分为并列分句，前面两句中的并列连词分别为 but 和 yet，后面两句中没有并列连词。

一个句子虽然是并列句，但如果并列的分句非常简短，则各分句之间可不用逗号分隔；一个句子若有由两个并列谓语组成的复合谓语，则这两个并列谓语间也可不用逗号分隔，这种句子实际上是简单句而不是并列句。例如：

• The survey was completed and we left the lab.（并列句）

• The product distribution results were obtained in sodium hydroxide and are listed in Figs. 5-8 and Table 10.（并列谓语）

• Heat, light, electricity, and sound are different forms of energy and can be changed from one form into another.（并列谓语）

3）在分词短语做状语的句子中，使用逗号分隔分词短语和句子。分词短语可放在句首、句末或句中，句子的主语和分词的逻辑主语相同。例如：

• While burning, fuel oil gives out heat energy.

• While using these high-precision instruments, we must be very careful.

• Heating water, you can change it into steam.

• On cooling, a crystalline phase may develop in coexistence with an amorphous phase.

• The computer works very fast, handling millions of data with the speed of light.

• We consulted many dictionaries, searching for a correct answer to the question.

• Once installed, this heater operates automatically.

• Compared with other products, the price of ours is very competitive.

• Considered from this point of view, the question under discussion is of great importance.

• Complicated in design and theory, the machine is not easy to manipulate.

• Held twice a year, the Guangzhou Fair is a mirror of Chinese economy.

4）在分词独立结构做状语的句子中，使用逗号分隔分词独立结构和句子。分词独立结构放在句首一般表示时间、原因或条件，放在句末一般表示附加说明或伴随、陪衬的动作。例如：

• With the experiments carried out, we started new investigations.

• The day's writing and editing being finished, I became relaxed and played a while.

• Christmas Day being a holiday, the

shops were all closed.

● Time permitting, we shall do the experiment tomorrow.

● Machine tools are built in various, their general theroy and construcion being the same.

● The war was over, without a shot being fired.

5）在分词短语（主要指现在分词短语）做插入语的句子中，使用逗号分隔插入语和句中其他成分。插入语表示对整个句子内容的态度或看法，通常放在句首，有时也可放在句中或句末。这种分词短语结构已成为固定短语，常见的有 all things considered, beginning with…, considering…, generally（frankly、strictly、roughly、seriously）speaking, judging from（by）…, speaking of…, talking of…, talking…into consideration 等。例如：

● Judging from the appearance, the machine must be of good quality.

● All things considered, this car is better than that one.

6）在有过渡语或插入语的句子中，使用逗号分隔过渡语或插入语和句中其他成分。过渡语起桥梁的作用，插入语一般对前面一句做附加解释，通常放在句首、句中或句末。充当或引导过渡语、插入语的常见词语有：accordingly、after all、also、as a result、as a matter of fact、at the same time、basically、besides、by the way、consequently、e. g. 、even so、finally、for example、fortunately、furthermore、hence、however、i. e. 、in addition、in conclusion、indeed、in effect、in fact、in essence、in general、in other words、instead、in summary、in the first place、in the meantime、likewise、moreover、namely、nevertheless、of course、on the contrary、on the other hand、then、that is、therefore、thus、what is more、too 等。例如：

● These oxides are more stable in organic solvents（e. g. , ketones, esters, and ethers）than previously believed.

● Many antibiotics, for example, penicillins, cephalosporins, and vancomycin, interfere with bacterial peptidogly can construction.

● However, these numerical techniques are computationally intensive.

● In addition, the wavelet network learns much faster than BP network.

● The direct displacement for parallel mechanisms is complex while the inverse displacement is, in general, simple.

● The new derivatives obtained with the simpler procedure, that is, reaction with organocuprates, were evaluated for antitumor activity.

● Basically, there are three types of locomotion mechanisms, wheeled, tracked, and legged styles, and many researchers have studied these mechanisms.（副词做过渡语或插入语时也可不用逗号分隔）

● Two steel plates, 320 mm in length and 200 mm in width of 12 mm thickness, were butt welded with chamfer in V.（画线部分为插入语）

● Several individual flows, each thicker than 25 m, have been traced for more than 160 km.（画线部分为插入语）

● Beauty, in its largest and profoundest sense, is one expression for the universe.（Ralph W. Emerson）（画线部分为插入语）

注意：插入语是句子独立成分的一种（另外两种是感叹语、呼语，一般不会在科技论文中出现），与句中其他成分没有语法关系，用逗号与其他成分隔开，但不能脱离句子而独立存在，词、短语或固定词组均可做插入语。

7）在有对比关系的一组单词、短语或独立句子的句子中，使用逗号分隔各个单词，或各个短语，或各个独立句子。例如：

• It is orange, not red.（分隔单词）

• Another approach, called systematic mapping, is more broad-brush.（分隔短语）

• The greater the risks are, the greater the probable gain from the treatment will be.（分隔句子）

• Potassium compounds such as KCl are strong electrolytes, other potassium compounds are weak electrolytes.（分隔句子）

• One part is the weir and groove ($r > R_g$), the other is the dam ($r < R_g$).（分隔句子）

• It is easy to draw a conclusion that the smaller the value of the max ΔF is, the more robust a solution is.（分隔同位语从句中的两个句子）

8）在复合句中，使用逗号分隔从句和主句。例如：

• After all ants finish their tours, the pheromone trails of the best route are updated following Eq.（8）.

• Where data are inaccurate or insufficient, results deviate from what is expected.

• Although 40 different P450 enzymes have been identified, only six are responsible for the processing of carcinogens.

以上三句中，画线部分为从句。

9）在复合句中，使用逗号分隔非限定性从句和句子，或非限定性同位语和句子。例如：

• In Eq.（9）, the resultant force matrix $\boldsymbol{F}_r$ comprises of the forces acting on the cylinders, which include the forces produced by the pressures in the cylinders（pA）, the friction forces（F）, and the equivalent loads of the specimen（including mass of the test stand）acting on each cylinder（M）.（画线部分为非限定性从句）

• Isaac Newton, a British scientist, who lived over 300 years ago, said he saw further than others, because he stood on the shoulders of giants.（第一处画线部分为 Isaac Newton 的非限定性同位语，第二处画线部分为非限定性从句。）

注意：对限定性从句或限定性同位语，一般不用逗号分隔。例如：

• The traditional linear dynamic analysis is based on the spring-mass-damp model which is shown in Fig. 1.（画线部分为 the spring-mass-damp model 的限定性从句）

• The book *The Gene Industry* will be released next week.（画线部分为 The book 的限定性同位语）

• This automobile is running at speed of 120 miles an hour.（画线部分为 speed 的限定性同位语）

10）使用逗号分隔以 such as 或 including 引导的非限定性短语。例如：

• $\eta_{ij}(g)$ is the heuristic pheromone on route（i, j）at iteration g, which is calculated by some heuristics, such as earliest due date（EDD）heuristic, and so on.

• Hydrogen-bonded complexes, including proton-bound dimers, are well-known species.

注意：对 such as 或 including 引导的限定性短语，不用逗号分隔。例如：

• Potassium compounds such as KCl are strong electrolytes, other potassium compounds are weak electrolytes.

• A virtual node including parallel nodes with different subscript represents the identical parallel machines.

11）在有几个并列形容词分别修饰同一名词的句子中，若调换这些形容词的顺序并不影响句子的意思，则用逗号分隔这些形容词。例如：

- Sample preparation is a repetitious, labor-intensive task.
- A powerful, versatile tool for particle sizing is quasi-elastic light scattering.

注意：若调换形容词的顺序影响句子意思，则不用逗号分隔这些形容词。例如：

- Polyethylene is an important industrial polymer.
- The rapid intra molecular reaction course leads to ring formation.

12）在介词短语做状语的句子中，使用逗号分隔介词短语和句子。例如：

- In the 1940s, the model of martensitic nucleation based on components fluctuation was presented by Fisher who thought the carbon-poor zone in steel could be the nucleation site of martensite.
- For nearly four decades, the development of semiconductor industry has been adhering to Moore's Law, which is found by Moore in 1965, and expressed as that the number of components per chip doubles every 18–24 months.
- From the late 1980s to the 1990s, numerical calculation model was adopted to research on adaptive control, by which the second generation welding quality controller for directly calculating the nugget diameter was developed.
- For the complexity of the sealing ring model, the deformation is usually calculated by finite element method (FEM).

13）在有一系列以数字或字母标识的单词或短语的句子中，通常用逗号分隔这些单词或短语。例如：

- Damage resulted from (1) vibration, (2) ground cracking, and (3) subsidence, etc..
- For each solution we calculate the following two entities: (1) domination count n_p, the number of solutions which dominate the solution p, and (2) S_p, a set of solutions that the solution p dominates.
- At the end of its run, the GA provides the optimal platform settings and product family design solutions with satisfying performance, where the results from the optimization include a) which variables should be made common (i. e., platform variables), b) the number of common values on each platform variable for multiple-platform design, c) the values that platform variables should take, d) the values that the remaining unique variables should take.

在这类表达中，有时也可用分号分隔（若标识的是句子，则多用句号分隔）。例如：

- The design of GNSGA-II for Vehicle Routing Problem in Distribution contains six steps: ①Coding; ②Initializing population; ③Fitness; ④Selection; ⑤Crossover; and ⑥Mutation.

14）在连续出现两个独立数字或符号（相邻但无关联）的句子中，用逗号分隔这两个数字或符号。例如：

- By the end of 1935, 1,000 experiments had been completed.
- During 2000, $876 000 worth of sales was financed through this plan.
- By October 10, 2019, 150 universities had submitted online reports for this project.（将 150 改为数量名词 one hundred and fifty 更恰当。）
- The vibration model in the nodes is written as ${}^iA_j$, i on up left means the number of substructures, j means the number of nodes.

注意：当两个独立的数字做定语修饰同一个名词时，可以考虑将第一个数字用数量名词的形式表示，而第二个数仍用数字的形式表示。例如：

● Each package contains twelve 2-inch nails.

● Be sure to order twenty-five 60 W bulbs for the lamps in the hall.

15）在地名或机构名的表达中，用逗号分隔其中不同级次的组成部分。例如：

● The specimens of species newly identified were deposited in the museum in Cairo，Egypt.

● K J Chen，P Ji. A mixed integer programming model for integrating MRP and job shop scheduling. In *Proceedings of the Fourth International Conference on e-Engineering and Digital Enterprise Technology*，Leeds，UK，2004：145–150.（参考文献著录）

● Department of Mechanical Engineering，South China University of Technology，Guangzhou，China.

16）在含有用“et al.”（或 et al）表示人名省略的句子中，如果不是处于句末，则可用逗号分隔它与其后面的部分（其前面可以加逗号，也可以不加）。例如：

● In 1978，Jacobson et al.，investigated mathematical models to predict man’s comfort response in different automobiles and environments.（“et al.”后的逗号也可去掉。）

● J C Bean，J R Birge，J Mittenthal，et al. Matchup scheduling with multiple resources，release dates and disruptions. *Operations Research*，1991，39（3）：470–483.（参考文献著录）

17）在基于“顺序编码制”的参考文献引用的表述中，如果所引文献的序号不连续，则可用逗号分隔这些序号。

● This method can be used in visual tracking of welding seam[10,19].

● Ar + H_2 become not uniform after entering arc space and the density of [H_2] in center is higher than the density in brim of arc column[2,13-15].

● For more detailed information，refer to Refs. [8，11，15].

注意：如果有连续的参考文献序号，则用短破折号来分隔（参见本小节（17）第二个例句（引文上标［2，13–15］）和后面 5.4 节“破折号”中的有关内容）。

18）在基于“著者-出版年制”的参考文献引用的表述中，用逗号分隔括号内所引文献的作者与年份。例如：

● The mycorrhizal fungus supplies the orchid with organic nitrogen（N）（Cameron et al.，2006）and a further study has demonstrated P transfer to juvenile protocorms（Smith，1966）.

19）在以“月、日、年”为次序排列的日期表达中，可用逗号分隔表示“日”与“年”的数字，但在以“日、月、年”为次序排列的日期表达中，不使用逗号。例如：August 8，2008；June 18–22，2019；8 August 2008；18–22 June 2019 等。

5.2 分号

分号（semicolon，“；”）通常用来分隔没有连接词连接的、语义关系密切的分句，这些分句因语义关系密切而组织为一个句子。分号还可用来替代逗号，分隔冗长、复杂或含有逗号的分句。分号主要用在以下场合：

1）在包含一系列其中含有逗号的单词、短语或数字的句子中，使用分号来分隔这些单词、短语或数字。例如：

● The compounds studied were methyl ethyl ketone；sodium benzoate；and acetic，benzoic，and cinnamic acids.

● The order of deposition was quartz and pyrite；massive galena，sphalerite，and pyrite；brown carbonates and quartz；and small amounts of all those named，together with fluorite, barite, calcite, and kaolin.

• Much of the unit is red, pink, or gray; medium to coarse grained; and equigranular or slightly porphyritic.

2）在包含由连接副词连接或表示列举、解释的引导词引导的独立分句（引导词之后的列举、解释语句中含有逗号或构成另外相对完整的意思）的句子中，常用分号来分隔这些独立分句。这样的连接副词或表示列举和解释的引导词语通常有：accordingly、besides、but、consequently、for example（e. g.）、for instance、furthermore、however、hence、indeed、in fact、in other words、likewise、moreover、namely（viz.）、nevertheless、notwithstanding、otherwise、on the contrary、so、still、then、therefore、thus、yet、that is（i. e.）、that is to say 等。例如：

• Numerical method solves the direct displacement using any of the available numerical techniques; however, these numerical techniques are computationally intensive.

• By adjusting the magnet field generated by adjustable magnetic poles, the adjustable function of the main flux is accomplished; therefore, the assistant control of engine for optimizing operating performance can also be achieved.

• The efficiency of the cross-coupling depends on the nature of X in RX; thus, the reaction is performed at room temperature by slow addition of the ester.

• The growth of a digital organism's wisdom is basically from bottom up; that is to say, the digital life will evolve wisdom by itself.（that is to say 引导的独立分句有相对完整的意思）

• I want to write a series of books; that is to say, I want to develop myself and improve students' writing level.

注意：如果引导词语之后的列举、解释语句中不含有逗号且没有构成另外相对完整的意思，则引导词语前后通常都用逗号；当列举或解释语作为插入成分时也不用分号隔开。例如：

• I want to write a series of books, that is to say, to develop myself and improve students' writing level.

3）在包含没有连接词（如 and、but、or、nor、yet、for、so 等）连接的独立分句的句子中，使用分号来分隔这些独立分句。例如：

• Computers were first developed in the 1940s; they have had a profound impact on our life today.

• A rotating feed machine is added to the spindle and the bearing becomes a complex one; its supporting rigidity and damp are also changed.

• Interface 1 between two axes is connected by a coupling; interface 2 is connected on taper faces.

• The participants in the first study were paid; those in the second were unpaid.

• In part A of Fig. 3（a）, the joystick doesn't move; in part B, moving the joystick grasps the tire; in part C, the tire is grasped; and in part D, the tire is unlocked.

• The concussion frequency of induction heating equipment is 90 kHz, current density 5.8×10^{7} A/m^{2}; relative permeability of aluminum, $u_{\mathrm{r}}=u/u_{0}=8$; electrical resistivity of Al-Si alloy, 2.1×10^{-7} Ω·m; density of ZL112Y alloy, 2 740 kg/m^{3}.

4）在对数学式中的符号进行解释的语句中，使用分号来分隔这些对应于不同符号的解释语。例如：

C—Number of customers;

w_i—Demand of each customer（$i=1$,

2，…，C)；

W—Capacity of vehicles；

n_k—Number of customers that vehicle k served.

此例也可表述为：

C is number of customers；w_i is demand of each customer ($i = 1, 2, \cdots, C$)；W is capacity of vehicles；n_k is number of customers that vehicle k served.

5）在有一系列以数字或字母标识的单词或短语的句子中，有时也可用分号来分隔这些单词或短语。例如：

- Yet it has been criticized mainly for 1) $O(MN^3)$ computational complexity (where M is the number of objectives and N is the population size)；2) nonelitism approach；3) the need for specifying a sharing parameter.

- The four basic arrangements are as follows：(a) liquid supply unit；(b) balance subsystem of storage hydraulic pressure；(c) injection structure；(d) liquid return unit.

- LARKS possess three properties that are consistent with their functioning as adhesive elements in protein gels formed from LCDs：(i) high aqueous solubility contributed by their high proportion of hydrophilic residues：serine，glutamine，and asparagine；(ii) flexibility ensured by their high glycine content；and (iii) multiple interaction motifs per chain (Fig. 3B)，endowing them with multivalency，enabling them to entangle，forming networks as found in gels (Fig. 2).

5.3 冒号

冒号（colon，“：”）属于句内标点，主要用来提示下文，或引出下文进行解释、说明、证明、定义或补充，也可用作特殊表达中的标识符。冒号主要用在以下场合：

1）引导前文所预期的解释、说明、引语、事项列举或详细信息，所引导的部分可以是几个词、短语或句子（特殊情况下也可以只有一个词、短语或句子），或若干词、短语、句子的组合，其中还可嵌套简单或复杂的修饰成分（修饰成分可以是词、短语或句子）。例如：

- For the HER2 data shown，the cancers were grouped into ten cancer-type categories：biliary，bladder，breast，cervical，colorectal，endometrial，gastro-oesophageal，lung，ovarian or other (for all other cancer types).（冒号后为 10 个词）

- These authors contributed equally：Ian B. Perry，Thomas F. Brewer.（冒号后为 2 个人名）

- This prevented bodies in the inner Solar System from accumulating large amounts of water ice，explaining why such bodies are mostly dry，and maintained an isotopic dichotomy between two types of meteorite：ordinary and carbonaceous chondrites.（冒号后为 1 个短语，由两个形容词加 1 个中心词构成，形容词做修饰语）

- Most human solid tumours exhibit one of three distinct immunological phenotypes：immune inflamed，immune excluded，or immune desert.（冒号后为 3 个短语）

- The composition of the alloy is as follows：Si 9.86%，Cu 3.44%，Fe 1.29%，and aluminum the rest.（冒号后为 4 个短语，相当于 4 个分句）

- A bleaching record in our analysis consists of three elements：the location，from 1 to 100；the year；and the binary presence or absence of bleaching.（冒号后分别为词、短语、词、短语）

- She first examined the “manuscript，” and asked：*how the copyright of the illustrations in the book was considered*?（冒号前面为 1 个疑问词，后面为 1 个句子，用冒号引出疑问词所问的具体问题）

● Furthermore, such analyses are labour-intensive and expensive: in medical fields, systematic reviews generally take about a year to conduct and can cost between US \$30,000 and \$300,000 each.（冒号后是 1 个有 2 个谓语的较长的单句）

● So it must be improved in design in two ways: one is to increase the spindle diameter, the other is to add supporting on axis.（冒号后为 2 个单句）

● General psychological insights offer an explanation: People may judge risk to be low without available personal experiences, may be less careful than expected when not observed, and may falter without an injunction from authority.（冒号后为 3 个单句）

● This structure (Fig. 2) is the same as the previous one, but the values of the following variables are different: $S = 35$, $d_0 = 15$, $\delta_p = 1$, $h_f = 10$.（冒号后为 4 个式子，相当于 4 个单句）

● Although our results are useful in a variety of contexts, their potential impact centres around a more unifying aim: catalysing action to narrow gaps in opportunity by improving accessibility for remote populations and/or reducing disparities between populations with differing degrees of connectivity to cities.（冒号后为 1 个有多重修饰成分的短语）

● However, these models have difficulty in explaining the diverse composition of objects in the Solar System: if all such bodies grew by accretion from the same flow of pebbles, then why do they have different compositions?（冒号后为 1 个条件状语从句）

● We assigned cellular function to these 400 proteins based on their UniProt annotations (Fig. 3C): 16% are DNA binding, 17% are RNA binding, and 4% are nucleotide binding, consistent with reports of nucleotide binding proteins in membraneless organelles.（冒号后为 3 个单句和 1 个对这些单句的较长的补充性修饰语）

2）用于有明显引出列举事项属性的词或短语（如 as follows、the following、including、such as、the following）后面，引出所列举的各个事项，列举事项较多、复杂时，各列举部分前可以加数字编号。例如：

● Previous discoveries include the following: LCDs can "functionally aggregate" (31); proteins with LCDs typically form more protein-protein interactions (32, 33); and proteins can interact homotypically and heteroty-pically through LCDs (1, 5, 34).

● Variation to the operational function is equal to do the same to the resource cell, including: the changing of number of production equipment and of the company.

● In systematic reviews, investigators generally pose a focused question, such as: 'Is surgery an effective treatment for knee osteoarthritis?'

● With this in mind, we generated all 144 possible two-cell circuit topologies according to the following interactions (Figure 3B): (1) three possibilities for cross-regulation (positive, negative, or absent); (2) two possibilities for production of growth factors: each cell type can or cannot produce a growth factor for its own growth and survival; and (3) two possibilities for internalization of growth factors: each cell type can or cannot remove its growth factor by receptor-mediated endocytosis.

● Color values represent normalized mean

accessibility of peaks overlapping known enhancers (top: erythroid and erythroid progenitor, middle: lymphoid and lymphoid progenitor, bottom: myeloid and myeloid progenitor).

3）表示数字比或量比（比值、相除、比例、比率）。例如：

- A 50:50 exchange rate in blood leukocytes was observed 1 week after surgery (fig. S2A), and lung and intestinal CD4T cells showed exchange rates of 50:50 and 30:70 to 40:60, respectively, 2 months after surgery (fig. S2B).（4组数字比）
- Actin and tubulin antibodies came from Sigma Aldrich and were used at 1:5,000 in 5% milk.（数字比）
- In brief, ~1,000 cells were plated in 10 μl of 1:1 matrigel to culture media in 96 well angiogenesis plates and allowed to solidify for 30 min at 37 degrees before 70 μl of culture media was added.（数字比）
- In both model and experiment, CSF1 addition mainly affected macrophage number and MP:FB ratio, whereas PDGFB addition mainly affected fibroblast number, with all effects eventually returning to baseline.（量比）

还可以用数字比的形式表示时间。例如：23:20 p. m.（或23:20 PM）；8:30 a. m.（或8:30 AM）。

4）作为标识符用在一些特殊标注或特殊表达中。例如：

- Reference architecture for holonic manufacturing systems: PROSA.（分隔主副题名）
- Gene therapy: The power of persistence.（分隔主副题名）
- Michigan: University of Michigan, 2016.（分隔出版地和出版机构）
- Chinese Journal of Laser, 2004, 31 (4): 495–498.（分隔期刊卷、期与页码范围）
- **Keywords**: laser quench, laser shock wave, microstructure, martensite transformation.（分隔Keywords与其后的具体关键词）
- Tel: +86-10-88379056.（分隔电话标志词Tel与电话号码）
- e-mail: dmacmill@princeton. edu.（分隔邮件标志词与邮件地址）
- https://doi. org/10. 1038/s41586-018-0366-x.（分隔网址标志词https与网址）
- Received: 20 April 2018; Revised: 10 July 2018; Accepted: 6 June 2018; Published online 1 August 2018.（分隔日期类别标志词如Received、Revised、Accepted等与日期）
- Nd: YAG laser.（分隔标识、型号或编号等的各个组成部分）
- Note: all in vivo studies must report how sample size was determined and whether blinding and randomization were used.（分隔特别词与特别词所强调的内容）

5）分隔动词（如be）或前置词（如as）与其受词。例如：

- The device numbers and associated colors are: 1, black; 2, green; 3, purple; 4, red; and 5, blue.
- The parameters used here are: $a = 0.6$ m, $b = 0.2$ m, $\rho = 7.80 \times 10^3$ kg/m^3.
- Cite as: F. Tian et al., Science 10.1126/science.aat7932 (2018).
- In systematic reviews, investigators generally pose a focused question, such as: "Is surgery an effective treatment for knee osteoarthritis?"

这类句子中的冒号完全可去掉，去掉后便形成完整的句子。这里之所以将冒号插在动词（或前置词）与其受词之间，虽然"破坏"了句子的完整性，但能起到强调和列举作用，有一定的修辞效果。但是，如果不为这种修辞效果，或者动词（或前置词）后面的项较为简单特别是只有一项时，则完全不必用冒号进行这种分隔。例如以下两句中的冒号冗余，去掉可能更妥当：

• The diameter increment, depth of angular distortion and grade of curvature are: three different concepts.

• The research on the control mathematic models is not in accordance with: the requirement of technique developments.

5.4　破折号

破折号（dash）分为短破折号（en dash, "–"）和长破折号（em dash, "—"）。前者的长度相当于英语字母 N 的宽度，约为连字符"-"（俗称小短横）的两倍；后者的长度相当于英语字母 M 的宽度，约为短破折号的两倍。在同一句子中，通常最多只能用两个成对或一个单独的破折号。破折号在句中所表示的停顿比逗号明显。

1. 短破折号的主要使用场合

1）用于组成术语的两个同等重要的词语之间（术语也可以用符号表示），与 and、to 或 versus 同义。例如：temperature–time curve；cost–benefit analysis；nickel–cadmium battery；vapor–liquid equilibrium；v–f_s characteristics；Al–Si alloy；austenite–martensite。

注意：表不同颜色的组合要用连字符连接而不用短破折号，如 blue-green，red-yellow 等。

2）用于表示由两个数字、时间或两个字母等组成的区间，与 to 或 through 同义。例如：Figs. 2–5；Eqs.（4）–（7）；Tables 3–6；100–150 m/s^2；2–6 h；parts B–E；Extended Data Figs 1–3，Supplementary Information。

注意：数字由某种符号（如正号、负号，负号在形式上与短破折号没什么区别）修饰时，表示区间应该用 to 或 through，或其他形式，但不宜用短破折号或浪纹线。例如：30 to +100 K；−145 to −30 ℃；−500 to 800；5 to >400 mL；<20 to 25 mg；$e_n = [-5\times0.59, +5\times0.59]\approx[-3, +3]$V（不表示为 $e_n = -5\times0.59 - +5\times0.59 \approx -3 - +3$ V，也不宜表示为 $e_n = -5\times0.59 \sim +5\times0.59 \approx -3 \sim +3$ V）。另外，用"from … to …""between… and…"等连接两个词语时，不能用短破折号替代其中的 to 或 and。例如：from 1 500 to 2 000 mL（不写为 from 1 500–2 000 mL）；between 8 and 12 days（不写为 between 8 – 12 days）；with temperatures of −15 to 35 ℃（不写为 with temperatures of −15–35 ℃）。

3）用于由两个同等重要的人名所组成的复合性修饰语中。例如：Jalm–Teller theory；Franck–Condon factor；Fisher–Johns hypothesis；Beer–Lambert law；Lineweaver–Burk method；Diels–Alder reaction；Bose–Einstein statistics；Garofalo–Arrheninus model。

复合性修饰语中的一部分可同时含有连字符（-）。例如：Columbia-Presbyteran–Brigham cases（其中 Columbia-Presbyteran 和 Brigham 为两所医疗机构）。

4）用于表示几个连续参考文献序号的引用，或文献著录中引文页码范围的著录。例如：

• For more detailed information, refer to Refs. [8–12].

• Although VITOR, et al.[3-5], have reported the growth of self-supported diamond tubes of different internal diameters and different external diameter to wall thickness aspect ratio, the diameter was only limited to 600 μm and uniform thickness could not be ensured.

• Tonghai Wu, Weigang Wang, Jiaoyi Wu, et al. Improvement on on-line Ferrograph image identification [J]. *Chinese Journal of*

Mechanical Engineering, 2010, 23 (1): 1-6.

5）用于表示不同组分的溶液或化学键。例如：hexane-benzene solvent; CH_3-CH_2-CH_2-CH_2-CH_3。

6）用于表示编号、型号等。例如：DAQ-2010（一种数据收集卡）；HAW-4、TPC-4（两种海底光缆）。

2. 长破折号的主要使用场合

1）一对破折号用在句中非限定性修饰语（词、短语、句子或其组合）的前后，相当于替代逗号。其作用通常包括：对语句作解释、说明或总结；表示作者的态度和看法；用来强调，增强表达力；引起读者注意；转移话题，说明事由；承上启下，使语句衔接更加紧密。例如：

- Another attribute of the Mowry Shale—a diagnostic one, and an unmistakable clue to the identity of the formation—is the presence of countless well-preserved fish scales found with little effort on nearly every outcrop.
- At some point—determined by how the virus was programmed—the virus attacks.
- However, the granularity of the transcriptional assessment—factors such as sequencing quality and which kinds of RNA are analysed—is a key parameter in delineating cell types.
- Many countries have a long history of subsidizing fossil fuels, and it seems logical that removing these subsidies—as the G20 group of nations has agreed to do—would help them to achieve their Paris climate commitments.
- In addition, the three broad groups—rather than being independent compartments, as typically framed within the ecosystem services approach—explicitly overlap.

注意：在可用其他标点符号清楚地表达时，尽量少用破折号分隔非限定性修饰语。例如：

- Knauth, not Stevens, obtained good correlation of results and calculations.
- Knauth—not Stevens—obtained good correlation of results and calculations.（不宜）
- The stress caused by the friction force σ_f, which is shown in Fig. 5, can be expressed as $\sigma_f = \sigma_1 - \sigma_2$.
- The stress caused by the friction force σ_f—which is shown in Fig. 5—can be expressed as $\sigma_f = \sigma_1 - \sigma_2$.（不宜）

2）一对破折号用在插入语的前后，相当于替代逗号。插入语也称独立成分，是插在句中的词、短语、从句或其组合，常用逗号或破折号隔开，与句子其他部分无语法关系，除了具有非限定性修饰语所具有的那些作用外，还有举例、列示之类的作用。例如：

- These 2 participants—1 from the first group, 1 from the second—were tested separately.
- In comparison, the success of the approach used in our study—notifying clinicians of a single fatal overdose—may have a number of explanations.
- However, by being apprised of studies that examine how a particular intervention has worked for an order—for birds in general, say—practitioners can better weigh up the chances of success for their intended programme.
- Two opposing—although not mutually exclusive—scenarios account for the generation of distinct kinds of neuron across the nervous system, and in the cerebral cortex in particular.
- We think that in fields in which data are sparse or patchily distributed, or where studies vary greatly in design and generalizability—as is the case in biodiversity conservation, international development and education, for example—a different approach might often be more appropriate.

• Although subdivision into internally consistent systems of categories is common in many local knowledge systems, a universally applicable classification—such as the one proposed in the generalizing perspective on NCP (table S1)—is not currently available and may be inappropriate because of cultural incommensurability and resistance to universal perspectives on human-nature relations.

3）一对破折号用在非限定性同位语的前后，相当于替代逗号或圆括号，以表示、突出或强调同位语，使句义更加清晰。非限定性同位语常由逗号隔开，必要时才用破折号。例如：

• All three experimental parameters—temperature, time, and concentration—were strictly followed.

• Aerosols—solid and liquids—also are carriers of sulfur, nitrogen, and hydrocarbons.

• Initial studies demonstrated that the maximum sensitivity of plasma DNA-based tests—liquid biopsies—was limited for localized cancers.

• In the United States, you must drive with the headlights—the large front lights of the car—on at night, whether you are driving in the city or not.

• The approach we're advocating—subject-wide evidence synthesis—combines elements of systematic reviewing and mapping, along with other techniques, to provide an industrial-scale, cost-effective way to synthesize evidence.

4）用来引导对上文陈述的总结，说明内容，解释原因，概括事项，列举示例。例如：

• Whether we locate meaning in the text, in the act of reading, or in some collaboration between reader and text—whatever our predilection, let us not generate a straitjacket from it.

• The Japanese beetle, the starling, the gypsy moth—these pests all came from abroad.

• Several UK government departments have published Areas of Research Interest (ARI; see, for example, ref. 7)—topics on which synthesized and new evidence would be most welcome.

• Only a select set of genes is differentially expressed between the three regions—a limited level of premitotic diversity that is consistent with the postmitotic model.

• In the second study, Nowakowski et al. focused on excitatory neurons that produce the neurotransmitter glutamate, in two neocortical areas in human fetuses—the prefrontal cortex and the primary visual cortex, which are involved in behavioural planning and in vision, respectively.

• They are typically physically consumed in the process of being experienced—for example, when organisms are transformed into food, energy, or materials for ornamental purposes.

5）用来表示叙述突然停顿、转折或有意中断一下，以突出强调或引起读者注意。例如：

• The human race has survived, and the planet seems to have replenished itself—there are fish, ocean, forests—but what kind of society exists in 2195?

• A critically unanswered question remained from these studies to pave a path toward therapeutic potential—will it work in vivo?

• In a word, the spirit of the whole country may be described as—self-reliance and arduous struggle.（总之，整个国家的精神可以说是——自力更生，艰苦奋斗。）

• But regardless of the outcomes of the assessments, the consideration of different knowledge systems—and the fact that generalizing, context-specific, and mixed perspectives are considered as equally useful—matters in terms of making IPBES procedures and outcomes more equitable.

6）分隔句中的多个成分。例如：

• Dr. Fitzpatrick of the USGS will "eyeball" a section of core sample to check the alternating cloudy and clear bands that represent deposits of summer snow—cloudy—and winter snow—clear.

• Many NCP may be perceived as benefits or detriments depending on the cultural, socioeconomic, temporal, or spatial context. For example, some carnivores are recognized—even by the same people—as beneficial for control of wild ungulates but as harmful because they may attack livestock.

7）表示直接引语的来源。例如：

• Publication of this letter does not indicate that it represents a policy of the American Chemical Society. —The Editor.（编辑的声明）

8）用来分隔主副标题。例如：

• Prediction of Right-Side Curves in Forming Limit Diagram of a Sheet Metal—Part Ⅰ: Predicting Fundamentals

• Prediction of Right-Side Curves in Forming Limit Diagram of a Sheet Metal—Part Ⅱ: Prediction Method

5.5 连字符

连字符（hyphen，"-"）又称连接号，其长度约为半个英语字母的宽度。不同词典或语法书对它使用的规定在某些细节方面不尽相同，写作时应勤查词典，力求论文中有连字符的复合词符合专业领域及相应出版物的表达习惯、规定。连字符与破折号在功能上的区别主要在于：连字符主要起连接作用，多用于复合词中，破折号主要起分隔作用。

下面列出连字符使用的一些通用规则。

1）连接单词与前缀。这种构词方式有以下七种情况：

① 前缀加普通词语。例如：semi-solid；super-plastic；self-configuration。

② 前缀加专有名词（或形容词，首字母大写）。例如：pre-Columbian；post-Copernican；non-Gaussian。

③ 前缀的尾字母与后面所连接的词的首字母重复。例如：anti-infective；meta-analysis；co-ordination；co-operate（多见于英式英语）。

④ 前缀加含有前缀的词。例如：mid-infrared；post-reorganization；bi-univalent。

⑤ 前缀加含有连字符的词。例如：non-radiation-caused effects；non-tumor-bearing organ。

⑥ 有前缀的化学术语。例如：non-alkane；non-hydrogen bonding；non-phenyl atoms。

⑦ 有前缀的数字（表年代）。例如：pre-1900s；post-1800s。

2）连接单词与后缀。这种构词方式主要有以下四种情况：

① 后缀的首字母与其前面单词的尾字母重复。例如：gel-like；shell-like；bell-like。

② 含like、wide等后缀的多音节词或含like、wide等后缀的已含有连字符的复合词。例如：resonance-like；radical-like；university-wide；rare-earth-like；transition-metal-like。

③ 含有后缀的数字。例如：6-fold；35-fold；4.8-fold。

④ 含有后缀的专有名词。例如：Kennedy-like；Claisen-type。

3）用于区别易混淆或具有不同词性的

单词或短语。例如：un-ionize，unionize；re-collect，recollect；re-form，reform；shut-down，shutdown，shut down。

4）用于构成复合词。这种构词方式主要有以下六种情况：

① 由几个单词组成、含义上需紧密配合使用的修饰性复合词。例如：Chinese-language；real-time；high-alumina；load-to-weight；slow-moving；variable-gain；Parent-1；six-degree-of-freedom。

② 以 better、best、ever、ill、lower、little、still、well 等副词开头的复合形容词。例如：best-known；ill-informed。

③ 包含名词、副词、形容词、分词的复合形容词。例如：machine-made；well-made；high-powered；state-owned；well-known；air-equilibrated；fluorescence-quenching；ion-promoted；steam-distilled。

④ 变换有几个部分的复合性形容词。例如：sodium-and potassium-conserving drugs；high-，medium-，and low-frequency measurements；second-and third-order reactions。

⑤ 由 east、south、west、north 中的若干词所组成的修饰语。例如：north-northeast；south-southeast。

⑥ 某些姓名中的复姓或复名。例如：Maier-Speredelozzi V；Yip-Hoi D；Ait-kadi D；Yip-Hoi Derek M；John Edward Lennard-Jones。

5）用于由数字、字母或元素符号，与名词或形容词组成的复合性修饰语。这种构词方式主要有以下四种情况：

① 含有数字的修饰语。例如：21st-century development；early-thirteenth-century architecture；six-coordinate system；one-way operation；three-dimensional model；two-layer structures；five-and nine-point finite-difference grids。

② 表示年龄的修饰语。例如：a 60-year-old scientist。

③ 由数值和单位共同组成的修饰语。例如：a 3-g dose；a 5-to 10-m-thick unit；the 20-km circumference；3–5–h sampling time（a 3-to 5-h sampling time）。

注意：如果数值或单位由多个部分组成或单位中含有“°”“′”“″”“%”，则不必加连字符。例如：$1.2\times10^{-6}\ mm^{-1}$ peak；a 100 ℃ water；90°angle；35% increase。

④ 由单一数字、字母或元素符号，与名词或形容词所组成的修饰语。例如：4-position；^{14}C-labeling；K-Ar age；O-ring；s-orbital；x-axis；U-Pb ratio；π-electron；α-helix；γ-ray；X-ray diffraction。

注意：表示同位素之比不用连字符而用斜线。例如：$^{207}Pb/^{206}Pb$；$^{40}Ar/^{39}Ar$。

6）用于用全拼单词所表示的分数的分子与分母之间。例如：one-half；one-third；one-fourth；thirty-nine hundredths；five-fourths；three-tenths。

7）用于 21～99 间的十位数和个位数之间。例如：ninety-nine；forty-one；sixty-six。

8）用于表示原子核子数（质量数）。例如：C-12（也可表示为^{12}C）；iodine-127。

9）用于文字录入、排版时同一单词的拆分转行。拆分转行是指一个英语单词在上行末排不下时分拆一部分移至下行之首。这种现象在英语科技论文中经常出现，容易出错。英语单词拆分转行的一般原则有以下八个方面：

① 尽量避免拆分单词，可以通过调整词间距（必要时也可调整同一词中的字符间距）、段落对齐方式等方法排版。

② 按英语词典所标示的音节拆分，并遵循单词的词源学规律，使得转接部分看起来像一个独立的单词。例如：information 可拆分为 infor-//mation，不能拆分为 informa-//tion；pathologic 可拆分为 path-//ologic，不

能拆分为 patho-//logic。

③ 多音节词（包括双音节词）一般按音节来拆分，双辅音（重叠辅音除外）或双元音不能拆开；单音节词（如 through、brought、plough 等）和较短的双音节词（如 also、into、away、oval 等）一般不拆分。

④ 相邻的两个音节之间有两个辅音字母时，转行应在两个辅音字母之间进行。例如：commune 拆分为 com-//mune；English 拆分为 Eng-//lish；doctor 拆分为 doc-//tor。

⑤ 派生词的转行要根据构词法来进行，即在词根和词缀之间转行。例如：illega 可拆分为 il-//legal；impatient 可拆分为 im-//patient；careless 可拆分为 care-//less；selfish 可拆分为 self-//ish。

⑥ 含有连字符的复合词应在连字符所在位置拆分，以避免出现更多的连字符。例如：cost-benefit analysis 拆分为 cost-//benefit analysis；well-known 拆分为 well-//known。

⑦ 长的化学名称或术语拆分后每行的字母不应少于 4 个，而且不能在描述性前缀的连字符处拆分。例如：2-acetylaminofluorene 可拆分为 2-acetyl-//aminofluorene，不可拆分为 2-//acetylaminofluorene。

⑧ 尽量避免缩写人名的转行或隔页转行；转行时不得在上行末或下行首只留一个字母；转行后的连字符应放在上行之末。

5.6　圆括号

圆括号（brackets 或 mark of parentheses，“（）”）主要用于涵括行文中相对独立的部分（如补充或说明材料、解释语及事后想法、建议等）。圆括号须成对使用，与成对出现的逗号和成对出现的破折号的作用相似，但在表示强调程度方面略有差异：圆括号较弱，逗号中性，破折号较强。圆括号主要用在以下七种场合。

1）表示补充信息，起解释、说明的作用，所括内容可以是单词、短语、句子（简单句、并列句、复合句），或数字、符号、式子，或几种要素的组合等。例如：

- The high transient energy is supplied by a series of accumulators (see Fig. 3).
- The dynamic rigidity declines when the ram is extending to the front (back constraint).
- The testing capacity of the shock machine is under 5 000 kg (including fixture).
- The inside temperature is the average temperature of 8 inner corners, and the outside temperature is the average temperature of the center of each exterior surface of the refrigeratory (totally 5 centers, except for the ground).
- With the system, the grinding wheel (or disk milling tool) axial cross-section that corresponds to the three-arc flute cross section can be conveniently simulated.
- Some secondary martensites induced by laser shock wave are formed on the base of the martensites induced by laser quench (such as the martensites arranged in cross direction).
- The use of native functional groups (for example, carboxylic acids, alkenes and alcohols) has improved the overall efficiency of such transformations by expanding the range of potential feedstocks.
- Suppose that $y = f(s, z, x)$ denotes the vector of responses for a particular set of factors, where s, z and x are vectors of the signal, noise (for instance, the tolerance of design parameters), and control factors (design variables), respectively.
- As the lengths of the links L_i ($i = 1, 2, \cdots, 6$) change, the movable platform will

move in all 6-DOF including translation motions (x,y,z) and rotation motions (φ,θ,ψ).

• After six openings (1-1, 1-2, 1-3, 1-4, 1-5, 1-6), six sloping pilot oil pipelines (2-1, 2-2, 2-3, 2-4, 2-5, 2-6) are arranged.

• Three scenarios are considered in the simulations: off-centered load along axis r (a = 50 mm), off-centered load along axis p (b = 50 mm), and off-centered load along axis r and axis p (a = b = 100 mm).

• Milligram morphine equivalents in prescriptions filled by patients of letter recipients versus controls decreased by 9.7% (95% confidence interval: 6.2 to 13.2%; P < 0.001) over 3 months after intervention.

所括内容为完整的句子时，若注释句子的局部，则括号内句子的首字母小写（量符号除外）；若注释整个句子，则括号内句子的首字母大写（量符号除外），句末使用标点，且连同括号放在整个所注释句子末尾的标点之后。例如：

• If the distance function d (x_i, x_j) < L (where L is the dynamic distance function), we regard that the individual x_i is similar to x_j, then compare their fitness.

• From this result, the transfer function of the joystick was estimated as one of the first order lag systems (T_m is 0.125 s, K_m is 0.18 N^{-1}) with a time lag element (L_m is 0.08 s).

• From Fig. 1 it is seen that the viscosity number is 4.12 × 10^{-2} N/(m • s) at 590 ℃ (the volume fraction of solid of the alloy is 20% at this temperature).

• K9 optical glass is used as confinement medium, one of whose sides connected with the sample was coated with black paint 86-1. (The thickness is about 0.025 mm.)

• The controller, servovalve, actuator, and test specimen models presented in the last three sections can now be assembled to give a simulation to verify the performance of the proposed control strategy. (Experimental verification is planned for the future.)

2）给出所用词语的缩略语或符号表达，或给出所用缩略语的全写。例如：

• Finite elements analysis (FEA) and modal synthesis analysis (MSA) are used to calculate the vibration state of 5-axis machine tool.

• The direct functionalization of carbon–hydrogen (C–H) bonds—the most abundant moiety in organic molecules—represents a more ideal approach to molecular construction.

• The world's first remote control system was a mechanical master-slave manipulator called ANL (Argonne National Laboratory) Model M1 developed by GOERTZ.

3）表示起序号作用的数字或字母（如式号、分图序等）。例如：

• However, we note two limitations of this method: (1) its performance is much less reliable on tissues with large class imbalance (dominated by a single-cell type, e.g., thymus) and (2) it does not handle cases where a cell type appears in one dataset but not the other.

• The input data for this design method are as follows: (a) meridional contour; (b) angular momentum distribution; (c) flow rate and rotational speed; and (d) blade number and thickness distribution, where the angular momentum distribution is very important.

• The fluid film thickness in the non-grooved area $h_1(r)$ and that in the grooved area $h_2(r)$ can be defined by Eqs. (1) and (2), respectively.

• Greedy algorithms have the following

sequence of steps:

(1) Calculate the initial population (see Fig. 2 (a)).

(2) If the program can go to a further step, calculate a sub-solution of the feasible solution according to feasible strategy.

(3) Combine all the sub-solutions to make a feasible solution.

4）表示复合单位中需要用圆括号括起的部分，或标目［即“量名称和（或）量符号/单位”］中复合单位的整个括起部分。例如：

• $W/(m^2 \cdot K)$; $C/(kg \cdot s)$; $m^3/(Pa \cdot s)$; $m^2/(sr \cdot J)$。

• Rotational speed $n/(r \cdot min^{-1})$; Coefficient of heat conductivity $\lambda_m/(W \cdot m^{-1} \cdot K^{-1})$。

5）表示数学式、化学式中需要用圆括号括起的部分。例如：

• $\frac{\sqrt{a}}{2\pi}\int_{-\infty}^{+\infty} s(\omega)\varphi^*(a\omega)\exp(j\omega b)\,d\omega$; $\theta(n) = \arctan(C_{si}(n)/C_{sr}(n))$。

• $CH_3(CH_2)_{10}CH_3$; $Fe(CN)_2 + 4NaCN \longrightarrow Na_4Fe(CN)_6$。

6）注释引文、引语的出处。例如：

• Saying and doing are two different things. (John Heywood)

• Before everything else, getting ready is the secret of success. (Henry Ford)

• Thinking is the talking of the soul with itself. (Plato)

7）标注参考文献著录中的信息项，包括期刊年卷期标志中的年份或期号，报纸的版次，电子文献更新日期、引用日期及非公元纪年、出版社信息等。例如：

• Crowley, P., Chalmers, I. & Keirse, M. J. *BJOG Int. J. Obstet. Gynaecol.* **97**, 11–25 (1990).

• Quinlan, A. R., and Hall, I. M. (2010). BEDTools: a flexible suite of utilities for comparing genomic features. Bioinformatics 26, 841–842.

• Tenopir, C. Online databases: quality control. *Library Journal*, 113 (3): 124–125 (1987).

• Turcotte, D. L. *Fractals and Chaos in Geology and Geophysics*. (Cambridge University Press, New York, 1992) (1998-09-23). http://www.seg.org/reviews//mccorm30.html.

5.7 方括号

方括号（square brackets，“[ ]”）主要用来标明行文中注释性的语句。方括号主要用在以下场合。

1）用在圆括号内的插入语或解释、注释语中，或用于表示含有圆括号的附加信息（或补充信息），使附加成分的层次更容易区分。例如：

• The auction of the Last Emperor's possessions (including jewelry, furs, furniture, movie costumes [from 1945 through 1975], china, silverware, and other memorabilia) is scheduled for July 17 and 18.

• In contrast, other clusters include cells from many tissues (e.g., cluster 4 derives from lung [44%], spleen [44%], bone marrow [5%], large intestine [2%], and others), and some tissues are distributed across many clusters (e.g., whole brain contributes to clusters 8 [34%], 5 [17%], 15 [13%], 21 [11%], and others) (Figures 2A, S2A, and S2B).

• However, primary refractoriness and acquired resistance after a period of response are major problems with checkpoint blockade therapy [reviewed in (38)].

• ΔR versus V_c for the 100-nm-wide gate on device 1 at $V_{tg} = 3.5$ V [white dashed line in (C)]

- Acetic anhydride [$(CH_3CO)_2O$].

2）用在直接引语或其他语句中加入的插入语或解释说明。例如：

- As Sir William Lawrence Bragg said, "The important thing in science is not so much to obtain new facts as to discover new ways [italics added] of thinking about them."
- The results of an analysis by Neyman, Scott, and Smith [Science, 163: 1445-1449] of a carefully conducted experiment were released to the press.
- However, in contrast to most reports on superconductors [for example, (41)], our study describes a beneficial effect from the light-induced lattice expansion.
- To check the above criteria in monolayer WTe_2, we fabricated devices with the structure depicted in Fig. 1A [see (26) and figs. S1 and S2].

3）表示物质或材料的化学式、浓度、剂量等。例如：$[W_{10}O_{32}]^{4-}$；$[Na^+]$；$[HCO_3^-]$；[%ID]。

4）用于式中需要用方括号括起的部分。例如：$\varphi[x(t)]=[x'(t)]^2-x(t)x''(t)$；$\theta_0\notin[\min(\theta_{\min},\theta_{\min}+\bar{\theta}-\psi)\pm 2k\pi, \max(\theta_{\max},\theta_{\max}+\bar{\theta}+\psi)\pm 2k\pi]$；$[Co(CO)_4]_2$；$[\mu_f(\boldsymbol{x},\boldsymbol{s}),\sigma_f^2(\boldsymbol{x},\boldsymbol{s})]$等。

5）用于参考文献标引（标注）、著录（包括文献序号、文献类型标志、电子文献的引用日期以及自拟的信息等）。例如：

- According to Ref. [15], we presented a new method to solve this difficult problem.
- There is limited research in this field[9-11].
- TURCOTTEDL. Fractals and chaos in geology and geophysics [M/OL]. New York: Cambridge University Press, 1992 [1998-09-23]. http://www.seg.org/reviews/mccorm30.html.
- T. F. Watson, S. G. J. Philips, E. Kawakami, D. R. Ward, P. Scarlino, M. Veldhorst, D. E. Savage, M. G. Lagally, M. Friesen, S. N. Coppersmith, M. A. Eriksson, L. M. K. Vandersypen, arXiv: 1708.04214 [cond-mat. mes-hall] (14 August 2017).

5.8 引号

引号（quotation marks）分为双引号（" "）和单引号（' '），主要用于表示需要着重论述的对象以及具有特殊含义的词语（也可用斜体、黑体或加粗体、下划线等形式来表示）。引号主要用在以下场合。

1）表示正文中某部分段落篇章的总结性用语，相当于标题，起总括作用，用以引起读者的注意。例如：

- "*Our team of researchers has searched every issue of nearly* 250 *journals for tests of some conservation intervention.*"
- "*In* 2017, *the Conservation Evidence website had* 15,000–25,000 *page views each month.*"

2）表示直接引语。直接引语后面的标点符号若是直接引语的一部分，则放在引号的内侧，否则就放在引号的外侧。例如：

- A computer program of ray-testing approach was implemented by using the commercial CAD/CAM system "CAXA Manufacturing Engineer".
- As for China, a batch of scientific researchers were organized to study utilizing transfer function method to calculate air conditioning load, and got a series of achievements, which were represented concentrated in "air conditioning technology" (1983).
- Ralph Waldo Emerson said, "The reward of a thing well done is to have done it."

- "It is a chronic problem masquerading as an acute one, but chronic problems aren't sexy," said Finkelstein. "Computer systems fail all the time, and the world doesn't come to end."

- The joystick ("SideWinder Force Feedback 2", Microsoft Co., Ltd.) can be operated to the *X*-axis and *Y*-axis directions.

注意：对于大段的直接引语，也可以采取另起段排版、两端或一端缩进的方式，或采取其他方式，具体方式视目标期刊的规定以及行文表达的实际情况而定。

3）表示出版物的各级标题。例如：

- A complete description of the oils is given in the section "Flavonoids in Citrus Peel Oils", and other references are listed in the bibliography.

- In the article, "Product Gene Representation and Acquisition Method Based on Population of Product Cases", Ligang Tai, et al., proposed a new methodology of product gene representation and acquisition from a population of product cases.

4）表示强调、突出或表达某种特殊的含义。例如：

- $\hat{\psi}(\boldsymbol{\omega})$ is the Fourier transformation of $\psi(\boldsymbol{x})$, $\psi(\boldsymbol{x})$ is called "mother wavelet".

- The UNDEX environment is very complex, composed of a "kick" from the incident shock wave followed by the effects of cavitations, bubble pulse, and structural whipping.

- The GA is used to train the PID neural network for control applications according to the performance index of the received error signal and evaluate the "goodness" of the control actions.

- The other pressure change from the valve's inlet to its outlet can be calculated by a model of "pressure and flow of valve" as below.

- The model provides a safe and easily controllable way to perform a "virtual testing" before starting potentially destructive tests on specimen and to predict performance of the system.

- To be consistent with data used in previous accessibility mapping research, we selected the 'high-density centres' variant of the GHS dataset, which is defined as "contiguous cells with a density of at least 1,500 inhabitants per km^2 or a density of built-up greater than 50% and a minimum of 50,000 inhabitants.

- Now that our rations of food, particularly of meat and wheaten bread, have been so appreciably reduced, the necessity of arranging our diet so as to ensure a sufficient supply of those elusive substances, the so-called "vitamines", is more important than ever.

- The natural sciences, and ecology in particular, were used to define "ecological production functions" to determine the supply of services, conceptualized as flows stemming from ecosystems (stocks of natural capital).

5）表示用于举例或解释的单词、短语、名称字符串或某种符号等。例如：

- Colbaugh, et al., discussed different resolutions for actuator redundancy and categorized them into two approaches, the "direct inversion" and "indirect inversion".

- The change of pressure of conical and converging pipelines and openings that are the key parts of this actuator can be simulated by models of "tube flow" and "converging opening".

- The term "silt" refers to unconsolidated rock particles finer than sand and coarser than clay.

● The markers "O" are the results from the measurement.

● The liquid used is usually machine oil; so we will use one word, "oil", to represent any available liquid as a medium of transition of force in the following description.

● As an advanced form of tele-operation, the concept of "tele-presence" was proposed by MINSKY.

● In Windows operation system, the function of "AfxBeginThread" can be used to create and start a thread, and the functions of "SuspendThread" and "ResumeThread" can be used to suspend and resume the thread.

● As a result of some recent work, McCollum and Davis concluded that two distinct types of vitamine exist, the "fat-soluble A" and the "water-soluble B".

按中国的习惯，使用引号时应优先使用双引号，在双引号内如果还需引用，则再使用单引号。但是，按国际习惯，英语论文中直接使用单引号的现象也较为普遍。例如：

● Generally, the prostrate leaves had a higher total photosynthesis rate than the 'erect' leaves.

● Alternatively, a practitioner asking, 'What can be done to conserve seabirds?' might want to read about all 48 interventions pertaining to the conservation of seabirds.

● In this paper, we use 'dormancy strength (weak-strong)' for a general description of a seed bath or taxa, and 'degree of dormancy (low-high)' for describing any specific moment on the continuous scale.

5.9　斜线号

斜线号（slash、virgule、solidus 或 shill，"/"）的主要作用是分隔供选用的词语，前后一般不留空。斜线号主要用在以下场合。

1）通常用在两个对立、两者选一个或几个并存的词语之间，而且这些词语是被当作一个词语来看待的。例如：

● It shows the feedforward/feedback control block diagram of the damping system.

● Tele-presence enables a human operator to remotely perform tasks with dexterity, making the user feel that she/he is present in the remote location.

● The Google roads data provided information critical for maintaining connectivity in areas where OSM coverage was sparse and/or fragmented owing to its piecemeal data collection approach.

● For box plots, centre mark is median, whiskers are minimum/maximum excluding outliers, and circles are outliers.

● In Eq. (7), the parameters f_e/f_c that are summation of friction, inertial force and weight of piston are called expanding/contracting motions threshold of driving forces.

● The concept of "tool-path loop tree" (TPL-tree) providing the information on the parent/child relationships among the tool-path loops (TPLs) is presented.

● However, the experienced designer may not always be available and/or may only have experience in a small number of turbine blade types.

● The grade C/gravel road/2.56×10^{-4} m^3 may be chosen as the test road.

2）表示除号或带有比值关系的"数字""常量""量名称""量符号""单位符号"及"量符号和单位符号"等（表示复杂的比值关系时，可以用中圆点和括号）。例如：$-\pi/2$；the efficacy/toxicity ratios of

PI3K inhibitors；$\partial \boldsymbol{I}/\partial \boldsymbol{P}_i=0$；$\lambda=|\sigma-\sigma_0|/\sigma_f$；optimum mass ratio（$m_2/m_1$）；$f_e/f_c$；about 2.0 GW/cm^2 in density；Oil density $\rho/(kg \cdot m^{-3})$；Extension coefficient δ/%；Flow of valve opened fully，m^3/s；2 kN•s/m；maximum power of 48 kW/5500 $r \cdot min^{-1}$；$([W_{10}O_{32}]^{5-}/[W_{10}O_{32}]^{6-})=-1.52$ V。

3）表示各种编号。例如：①DOI：10.3901/CJME.2016.06.001；②Paper No.97-DETC/DAC3978；③ GB/T 7714—2015；④Test standard BV043/85 等。

4）用于参考文献著录，“/”主要用在合期的期号间以及文献载体标志前，“//”主要用在专著或连续出版物中的析出文献的出处项前。例如：

- A Hopkinson. UNIMARC and metadata：Dublin Core［EB/OL］.［1999-12-08］. http://www.ifla.org/IV/ifla64/138-161e.htm.
- Jinhua Xu，D W C Ho. Adaptive wavelet networks for nonlinear system identification［C］//Proceeding of the American Control Conference，San Diego，California，USA，June 2-4，1999：3472-3473.

5.10　撇号

撇号（apostrophe，“’”）主要用来表示单词的所有格或构成缩写形式。撇号表示所有格的使用规则大体上有以下六个方面。

1）表示单数名词的所有格，在其后面加一个撇号和一个 s。例如：alloy’s temperature；Biological Diversity’s strategic plan；container’s outlet；doctor’s degree；Europe’s railways；machine tool’s static and dynamic rigidity；nature’s contributions；season’s greeting；the HITACHI’s TEM；the world’s first remote control system；valve’s inlet；wave’s opening；ZL112Y’s liquidus。

2）表示以 s 结尾的复数名词的所有格，只在其后加一个撇号。例如：two days’ paid vacation；martensites’ growth；machine tools’ quality；drivers’ operating demands；French railways’ TGV；testees’ contact pressure or electromyography；bidders’ private information。

3）表示不以 s 结尾的不规则复数名词的所有格，在其后加一个撇号和一个 s。例如：women’s studies；children’s toys；people’s quality of life。

4）表示系列名词的所有格，如果它们共享所有权，只加一个撇号和一个 s；如果所有权是分开的，则在每个名词的后面加一个撇号和一个 s。例如：Palmer and Golton’s book on European history（Palmer 和 Golton 合著关于欧洲历史的论著，指一部书）；Palmer’s and Golton’s books on European history（Palmer 和 Golton 各自关于欧洲历史的论著，指两部书）。

5）表示单数专有名词的所有格，在其后加一个撇号和一个 s。例如：Bernoulli’s theorem；Descartes’s philosophy；Marx’s precepts；Newton’s second law；Taylor’s series expansions 等。

6）表示复数专有名词的所有格，在其后加一个撇号。例如：the Dickenses’ economic woes（Dickens 全家的经济困难）。

撇号还可以构成缩写，如 can’t（can not），wouldn’t（would not），it’s（it is），isn’t（is not），doesn’t（does not）等。

撇号也可以表示时间，如 1990’s（1990’s 与 1990s 表意相同）；也可以构成地名、会议名等，如 Xi’an，ICME’2000 等。

5.11　省略号

省略号（ellipsis，“…”）表示句中某一（些）成分或语句被有意省略，这样既可避免把一段材料中不必要的、不甚相关的、不愿写出的或按某种语境无须写出的部分写出来，又可把一段话中未表达出来的部分在形式上以省略的形式展示出来。省略号还可用在数学式中来表示省去的部分。

省略号的形式为三个连续的句点，通常与句号、逗号等在同一水平线上，即处于底平齐的位置，如果用在句末，则加上句号就成了四个句点，但省略号后面的这个句点通常可以省去。数学式中的省略号与数学运算符和（或）关系符连用时，应将省略号提至与其前、后运算符和（或）关系符的中轴线同一高度的位置（即上下居中）。例如：

- The British Home Office seems to be modifying slightly its attitude to the tests by which motorists in Britain can now be convicted of driving under the influence of alcohol. A recent paper ... by Professor J. B. Payne ... suggests that the methods used ... are far from accurate...So far the Home Office has not been forced to act, because no motorist accused of driving under the influence of drink has quoted Professor Payne's work in his defence ... Originally ... police surgeons were advised to take small samples of capillary blood for use in the test, although it was also open to them to take venous blood if they preferred. The work at the Royal College of Surgeons suggests that the latter is likely to give more accurate results ... The Home Office has now sent a circular to police authorities pointing out that it is within their discretion to take venous rather than capillary blood ... The circular also points out that the motorist has the right to keep a sample of his own blood for independent analysis. (**From *Nature* 23 March 1968**)
- The NCP approach aims at ... products that are ... more likely to be incorporated into policy and practice.
- Assign to groups based on even or odd number of groups (to create even distribution): (i) Odd # of groups, in straight sequential (1, 2, 3, 4, 5, 1, 2, 3, 4, 5 ...) and (ii) Even # of groups, in snaking-sequential (1, 2, 3, 4, 4, 3, 2, 1...).
- **Authors** Darren A. Cusanovich, Andrew J. Hill, Delasa Aghamirzaie, ..., Christine M. Disteche, Cole Trapnell, Jay Shendure
- 1, 2, ..., C indicate customers and 0 indicates distribution center.（其中，省略号也可排为上下居中的形式）
- According to fuzzy inference system (FIS), there are input-output relationship matrices $\boldsymbol{R}_1$, $\boldsymbol{R}_2$, ..., $\boldsymbol{R}_n$ corresponding to fuzzy logic control rules... .（其中，第一个省略号也可排为上下居中的形式）
- $K(x_i, x_j) = \tan[k(x_i, x_j) - \theta], i, j = 1, 2, \cdots, N.$
- $y_k = a_1 y_{k-1} + a_2 y_{k-2} + \cdots + a_p y_{k-p} + b_1 u_{k-1} + \cdots + b_q u_{k-q}.$
- $h(j) = b_1(\alpha_1^{j-1} + \alpha_1^{j-2}\alpha_2 + \alpha_1^{j-3}\alpha_2^2 + \cdots + \alpha_2^{j-1})(j = 1, 2, \cdots).$

根据表达的需要，省略号有时也可排为竖排的形式，即“⋮”。

5.12 句号

句号（period 或 full stop，“.”）也称句点，是句末点号的一种，主要用来表示一个陈述性句子或短语的结束。句号主要用在以下场合。

1）用于完整的陈述性句子或短语的结束。例如：

- A wavelet network suiting to approach multi-input and multi-output system is constructed.
- Yes, for the most part. Our manufactures

have increased their output since the new increased program was established.

注意：陈述句以含有缩写点的缩写符号结束时，末尾可以不加句号。例如：

- Most of these products were manufactured in the U. S. A.

- A technique for constrained B-spline curve and surface fitting was developed by ROGERS, et al.

2）用于一些非科技术语或拉丁语的缩写。例如：e. g. ; i. e. ; op. cit. ; et seq. ; s. t. ; Mr. ; Ph. D. ; No. ; Thomas A. Smith, Jr. ; GOLUB G. H. 等。

注意：大多数计量单位的缩写的后面不加句点，但缩写后的单位如果容易与其他单词混淆，则应该加句点。例如：inch（es）的缩写为“in.”；foot 的缩写为“ft.”；cubic meter 的缩写为“cu. m.”。

3）在有一系列以数字或字母标识（或编号）的语句中，可用句号分隔这些语句。例如：

- The results show the following：(1) For rolling-sliding case, the thermal stress in the thin layer near the contact patch due to the friction temperature rise is severe. The higher rolling speed leads to the lower friction temperature rise and thermal stress in the wheel. (2) For sliding case, the friction temperature and thermal stress of the wheel rise quickly in the initial sliding stage, and then get into a steady state gradually.

4）用作小数点、提纲（目录、标题）排序数字编号中的分隔符。例如：

- Of the customers responding to our survey, only 34. 7 percent rated our service as “Excellent”.

- 5. 1 Profile curve from the engine cover of a car

- 2. Sealing of surfaces prevents excessive moisture absorption

5.13 问号

问号（question mark，“?”）是句末点号的一种。用在疑问句的后面表示疑问语气，用在反问句的后面表示反问语气，用于括号内则表示怀疑语气。问号主要用在以下场合。

1）用于疑问句的末尾，直接提出问题或表示疑问。例如：

- What causes the moon to rise in the east and set in the west?

- Which is the more costly assimilatory structure to make—the leaf or the pitcher?

- Are *Nepenthes* species similar in nutrient status and limitation to other carnivorous plants?

- A critically unanswered question remained from these studies to pave a path toward therapeutic potential—will it work in vivo?

- Why might this stalling occur? The authors' tomographic reconstructions revealed numerous regions of electron density located between a poly (GA) ribbon and the site where the protein RAD23 binds to the proteasome.

- For example, conservationists might ask, 'How can we reduce fulmar by-catch at sea?'

- We first examined the “ceiling,” and asked: which cell type is constrained by carrying capacity?

- Alternatively, a practitioner asking, 'What can be done to conserve seabirds?' might want to read about all 48 interventions pertaining to the conservation of seabirds.

- For example, if a drug that targets a specific protein can treat a person with breast cancer who has a mutation in the gene encoding the protein, could the drug treat another patient who has a different mutation in that gene? And could it treat a person with a mutation in the same gene, but in a tumour that has developed in a different tissue?

• Several important issues were discussed at the conference last week: Which style trends will be popular during the next decade? What comfort demands will the public make on furniture manufacturers? How much will price influence consumer to purchase furniture?

2）用于文章标题的后面，表示疑问、探究或征询语气。例如：

• A death knell for relapsed leukemia?

• CDC25 phosphatases in cancer cells: key players? Good targets?

3）用于一般陈述句的末尾表示反问。例如：

• The policy was overturned. The loss of 10 species were not too much?

• The Conference of Bioengineering Technology in the Future will be launched in 2021?

4）用于疑问词带有逗号的陈述句的末尾表示强调。例如：

• The committee asked, why were so many species killed?

此句的正常表达是“The committee asked why so many species were killed.”，之所以末尾用了问号，并改变了语序，就是为了强调。

5）用于表达句中不确定或有疑问的部分。例如：

• Girolamo Fracastoro (1483?-1553) was in effect the father of the concept of infectious disease.

6）引用文献原文时，所引部分包含问号时，此问号不要省去。例如：

• Another approach, called systematic mapping, is more broad-brush. But this typically does not describe the findings of the research, and so cannot be used to answer questions about policy (see 'Review or map?').

5.14 叹号

叹号（exclamation mark，“!”）又称惊叹号，属于句末点号的一种，用在句子的末尾表示强调某种语气。叹号在科技论文中一般很少使用，偶尔也会出现，主要用在以下场合。

用于陈述句、祈使句或疑问句的末尾表达强调的语气。例如：

• Freud's “science” was pure metaphysics!

• By modifying the weakest component (spindle-tool part), the limit cutting depth is extended downwards 100%!

• Come, come, come! We all enjoy this beautiful music together.

• Isn't this new robot a product of artificial intelligence!

数学式中的“!”不是叹号，而是表示阶乘的数学符号。例如：$[(n+1)\times(m+1)]!$；100！等。

第6章 英语科技论文修辞实例

修辞就是对“辞”做“修”，即对语句进行修改，使其表达规范，或使其表达效果提升。使语句表达规范，就是消除语病，使语句符合语言表达要求，相当于语言的一般润色，对应修辞中的消极修辞（一般修辞）；使语句表达效果提升，就是对没有语病的语句重新表达，选用最佳词语组合，达到理想的表达效果，甚至达到语用的艺术化，相当于语言的高级润色，对应修辞中的积极修辞（高级修辞）。一篇论文有那么多语句，每句达到最佳不太可能，也无必要，尤其对于刚写成的初稿，存在这样那样的语病也很正常，况且语言效果的提升永无止境，下多少工夫去做都不为过。因此，论文初稿完成后，进行修辞是必要的。通过修辞，发现问题，进行词语替换、补充或删除，或进行句式调整、优化或关联，或修改语言使其符合逻辑，达到语言表达的规范；然后，做进一步的修辞，使用恰当的修辞手法，达到语言表达的提升。这样，最终才有可能成就高质量的论文。本章以英语科技论文中的一些常见问题语句作为实例进行分析和修辞，给出达到规范或得到提升的语句修改方案。

6.1 词法

6.1.1 代词或名词指代不明

代词或前面加定冠词的名词、名词短语所指代的对象必须是确指的。在前词没有出现或不明确时，不宜直接使用代词或加定冠词的名词、名词短语进行指代，也不要滥用代词而造成指代不明。

1）After **that**, applying the improved method to the master-slave system, we confirm the availability of the method by experiment.

此句中的代词 that 指代不明，属于冗词，可将其连同逗号一并删除。

句1）的参考修改方案：

✓ After applying the improved method to the master-slave system, we confirm the availability of the method by experiment.

2）Li and Azarm extended the sensitivity region analysis to multi-objective optimization. **The method** doesn't require presumed probability distribution of uncertain factors or gradient information of constraints.

此句中的 The method 指代不明，应在其前面句子的适当位置补出其前词，或给 method 加上适当的修饰语。

句2）的参考修改方案：

✓ Li and Azarm extended the sensitivity region analysis to multi-objective optimization, and presented **a relevant method. The method** doesn't require presumed probability distribution of uncertain factors or gradient information of constraints.

✓ Li and Azarm extended **the sensitivity region analysis method** to multi-objective optimization. **The method** doesn't require presumed probability distribution of uncertain factors or gradient information of constraints.

✓ Li and Azarm extended **the sensitivity region analysis** to multi-objective optimization. **The relevant method** doesn't require presumed probability distribution of uncertain factors or gradient information of constraints.

3) Though some studies have been done to improve the surface quality of the magnetic head to some extend, people still can not be satisfied with **it**.

此句中的代词 it 指代不明，改为 the result 稍好些，或改为 the research result。

句 3) 的参考修改方案：

✓ Though some studies have been done to improve the surface quality of the magnetic head to some extend, people still can not be satisfied with **the** (**research**) **result**.

6.1.2　冠词遗漏或位置不当

在需要用冠词时不要遗漏冠词，不需要用时不要误用，而且冠词的类别要正确，位置要合适。冠词使用的基本原则：用定冠词时，应做到确指，不确指时，就用不定冠词或不用冠词。定冠词用于表示独一无二的事物、形容词最高级等情况时较容易掌握，但用于特指时，容易遗漏。

(1) 不定冠词的遗漏

4) Faults in **rotating machine** are difficult to detect and identify, especially when the system is complex and nonlinear.

5) These membrane antenna structures generally involve **membrane surface**, support structures **and** a tensioning system.

6) Experimental results on **scale** model of **offshore** crane show the feasibility of the heave motion estimation method.

此三句中，句 4) 的 rotating machine 和句 5) 的 membrane surface 的前面均缺不定冠词 a，且句 5) 的 and 前面应加一个逗号；句 6) 的 scale 和 offshore 前面分别遗漏了不定冠词 a 和 an。

句 4) ~6) 的参考修改方案：

✓ Faults in **a rotating machine** are difficult to detect and identify, especially when the system is complex and nonlinear.

✓ These membrane antenna structures generally involve **a membrane surface**, support structures, **and** a tensioning system.

✓ Experimental results on **a scale** model of **an offshore** crane show the feasibility of the heave motion estimation method.

(2) 定冠词的遗漏

7) **Author** designed a new machine. **Machine** is operated with solar energy.

8) **Development** and a detailed comparison of these five methods are presented.

9) **Distribution** of residual stresses and its varying characteristics were also studied.

10) The calculated stresses of **hardening** of rotating disks were presented.

句 7) 的名词 Author 和 Machine 前缺少定冠词，未将“本论文的作者”和“前面提到的那台新设计的机器”之意表达出来。句 8) 的名词 Development、句 9) 的名词 Distribution 和句 10) 的名词或动名词 hardening 的前面均缺定冠词。

句 7) ~10) 的参考修改方案：

✓ **The author** designed a new machine. **The machine** is operated with solar energy.

✓ **The development** and detailed comparison of these five methods are presented.

✓ **The distribution** of residual stresses and its varying characteristics were also studied.

✓ The calculated stresses of **the hardening** of rotating disks were presented.

11) It shows the distribution of relative velocity on **surface** of **main and splitter blades**.

12) Molar mass M has long been considered one of **most important** factors among various

relevant parameters. **Parameter** Molar mass *M* was calculated with **following equation.**

13）$X^{(m,t)}$ is the value of **tth** neuron in the hidden（**mth**）layer.

句 11）有两处均缺定冠词，一处是名词 surface 前，另一处是名词短语 main and splitter blades 前。句群 12）有三处缺定冠词：第一处在 most important 前，用来构成形容词最高级；第二处在名词 Parameter 前，特指“前面提到的参数”；第三处在 following equation 前，用来构成固定短语。句 13）中 tth 和 mth 中的 t 和 m 均为变量，th 表示序数，它们前面均缺定冠词，而且表变量的 t 和 m 用斜体更规范。

句 11）~13）的参考修改方案：

✓ It shows the distribution of relative velocity on **the surface** of **the main and splitter blades.**

✓ Molar mass *M* has long been considered one of **the** most important factors among various relevant parameters. **The** parameter Molar mass *M* was calculated with **the following equation.**

✓ $X^{(m,t)}$ is the value of **the *t*th** neuron in the hidden（**the *m*th**）layer.（*t*th 和 *m*th 也可用 *t*-th 和 *m*-th 的形式。）

（3）冠词类别不当

14）An energy management strategy combining **the logic threshold approach** and **the instantaneous optimization algorithm** is proposed for the investigated PHEUB.

此句为某论文的摘要的第二个句子，该摘要的第一个句子中没有出现过 logic threshold approach 和 instantaneous optimization algorithm，因此，这句中在它们前面用定冠词 the 不当，应改为不定冠词。

句 14）的参考修改方案：

✓ An energy management strategy combining **a logic threshold approach** and **an instantaneous optimization algorithm** is proposed for the investigated PHEUB.

15）**A efficient** method for establishing **mathematical model** was given.

16）**A SFC**-enabled approach for processing SSL/TLS encrypted traffic in Future Enterprise Networks.

17）This paper reports **a X-ray** absorption method for on-line analysis of calcium chloride.

此三句的不定冠词使用不当，A（或 a）应改为 An（或 an），因为其后单词、缩写词或短语（efficient，SFC，X-ray）的首音发元音，尽管这些单词、缩写词或短语的首字母在形式上可能不是元音字母。另外，第一个句子的名词短语 mathematical model 的前面缺少定冠词。

句 15）~17）的参考修改方案：

✓ **An efficient** method for establishing **the mathematical model** was given.

✓ **An SFC**-enabled approach for processing SSL/TLS encrypted traffic in Future Enterprise Networks.

✓ This paper reports **an X-ray** absorption method for on-line analysis of calcium chloride.

6.1.3 名词单复数混淆

一些名词是不可数名词，或表达某种意思时是不可数名词，没有复数形式，若不小心，就容易在其后加 s，即把不可数名词当可数名词用，进而还会引起修饰语、谓语或其他成分出错。

18）**Literatures do** not support the need for removal of all bone and metal fragments.

19）However, there **have been few literatures** providing a comprehensive review and comparison of different membrane antenna structures.

此两句的Literatures do和have been few literatures错误。literature指文献、著作，是不可数名词，是单数，如果用复数，则可以用references代替。另外，后一句的时态为完成时，改为现在时更合适，用现在时可表示出一种实际状况。

句18）、19）的参考修改方案：

✓ **Literature does** not support the need for removal of all bone and metal fragments.

✓ However, there **is little literature** providing a comprehensive review and comparison of different membrane antenna structures.

20）Variation to the operational function is equal to do the same to the resource cell, including: the changing of number of production **equipments** and of the company.

此句的equipments错误，equipment表示设备、装备、器材，是不可数名词，没有复数形式，其结尾的s多余。另外，including后面的冒号也可去掉。

句20）的参考修改方案：

✓ Variation to the operational function is equal to do the same to the resource cell, including the changing of number of production **equipment** and of the company.

21）The demand for large antennas in future space missions has increasingly stimulated the development of deployable membrane antenna structures **due to** their light **weights** and small stowage **volumes**.

此句的weights和volumes错误，二者的表意分别是“重量”和“体积”，是不可数名词，没有复数形式。另外，按传统语法，due to主要用来引导表语，其同义词语owing to和because of主要引导状语，句中due to改为owing to或because of更好些。

句21）的参考修改方案：

✓ The demand for large antennas in future space missions has increasingly stimulated the development of deployable membrane antenna structures **owing to**（**because of**）their light **weight** and small stowage **volume**.

22）With an increasing demand for large-aperture（hundreds of square meters or more）space-borne antennas, deployable membrane antennas have been attracting **interests** in space research areas.

本句中interest的意思是“兴趣”，表达此意时是不可数名词，其结尾加s错误。

句22）的参考修改方案：

✓ With an increasing demand for large-aperture（hundreds of square meters or more）space-borne antennas, deployable membrane antennas have been attracting **interest** in space research areas.

23）Table 2 shows some typical performance parameters for these two kinds of polymer **films**.

本句结尾的名词film是“薄膜”之意，是不可数名词，结尾不加s。（它表示电影、影片时，是可数名词，但表示薄膜或胶片、软片、胶卷时是不可数名词）。

句23）的参考修改方案：

✓ Table 2 shows some typical performance parameters for these two kinds of polymer **film**.

24）Based on **analysis** mentioned above, **following** conclusions can be drawn.

本句中analysis（分析）是单数形式，按语义，应改为复数形式analyses，而且其前面及following前面均缺少定冠词。

句24）的参考修改方案：

✓ Based on **the analyses** mentioned above, **the following** conclusions can be drawn.

6.1.4 数词数字使用不当

可用一两个单词表达的整数，或位于句首的数字，一般用英语数词而不用阿拉伯数字，而且阿拉伯数字不宜用在句子或含有数字序号的句子的开头，也不宜用在表达全体信息的词语后面。

25）In 2011, JPL developed a 2.3 m × 2.6 m active phased array antenna [57], which reduced the number of membrane layers from **3** to **2**.

26）**10** algorithms are developed with Visual C #, and the correctness of them is verified through examples test.

27）**20 more** mathematical models are built in this test.

28）**All 6** studies concluded that the mean temperature should be 30 ℃.

第一句介词结构（from 3 to 2）中的数字3、2 宜改为英文数词 three、two，第二句开头的阿拉伯数字 10 应改为英语数词 Ten，第三句开头的 20 more 应改为 More than 20 或 Over 20，最后一句的阿拉伯数字 6 应改为英语数词 six。

句 25）~28）的参考修改方案：

✓ In 2011, JPL developed a 2.3 m × 2.6 m active phased array antenna [57], which reduced the number of membrane layers from **three** to **two**.

✓ **Ten** algorithms are developed with Visual C #, and the correctness of them is verified through examples test.

✓ **More than 20** (**Over 20**) mathematical models are built in this test.

✓ **All six** studies concluded that the mean temperature should be above 30 ℃.

29）The measuring accuracy varies from **3-6%** with the efficiency of **85-90%**.

此句中表示百分比范围（3%-6% 和 85%-90%）的前一个数字后缺少必要的百分号，从而可以理解为是从 3 到 0.06，从 85 到 0.9；另外，from 3-6% 的意思是从 3% 到 6%，即 from 3% to 6%，其中 from…to… 是一个常见的固定短语，应该使用此短语。

句 29）的参考修改方案：

✓ The measuring accuracy varies **from 3% to 6%** with the efficiency of **from 85% to 90%**.

6.1.5 介词使用不当

介词在准确表达语义方面发挥着非常重要的作用，同一句子的某一处使用不同的介词，句义就会有差异，介词使用不当甚至会引起逻辑不通。

30）Recent **advances of structure configurations**, tensioning system design **and** dynamic analysis of planar membrane antenna structures are investigated.

本句中使用介词 of，表示的意思是“结构构型（structure configurations）的进展（advances）”，而实际上“构型”本身是不能“进展”的，实际意思应是“在构型方面的进展”，因此使用 of 不妥，应改用 in。另外，用名词 structure 做名词 configurations 的修饰语不妥，应改为用形容词（structural）做修饰语；连词 and 连接三个名词短语（…advances, …design, …analysis），其前面应有逗号。

句 30）的参考修改方案：

✓ Recent **advances in structural configurations**, tensioning system design, and dynamic analysis for planar membrane antenna structures are investigated.

31）Through a review of large deployable

membrane antenna structures, guidance **to** space membrane antenna research and applications is provided.

名词 guidance 与 to 和 for 都可搭配，但与 for 搭配显得更正式，与 to 的搭配更常见的是动词不定式（guide to 也较为常见）。此句 guidance 后使用介词 to 不如 for 妥当。

句 31）的参考修改方案：

✓ Through a review of large deployable membrane antenna structures, guidance **for** space membrane antenna research and applications is provided.

32）Since 1970, a series of explorations **on** membrane antennas have been carried out in the United States, Europe, Japan, etc.

名词 exploration 与介词 on 和 of 都可搭配，表意相同，但与 of 搭配更常见、正式。

句 32）的参考修改方案：

✓ Since 1970, a series of explorations **of** membrane antennas have been carried out in the United States, Europe, Japan, etc.

33）In 2004, **Naboulsi, et al [21],** from the Air Force Institute of Technology, carried out a detailed study on the main structure of an inflatable antenna (including the membrane reflector, inflatable struts **and** torus) **by** software ABAQUS.

使用某某软件，应该用介词 with，即 with + 软件名称，一般不用 by。此句还有三处问题，缩写词 et al 后缺少表示缩写的小圆点，姓氏 Naboulsi 和引文序号［21］后的两处逗号多余，连词 and 前缺少逗号。

句 33）的参考修改方案：

✓ In 2004, **Naboulsi et al.**［**21**］from the Air Force Institute of Technology, carried out a detailed study on the main structure of an inflatable antenna (including the membrane reflector, inflatable struts, **and** torus) **with the** software ABAQUS.

34）**At ground**, the antenna is packed by **external mechanical load** above the glass transition temperature of SMP material.

本句逗号前面的句首状语，按句义是指在地上（某一位置或区域），相应的词语应为 On the ground，而不是 At ground。介词 at 可以与 ground 搭配，不过此时的 ground 通常是形容词，后面再跟名词，如 at ground level（在地面上），而不是直接跟名词。另外，external 前面遗漏了不定冠词。

句 34）的参考修改方案：

✓ **On the ground**, the antenna is packed by **an external mechanical load** above the glass transition temperature of SMP material.

35）**Due to** the electric field force, the metal-coated membrane is suspended on the front net and finally becomes a parabolic surface.

本句使用复合介词 due to，表示原因（状语），但按常规语法，due to 主要用来引导表语，其同义词语 because of 和 owing to 主要引导状语。

句 35）的参考修改方案：

✓ **Because of** (**owing to**) the electric field force, the metal-coated membrane is suspended on the front net and finally becomes a parabolic surface.

6.1.6　修饰语词性不当

修饰语有形容词和固定短语等，形容词是名词修饰语的主要形态。对于一个修饰性意思的表达，应优先用形容词而不是名词。

36）Although **structure characteristics** and application prospects of inflatable antennas were summarized in Refs. [12-14], and advances in parabolic membrane antennas over **past decades** were presented in Refs. [15-16], characteristic comparisons were not **concerned** among these membrane antenna structures.

本句名词 characteristics 的修饰语为名词 structure，应改为形容词 structural。另外，还有定冠词遗漏、谓语动词选用不当的问题。

句 36）的参考修改方案：

✓ Although **structural characteristics** and application prospects of inflatable antennas were summarized in Refs. [12-14], and advances in parabolic membrane antennas over **the past decades** were presented in Refs. [15-16], characteristic comparisons were not **considered** among these membrane antenna structures.

37）In 2001, **Greschik, et al [20],** from the University of Colorado, investigated the influence of membrane thickness, membrane materials, thermal effects, boundary perturbations **and** membrane wrinkles on surface accuracy, but the **geometry differences** of **membrane** reflector shape during its deformation process **was** not considered in the calculation.

本句名词 differences 的修饰语为名词 geometry，应改为形容词 geometric。另外，还有遗漏小圆点（表示缩写）、逗号和定冠词以及主谓不一致（指数的不一致）的问题。

句 37）的参考修改方案：

✓ In 2001, **Greschik et al. [20]** from the University of Colorado, investigated the influence of membrane thickness, membrane materials, thermal effects, boundary perturbations, **and** membrane wrinkles on surface accuracy, but the **geometric differences** of **the membrane** reflector shape during its deformation process **were** not considered in the calculation.

38）Table 3 lists the characteristics of **above-mentioned** planar membrane antenna **structure** configurations, and shows that planar membrane antenna frame shapes have changed from **circle** to horseshoe **shape**, and to **rectangle**.

此句有三个名词（structure, circle, rectangle）做修饰语，第一个名词修饰名词 configurations，后两个修饰名词 shape），应将这三个名词改为相应的形容词（structural, circular, rectangular）。另外，在 above 前面遗漏了定冠词。

句 38）的参考修改方案：

✓ Table 3 lists the characteristics of **the above-mentioned** planar membrane antenna **structural** configurations, and shows that planar membrane antenna frame shapes have changed from **circular** to horseshoe, and to **rectangular shape**.

39）Ship motions in **sea waves** negatively affect many offshore engineering operations, including oil-drilling technology in drill-ships, building and maintaining installations **as** oil rigs or windmills, and the launching and **recovering operations** of remotely operated vehicles (ROV).

此句中动名词 recovering 做名词 operations 的修饰语，比不上改用同义的名词 recovery。另外，sea waves 的意思是海浪，说船在海浪里运动（Ship motions in sea waves）比不上船在海里运动（Ship motions in sea）更准确；按表意，是在石油钻塔或风车（oil rigs or windmills）上建造及维修设施（building and maintaining installations），

表示“在……上”应该用介词 on 而不是 as。

句 39）的参考修改方案：

✓ Ship motions in **the sea** negatively affect many offshore engineering operations, including oil-drilling technology in drillships, building and maintaining installations **on** oil rigs or windmills, and the launching and **recovery** operations of remotely operated vehicles (ROV).

注意：能用名词做定语的通常不用动名词，能用形容词做定语的通常不用名词，即做定语的优先顺序为形容词→名词→动名词。例如：experimental accuracy 胜过 experiment results；measurement accuracy 胜过 measuring accuracy。

40）**The** β_1 is determined by the equation **as follows**:

41）$\boldsymbol{J}$ is a 4 × 3 matrix **as follows**:

42）$\boldsymbol{J}$ is defined **as following**:

句 40）、41）中的修饰语 as follows 不妥。该短语表示“如下”“如下所述”之意，相当于副词，主要用来修饰谓语动词。这两句中，as follows 位于名词 equation、matrix 后，表意是“如下方程”“如下矩阵”，用 as follows 修饰名词显然不妥，可改为用形容词 following 来修饰名词。另外，符号 β_1 自身已具备确指特性，无须再用一个定冠词来指定。句 42）的 as following 应改为 as follows，修饰谓语动词 defined。

句 40）~42）的参考修改方案：

✓ β_1 is determined by the **following** equation:

✓ $\boldsymbol{J}$ is a **following** 4 × 3 matrix:

✓ $\boldsymbol{J}$ is defined **as follows**:

43）Double-roller clamping spinning (DRCS) is a new **process to form thin-walled cylinder** with **complex surface flange**.

此句中的名词 process 的修饰语用动词不定式（to form），不如用介词结构（for forming，介词 + 动名词）更正式和常见，同时短语 thin-walled cylinder 和 complex surface flange 前均缺少不定冠词。

句 43）的参考修改方案：

✓ Double-roller clamping spinning (DRCS) is a new process **for forming a thin-walled cylinder** with **a complex surface flange.**

6.1.7　动词的名词形式使用不当

动词的名词形式（做主语、宾语）有时比不上直接用动词（做谓语）或不定式简洁、有效，即能用动词时尽量不用或少用其名词形式。动词的名词形式包括名词和动名词，如 measure、measurement、measuring，develop、development、developing，analyze、analysis、analyzing，research、research、researching 等。但有时巧妙使用动名词，可以将句子的谓语变为一个短语，即把从句变成短语，语句得到简化，表达更加有效。

44）Many scholars **carried out** the **researches** on the power spinning.

此句使用名词 researches 做宾语，不如直接用它做谓语，这样可以省去原谓语 carried out，表达简洁、明快。另外，时态用完成时更贴切。

句 44）的参考修改方案：

✓ Many scholars **have researched** the power spinning.

✓ Many scholars **have investigated** the power spinning.

✓ Many **researchers** have **investigated** power spinning.

45）**Measurement** of the thickness of that plastic sheet was **made**.

46）**Development** of a new generation of

intelligent manufacturing systems **has been finished**.

这两句分别用由动词 measure 和 develop 的名词形式 measurement 和 development 做主语，比不上直接用这两个动词直接做谓语简洁，当直接用动词做谓语时，可省去句中原来的谓语动词，表意直截了当，句子也能得到简化。

句 45）、46）的参考修改方案：

✓ The thickness of that plastic sheet was **measured**.

✓ A new generation of intelligent manufacturing systems **has been developed**.

47）Techniques are used **for identification and quantitation** of the experimental results.

此句中介词结构 for identification and quantitation 为介宾结构（介词 + 并列名词），不如用 to identify and quantify 的动词不定式结构（to + 并列动词），后一种结构更紧凑简单，表意更清晰明了。这就是说，句中用名词 identification、quantitation 不太妥当，可改用不定式形式。

句 47）的参考修改方案：

✓ Techniques are used **to identify and quantify** the experimental results.

48）**Test** and **analysis** of this method is **done** from the aspects of boundary cell type, error distribution and accuracy.

49）In this paper, both a finite element numerical **simulation** and experimental **research** on the double-roller clamping spinning process are **carried out**.

这两句分别用名词 test、analysis 和 simulation、research 做主语，比不上直接用其相应的动词做谓语更能突出动作。test、research 名词兼动词，analysis 的动词形式为 analyze；simulation 是动词 simulate 的名词形式。

句 48）、49）的参考修改方案：

✓ This method is **tested** and **analyzed** from the aspects of boundary cell type, error distribution and accuracy.

✓ In this paper, the double-roller clamping spinning process is numerically **simulated** and experimentally **researched** with a finite element method.

50）It is important to develop a **way in which** the effect of ground microgravity simulation systems on **experimental measurement results of** lightweight and flexible membrane antenna structures **will be minimized** in the future.

此句定语从句中的谓语动词 minimized，可以转变为动名词的形式充当介词 for 的宾语而形成介宾结构 for minimizing…，来做名词 way 的后置修饰语。考虑到 way 一词的专业性较差，可改为专业性较强的 method 一词。此外，还有定冠词遗漏（experimental 前）、介词（of）使用不当的问题。

句 50）的参考修改方案：

✓ It is important to develop a **method for minimizing** the effect of ground microgravity simulation systems on **the experimental measurement results for** lightweight and flexible membrane antenna structures in the future.

6.1.8 分词形式使用不当

一个动词在句中以分词的形式出现时，用现在分词还是过去分词，需要考察它与修饰对象之间的关系。通常，主动关系用现在分词，被动关系用过去分词。

51）For parabolic membrane antenna structures, there are five deploying and forming methods

including inflation, inflation-rigidization, **elastic ribs driving**, **SMP**-inflation **and** electrostatic forming.

此句中 including 的受词为其后 5 个并列的名词或名词短语，其中 elastic ribs driving 由名词短语 elastic ribs 加现在分词 driving 组成，driving 的逻辑主语是 elastic ribs。但仔细考量一下，elastic ribs 是动词 drive（驱动）的对象，因此用 driving 不妥，而应改用 driven，修改后 driven 为 elastic ribs 的后置修饰语，表意是“驱动的弹性肋”。另外，本句引自一篇综述，在其前面的语句中并未出现过 SMP 这一简称，因此在本句中直接使用这一简称不合适，应该改为先全写后简称；在最后一个连词 and 前缺少逗号。

句 51）的参考修改方案：

✓ For parabolic membrane antenna structures, there are five deploying and forming methods, including inflation, inflation-rigidization, **elastic ribs driven**, **shape memory polymer** (**SMP**)-inflation, and electrostatic forming.

52）Inflatable-rigidizable antennas avoid some **disadvantages facing** inflatable antennas, including **surface accuracy decrease**, antenna performance degradation **and requirement** for **continuous gas supplement**.

此句用现在分词 facing 修饰名词 disadvantages 不妥。动词 face 与名词 disadvantages 为动宾关系，即 disadvantages 是 face 的宾语，当 disadvantages 前置时就形成了被动关系，应该用过去分词 faced。另外，此句还存在修饰语位置和结构（surface accuracy decrease）不妥，定冠词（requirement 前）、不定冠词（continuous gas supplement 前）遗漏的问题。

句 52）的参考修改方案：

✓ Inflatable-rigidizable antennas avoid some **disadvantages faced by** inflatable antennas, including **decrease in surface accuracy**, antenna performance degradation, **and the requirement** for **a continuous gas supplement**.

6.2 时态

6.2.1 对所述内容的时间属性未准确把握

时态的基本属性就是时间，时间决定时态，只有准确把握内容的时间属性，才能用对时态。

53）Synthetic genetic approaches **revealed** potential interacting partners for any given target protein.

54）In comparison with traditional rigid antennas, membrane antennas **could** easily achieve **large scales** with **light weights**, **small stowage volumes** and **low costs**.

55）DNA microarrays **is** the established technology for measuring gene expression levels.

这三句均来自研究背景介绍或现状描述。句 53）的意思是“合成遗传方法揭示了任何特定目标蛋白的潜在相互作用伙伴”，所述内容是不受时间影响的普遍事实，用过去时不合适，应该用现在时。句 54）的意思是“与传统刚性天线相比，薄膜天线具有重量轻、装载体积小和成本低的优点”，所述内容也是不受时间影响的普遍事实，用过去时不合适，应该用现在时。此句还有名词单复数混淆的问题，即把不可数名词当可数名词用。另外，句首用了固定短语 In comparison with，语义上明显有对比意味，后面的形容词（large、light、small、 low）应该用比较级。句 55）的意思“DNA 微阵列长期以来一直是测量基因表达水平的成熟技术”，所述内容是对某种研究趋势的概述，含“到目前”“截至目前”的时间属性，适于用完成时，用现在时的表达效果会差一些。

句 53）~55）的参考修改方案：

✓ Synthetic genetic approaches (**often**) **reveal** potential interacting partners for any given target protein.

✓ In comparison with traditional rigid antennas, membrane antennas **can** easily achieve **larger scale** with **lighter weight**, **smaller stowage volume**, and **lower cost**.

✓ DNA microarrays **have long been** the established technology for measuring gene expression levels.

56) Only a few studies on the ALT under time-**depended** stress **are reported**.

57) The statistical model **has been** determined **during the plan design**, but it is expected that the test plan could achieve the best possible estimation accuracy, even if the wrong model is selected.

句 56）的主体意思是“仅有一些有关……的研究被报道了”，谓语动词 are reported 明显具有截至目前的时间属性，因此用现在时不太合适，应该用完成时。另外，做修饰语的过去分词 depended 可直接改为形容词 dependent。句 57）中第一分句结尾的时间状语 during the plan design（在……设计期间），明显带有过去时间的属性，因此用完成时不太好，应该用过去时。

句 56）、57）的参考修改方案：

✓ Only a few studies on the ALT under time-**dependent** stress **have been reported**.

✓ The statistical model **was** determined **during the plan design**, but it is expected that the test plan could achieve the best possible estimation accuracy, even if the wrong model is selected.

58) The results **showed** that pipeline steel itself **had** good strength and toughness, when hydrogen **was electrochemically introduced**, the yield strength and tensile strength of the material **would decrease**, and the performance deterioration **would intensify** with the increase of hydrogen charging time.

此句的主体意思是想说明（表述/描述）研究结果，旨在说明所得结果的规律性、事实性和科学性，“现在”的时间属性较为明显，因此用过去时不大合适，应改用一般现在时。

句 58）的参考修改方案：

✓ The results **show** that pipeline steel itself **has** good strength and toughness, when hydrogen **is electrochemically introduced**, the yield strength and tensile strength of the material **will decrease**, and the performance deterioration **will intensify** with the increase of hydrogen charging time.

注意：当研究结果明显带有过去时间的痕迹时，用过去时表述更合适。所谓过去时间的痕迹，是指研究结果只是当时的现象、结果，属于一定范围内的发现、现象，尚不能确认为自然规律、永恒真理。

关于研究结果表述的时态问题，在 6.2.3 节中还会提及。

6.2.2　对描述对象是论文还是研究未仔细区分

“研究”意味着本次研究工作的结束，过去的时间属性明显；而“论文”虽是对过去研究工作的记录，却随时可以被拿出来阅读，而阅读就意味着目前正在进行，甚至将要进行，现在的时间属性明显。因此，写作行文对描述对象是论文还是研究应仔细区分，这样才能用好时态。

59）**This paper focused** on the contribution of electron techniques to our understanding of cellular processes.

60）**This study discusses** the industrial application prospects of amphibian aircraft and the direction of technological development.

在叙述研究目的或主要研究活动时，描述“论文对象”时，多使用现在时（如 This paper presents...）；描述“研究对象”时，多使用过去时（如 This study presented...）。句 59）的主语为 This paper，谓语（focused）为过去时，不太妥当；句 60）的主语为 This study，谓语（discusses）为现在时，也不妥当。

句 59）、60）的参考修改方案：

✓ **This paper focuses** on the contribution of electron techniques to our understanding of cellular processes.

✓ **This study discussed** the industrial application prospects of amphibian aircraft and the direction of technological development.

61）**We describe** a new molecular approach to analyze the genetic diversity of complex microbial populations.

We 是此句的主语（类似的还有 the author、authors 和 the team 等），虽然形式上不是 this（the）study，但描述的对象是“研究”，用现在时不太妥当，改用过去时更合适。

句 61）的参考修改方案：

✓ **We described** a new molecular approach to analyze the genetic diversity of complex microbial populations.

本句也可改为被动语态的形式，是用现在时还是过去时，取决于描述对象是论文还是研究，例如：

✓ A new molecular approach **is described**（by this paper）to analyze the genetic diversity of complex microbial populations.

✓ A new molecular approach **was described**（by us，by the author 或 by authors）to analyze the genetic diversity of complex microbial populations.

这两句为被动语态，表示描述对象是论文还是研究的关键词语省略了，这样就没有区别了，即用现在时或过去时都是没有问题的（有时可通过上下文语境来确定），但用主动语态时，还是有区别的。这种细微差别只有经过仔细区分才能被体会出来。

62）The following subsection **will discuss** other parts of the bit-simulant interaction.

此句的主语（The following subsection）是论文中的层次标题，即“小节”，本质上属于论文范畴，用将来时不太妥当，用现在时更合适。

句 62）的参考修改方案：

✓ The following subsection **discusses** other parts of the bit-simulant interaction.

63）A typical planar membrane antenna structure **is mainly composed of** a deployable frame and a multi-layer flexible membrane, which **is supported** by the frame [40]. The multiple membrane layers **were deployed** to a planar structure with the deployment of **frame** and **maintained** the required surface flatness **by** the tensioning system between the membrane and frame. Fig. 7 shows the structural configurations of planar membrane array antennas developed in the USA between 1998 and 2008.

此句群由三个句子组成。句一是从论文（文献 [40]）的角度来描述的，用现在时（is mainly composed of，is supported）是合适的。句二承接了句一，自然也是以论文（文献［40］）为描述对象，但用过去时（were deployed，maintained）是不合适的，应改为现在时；另外，名词 frame 前遗漏了定冠词，介词 by 使用不当。句三是从图（Fig. 7）的角度来描述的，相当于以论文为描述对象，用现在时（shows）是合适的。

句 63）的句二参考修改方案：

✓ The multiple membrane layers **are deployed** to a planar structure with the deployment of **the frame** and **maintain** the required surface flatness **through** the tensioning system between the membrane and frame.（简单句，一个主语，一个由两个谓语动词组成的复合谓语，这两个谓语动词的语态不相同。）

✓ The multiple membrane layers **are deployed** to a planar structure with the deployment of **the frame**, and the required surface flatness **is maintained through** the tensioning system between the membrane and frame.（并列句，由 and 连接，前后句子的语态相同。）

6.2.3　表述现状、目的、结果的时态不妥

64）Ti_2AlN **was focused** by many researchers for its excellent properties, however the research on fabrication and properties of the Ti_2AlN film **was** little.（三元氮化物 Ti_2AlN 因其优良的性能而受到很多科研工作者的关注，但对其薄膜材料制备及性能的研究很少。）

以上整句用来表述研究的现状或背景，属于常识或客观事实，但前后分句都用了过去时，不太妥当，都改用现在时更合适。

句 64）的参考修改方案：

✓ Ti_2AlN **is focused** by many researchers for its excellent properties, however the research on fabrication and properties of the Ti_2AlN film **is** little.

再仔细琢磨一下，不难发现句 64）前一分句的谓语（was focused）还有截至目前的时间属性，用完成时更妥当。即改为：

✓ Ti_2AlN **has been focused** by many researchers for its excellent properties, however the research on fabrication and properties of the Ti_2AlN film is little.

65）**Our aim was** to explore the relationship between the response of antioxidant enzymes and temperature in the two invasive weeds.

此句的主语是 Our aim，谓语是 was，直接表述研究目的，但用过去时不太妥当，宜改为现在时，学术论文中叙述研究目的宜用现在时。

句 65）的参考修改方案：

✓ **Our aim is** to explore the relationship between the response of antioxidant enzymes and temperature in the two invasive weeds.

66）**The results showed** that the Ti_2AlN film which **possessed** better corrosion resistance and high temperature oxidation resistance compared with TiN film **can be obtained** under the optimized process.（结果表明在优化制备工艺下可以制备出与 TiN 薄膜相比，耐腐蚀性和抗高温氧化性更好的 Ti_2AlN 薄膜。）

此句的主语是 The results，谓语是 showed，直接表述研究结果，但用过去时不妥，改用现在时更好，学术论文中表述研究结果（含结论）宜用现在时。另外，以上研究结果后面部分有臆测意味，宜使用 may、should、could 等助动词。

句 66）的参考修改方案：

✓ **The results show** that the Ti_2AlN film

which **possesses** better corrosion resistance and high temperature oxidation resistance compared with TiN film **could be obtained** under the optimized process.

67) Ti_2AlN film with excellent corrosion resistance and high temperature oxidation resistance **could be fabricated** by multi-arc ion plating and vacuum annealing process, which **provided** the theoretical support for the study of Ti_2AlN film.（采用多弧离子镀技术和真空退火工艺可制备出耐腐蚀性和高温抗氧化性优良的 Ti_2AlN 薄膜，可为 Ti_2AlN 薄膜的研究提供理论支撑。）

此句表述研究的结论，属于结果的范畴，主句的谓语（could be fabricated）有臆测意味，可以用过去时，但从句的谓语（provided）用过去时不如用现在时好。

句 67）的参考修改方案：

✓ Ti_2AlN film with excellent corrosion resistance and high temperature oxidation resistance **could be fabricated** by multi-arc ion plating and vacuum annealing process, which **provides** the theoretical support for the study of Ti_2AlN film.

68) We **suggested** that climate instability in the early part of the last interglacial **may have delayed** the melting of the Saalean ice sheets in America and Eurasia, perhaps accounting for this discrepancy.

叙述结论或建议多用现在时，特定情况下可用臆测动词（如 think/suppose, believe, imagine/fancy, intend/mean, guess）或 may、should、could 等助动词。此句主句用过去时不妥，应改用现在时。

句 68）的参考修改方案：

✓ We **suggest** that climate instability in the early part of the last interglacial **may have delayed** the melting of the Saalean ice sheets in America and Eurasia, perhaps accounting for this discrepancy.

6.2.4　表述工作内容、过程的时态不妥

69) **In this study**, multi-arc ion plating and subsequent vacuum annealing **are used to prepare** the Ti_2AlN film, and the corrosion resistance and high temperature oxidation resistance of the film **are tested and analyzed.**（本研究采用多弧离子镀技术和真空退火工艺制备 Ti_2AlN 薄膜，还对薄膜的耐腐蚀性和高温抗氧化性进行测试及机理分析。）

此句以 In this study 开头，明显是研究导向的，表述本研究做了什么工作，属于具体工作内容（即使去掉开头的 In this study，句子表示“本研究做了什么工作”的主体内容也很明显），因此用现在时（are used to prepare; are tested and analyzed）不太合适，应改用过去时。但如果改用论文导向（In this paper）行文，则用现在时是可以的。

句 69）的参考修改方案：

✓ **In this study**, multi-arc ion plating and subsequent vacuum annealing **were used to prepare** the Ti_2AlN film, and the corrosion resistance and high temperature oxidation resistance of the film **were tested and analyzed.**

✓ **In this paper**, multi-arc ion plating and subsequent vacuum annealing **are used to prepare** the Ti_2AlN film, and the corrosion resistance and high temperature oxidation resistance of the film **are tested and analyzed.**

70) To investigate the mechanisms controlling flowering time, we **screen** for Arabidopsis mutants with late-flowering phenotypes. One mutant **is identified** with delayed flowering

time.

71) A plant that is highly susceptible to this fungus **is analyzed**.

这两例描述的是已经发生的具体行为和动作，用现在时不妥，应改用过去时。

句群 70) 和句 71) 的参考修改方案:

✓ To investigate the mechanisms controlling flowering time, we **screened** for Arabidopsis mutants with late-flowering phenotypes. One mutant **was identified** with delayed flowering time.

✓ A plant that is highly susceptible to this fungus **was analyzed**.

72) The questionnaires **are distributed** to high score users of 44 mainstream interactive q&a platforms in China through email addresses, and 220 questionnaires collected **are evaluated** with partial least square method.

此句描述的是所做的具体工作或某种方法的具体过程，用现在时不妥，应改用过去时。

句 72) 的参考修改方案:

✓ The questionnaires **were distributed** to high score users of 44 mainstream interactive q&a platforms in China through email addresses, and 220 questionnaires collected **were evaluated** with partial least square method.

6.3 标点

英语标点符号在类别和形式上基本同中文标点符号，但其间差别还是有的。主要差别有：英语的句号为句点，省略号为连续排列的三个句点；英文中没有顿号“、”和书名号“《》”；英文中有撇号“'”，中文中则没有。

6.3.1 缺少必要的标点

73) **Then** properties of membrane materials (including polyester film and polyimide film) for parabolic membrane antennas are compared.

此句副词 Then 与其后面的句子在语义上有停顿，其间缺少逗号，应加上。

句 73) 的参考修改方案:

✓ **Then**, properties of membrane materials (including polyester film and polyimide film) for parabolic membrane antennas are compared.

74) Additionally, for planar membrane antenna structures, **their** frame shapes have changed from **circle** to **rectangle**. **And** different tensioning systems have emerged successively, including single, **Miura-Natori**, double **and** multi-layer tensioning **system**.

连词 and 用来连接前后两个部分，当连接句子特别是较长的句子或用来连接多个成分时，它与前面的句子或成分之间在语义上有停顿，在其前面应加逗号。通常，and 不宜用在句子的开头，即一个句子不宜直接从 and 开始。另外，此句还有几处问题：their 与 planar membrane antenna structures 同指，因此多余；表达形状（shapes）应该用形容词 circular（圆形的）、rectangular（矩形的），而不是名词 circle（圆形）、rectangle 矩形）；Miura 和 Natori 是两个同等重要的词语或名称，组成一个整体即复合性修饰语（修饰 system），其间不是连接关系，因此用连字符不合适，而应该用短破折号；按语义，应该有多个系统（system），结尾的 system 不应用单数形式。

句 74) 的参考修改方案:

✓ Additionally, for planar membrane antenna structures, frame shapes have changed from **circular** to **rectangular**, **and** different tensioning systems have emerged successively, including single, **Miura-Natori**, double, **and** multi-layer tensioning **systems**.

75）Finally，future trends **of** large space membrane antenna structures are pointed out and technical problems are proposed **including** design and analysis of membrane structures，materials and processes，membrane packing，surface accuracy stability，and test and verification technology.

此句介词短语（including…）所述内容是针对句子的主语（future trends）来说的，并非针对谓语（are proposed），因此它同其前面的成分（are proposed）在语义上不大相关或紧密，即应该有停顿，也就是说 including 前面应该有逗号。另外，trends 后用介词 for 而不是 of 表示趋势更妥当。

句 75）的参考修改方案：

✓ Finally，future trends **for** large space membrane antenna structures are pointed out and technical problems are proposed，**including** design and analysis of membrane structures，materials and processes，membrane packing，surface accuracy stability，and test and verification technology.

6.3.2　标点符号错用

76）At present，there are **mainly** two kinds of space-borne membrane antenna structures，**parabolic and planar membrane antenna structures**.

此句后面部分 parabolic and planar membrane antenna structures 是前面所述两种结构（two kinds of…structures）的具体解释，其间用逗号不妥，应该用冒号。另外，"主要"这一意思侧重在 two kinds 而不在系动词 are，因此副词 mainly 宜改为形容词 main 而置于 kinds 前面。

句 76）的参考修改方案：

✓ At present，there are two **main** kinds of space-borne membrane antenna structures：**parabolic and planar membrane antenna structures**.

77）In 2012，**Li，et al**［50］，developed a 6m × 2m L/C dual-band，single-layer planar membrane antenna，but the deployment was not **involved** in this paper.

此句中作者姓氏 Li 与 et al 之间在语义上无停顿，其间的逗号多余，而且 et al 是一个缩写词语，其后表示缩写的小圆点不宜遗漏。另外，数字与单位之间宜加空，动词 involved 使用不当，应改为 considered。

句 77）的参考修改方案：

✓ In 2012，**Li et al.**［50］，developed a 6 m × 2 m L/C dual-band，single-layer planar membrane antenna，but the deployment was not **considered** in this paper.

6.3.3　中文顿号充当英语逗号

78）The gain and the power of this test are 20 dB、10 mW，respectively.

79）I_i（i = 1，2）is the moment of inertia of system，K_1 is the linearity torsional rigidity of system，K_2、K_3 **is** the nonlieaarity torsional rigidity of the system，θ_i（i = 1，2）、$\dot{\theta}_i$（i = 1，2）are rotational angle and speed respectively.

此两句中的 20 dB、10 mW，K_2、K_3，θ_i（i = 1，2）、$\dot{\theta}_i$（i = 1，2）中的顿号使用不当，应改用逗号。英语中没有顿号，词语间的停顿常用逗号，相当于中文中的顿号。另外，句 79）由多个分句组成，在最后一个分句前宜加一个逗号及 and。

句 78）、79）的参考修改方案：

✓ The gain and the power of this test are 20 dB，10 mW respectively.

✓ I_i（i = 1，2）is the moment of inertia of system，K_1 is the linearity torsional rigidity of system，K_2，K_3 are the nonlieaarity torsional rigidity of the system, and θ_i, $\dot{\theta}_i$ ($i=1$, 2) are rotational angle and speed respectively.

6.3.4 中文连接号、破折号充当英语破折号

80) The age dated by fossil ice wedges shows that the ancient aeolian sand deposited during a period of **27 ka ~ 10 ka BP.**

81) The threshold current of this test is **100 ~ 120 A**.

82) M J Hoeijmakers, A J Ferreira. The electrical variable transmission [J]. *IEEE Transactions on Industry Applications*, 2006, 42(4): **1092 ~ 1100**. (参考文献著录)

此三例用中文连接号"~"(浪纹线)表数值范围,不妥,应改用英语短破折号"–"。

句 80)~82)的参考修改方案:

✓ The age dated by fossil ice wedges shows that the ancient aeolian sand deposited during a period of **27 ka–10 ka BP**.

✓ The threshold current of this test is **100–120 A**.

✓ M J Hoeijmakers, J A Ferreira. The electrical variable transmission [J]. *IEEE Transactions on Industry Applications*, 2006, 42 (4): **1092–1100**. (参考文献著录)

注意:对以上例子中的英语短破折号"–",有些期刊使用英语连字符(又称连接号)"-"替代。

83) Where P'_e——Effective power flow transferred from the DRM to the SRM; η_{td}, η_{ts}——Working efficiency of the transducers on the DRM side and on the SRM side.

此例用中文破折号"——"表英语式注不妥,应改用英语长破折号"—"。

句 83)的参考修改方案:

✓ Where P'_e—Effective power flow transferred from the DRM to the SRM; η_{td}, η_{ts}—Working efficiency of the transducers on the DRM side and on the SRM side.

✓ Where P'_e is effective power flow transferred from the DRM to the SRM, η_{td}, η_{ts} are working efficiency of the transducers on the DRM side and on the SRM side.

6.3.5 连字符、短破折号混淆

84) A W Burton, A J Truscott, P E Wellstead. Analysis, modelling and control of an advanced automotive self-levelling suspension system [J]. Control Theory and Applications, 1995, 142 (2): **129-139**. (参考文献著录)

85) This paper presents the advances obtained at State Key Laboratory of Advanced Welding Production Technology in development of ultrasonic stress measurement[7-10], where the ultrasonic stress measurement experimental installation is established. (参考文献引用)

此两例中用连字符"-"表示数值范围不妥,应改用短破折号"–"。但是,目前国际期刊包括一些世界名刊,用连字符"-"表示数值范围有增多的趋势,使用多了,也就"正常"了。

句 84)、85)的参考修改方案:

✓ A W Burton, A J Truscott, P E Wellstead. Analysis, modelling and control of an advanced automotive self-levelling suspension system [J]. Control Theory and Applications, 1995, 142 (2): **129–139**. (参考文献著录)

✓ This paper presents the advances obtained at State Key Laboratory of Advanced Welding Production Technology in development of ultrasonic stress measurement[7–10], where the ultrasonic stress measurement experimental installation is established.

86) We proposed an improved method which

changes the definition of the **4–neighborhood** model.

87）The OSM database also provided the necessary information for assigning **country–and road–type–specific** speed data to road pixels.

此两句中的“4–neighborhood”和“country–and road–type–specific”为由某几种要素（如数字、单词）组成的短语，组成要素间误用了短破折号“–”，应改用连字符“-”。

句86）、87）的参考修改方案：

✓ We proposed an improved method which changes the definition of the **4-neighborhood** model.

✓ The OSM database also provided the necessary information for assigning **country-and road-type-specific** speed data to road pixels.

6.3.6 数学比例号充当英语冒号

88）A Tessier，P G C Camplell，M Bisson. Sequential extraction procedure for the speciation of particulate tracemetals［J］. *Analytical Chemistry*，1979，51(7)∶844–851.

此例为期刊析出文献著录，51（7）为析出文献（即所引用的论文）所在期刊的卷期号，844–851为起止页码，其间误用了比例号“∶”，应改用冒号“：”（用来提示页码）。

句88）的参考修改方案：

✓ A Tessier，P G C Camplell，M Bisson. Sequential extraction procedure for the speciation of particulate tracemetals［J］. *Analytical Chemistry*，1979，51(7)：844–851.

89）S Y Son. *Design principles and methodologies for reconfigurable machining systems*［D］. Michigan∶University of Michigan，2000.

此例为博士学位论文著录，“∶”前面的Michigan为所引博士论文的保藏单位所在的城市，后面的大学名称为保藏单位，其间误用了比例号，应改用冒号。

句89）的参考修改方案：

✓ S Y Son. *Design principles and methodologies for reconfigurable machining systems*［D］. Michigan：University of Michigan，2000.

90）In this paper，the main source data is still from LANDSAT TM digital tape except some additional DEM data from 1∶100 000 topographic map.

此句1∶100 000中用比例号是正确，但也可改用冒号。英语中的比例号通常用冒号，即冒号可以表示相除（相比）。

句90）的参考修改方案：

✓ In this paper，the main source data is still from LANDSAT TM digital tape except some additional DEM data from 1：100 000 topographic map.

6.3.7 中文书名号充当英语书名号

91）《Science》、《Cell》and《Nature》etc. are the authoritative journals in the world.

此句使用中文书名号及顿号是错误的。英语中没有书名号和顿号，书刊名一般用斜体形式表示，有时也可通过给书刊名加引号的方式来表示。

句91）的参考修改方案：

✓ *Science*，*Cell* and *Nature* etc. are the authoritative journals in the world.

✓“Science”，“Cell”and“Nature”etc. are the authoritative journals in the world.

6.4 选词

6.4.1 未用专业词语

92）The authors confirm this using growth

timescales for angrites, ureilites, Vesta and Mars that were inferred by nuclear chronometry (a dating technique that uses the decay of radioactive isotopes) and **heat** models.

heat（热）属日常用词，在学术论文中，特别是用来做形容词性修饰语时，应使用与其同义的专业词 thermal。

句 92）的参考修改方案：

✓ The authors confirm this using growth timescales for angrites, ureilites, Vesta and Mars that were inferred by nuclear chronometry (a dating technique that uses the decay of radioactive isotopes) and **thermal** models.

93）Such **movement** is possible whenever the electrochemical potentials of the two dots are aligned—i. e. , where it costs equal energy for an electron to be in either dot.

在学术领域如生理学或力学中，常用 motion 表示抽象的动作、行为，与静止相对，指位置的移动、条件的改变或系列运动的各个过程。它指动作或姿态时，与 movement 差不多，但 movement 多用于普通英语，指具体的动作、行为。因此，此句用 movement 不太合适，应改用 motion。

句 93）的参考修改方案：

✓ Such **motion** is possible whenever the electrochemical potentials of the two dots are aligned—i. e. , where it costs equal energy for an electron to be in either dot.

94）**The object** of this work is to develop a computer aided editing system that is able to find and correct writing mistakes.

学术论文中表示研究目的的常用词有 objective、purpose、aim 等，其中最正式的是 objective。虽然 object 也有“目的”之意，但在学术论文中很少用它来表示此意，而是多用来表“物体”之意。

句 94）的参考修改方案：

✓ **The objective** of this work is to develop a computer aided editing system that is able to find and correct writing mistakes.

✓ **The purpose** of this work is to develop a computer aided editing system that is able to find and correct writing mistakes.

✓ **The aim** of this work is to develop a computer aided editing system that is able to find and correct writing mistakes.

95）If **cable** winding **possibility** can be reduced and high voltage hazard can be eliminated, surface accuracy of electrostatic forming membrane antennas **will be further improved on the basis of** existing Astromesh antennas.

possibility 和 probability 都可表示可能性，但后者相对更专业一些，可以表示事件发生的概率，而前者仅仅是在感官上认为的可能，在学术论文中用 probability 更正式。（如果不考虑学术性，则除了一些固定短语，如 in all probability 不能互换以外，多数情况下这两个词可以等同。）另外，用 on the basis of 修饰谓语（will be further improved）不合适，应改用 based on；cable 前面缺少定冠词。

句 95）的参考修改方案：

✓ If **the cable** winding **probability** can be reduced and high voltage hazard can be eliminated, surface accuracy of electrostatic forming membrane antennas **will be further improved based on** existing Astromesh antennas.

96）In 2000, the two companies co-developed a **3-meter** Ka-band membrane **reflectarray antenna** [42-43], as shown in Fig. 7(b).

句中 reflectarray antenna 表示“反射阵

列天线”之意，是一个专业词语，其准确的英语名称应是 reflective array antenna。另外，meter 一词用在数字后面充当单位也不专业，可改用国际单位 m。

句 96）的参考修改方案：

✓ In 2000, the two companies co-developed a **3 m** Ka-band membrane **reflective array antenna** [42-43], as shown in Fig. 7 (b).

6.4.2　未用正式词语

97）Advances in Deploying and Forming **Ways of** Parabolic Membrane Antennas.

98）The expression of p16 gene in 80 cases of peripheral lung cancer tissues was detected through immunohistochemistry SP **way**.

此两句中的 Ways 和 way 用词不当，way 属于普通用词，指一般的方法、道路和方向等意思，而学术论文中表示“方法”的较正式的用词为 method。

句 97）、98）的参考修改方案：

✓ Advances in Deploying and Forming **Methods for** Parabolic Membrane Antennas.

✓ The expression of p16 gene in 80 cases of peripheral lung cancer tissues was detected by immunohistochemistry SP **method**.

99）The correlation of μ^{48}Ca values with planetary-body masses therefore also **becomes a positive correlation with** the timescale of the growth of such bodies.

表述一事物与另一事物之间正相关，不宜用句型“is (becomes) a positive correlation with…”，而应该用“is (becomes) positively correlated with…”，或“a positive correlation was found between … and …”，或 there is a positive correlation between…and…。表述负相关、逆相关的词语是 negatively correlated（或 inversely correlated），negative correlation（或 inverse correlation 和 reverse correlation）。

句 99）的参考修改方案：

✓ The correlation of μ^{48}Ca values with planetary-body masses therefore also **is** (**found**) **positively correlated with** the timescale of the growth of such bodies.

✓ The correlation of μ^{48}Ca values with planetary-body masses therefore also **becomes positively correlated with** the timescale of the growth of such bodies.

✓ **A positive correlation was found between** the correlation of μ^{48}Ca values with planetary-body masses **and** the timescale of the growth of such bodies.

✓ **There is a positive correlation between** the correlation of μ^{48}Ca values with planetary-body masses **and** the timescale of the growth of such bodies.

100）Doubling the area of the well doubled the cell number at which proliferation reached zero, suggesting that in this setting, carrying capacity **is positively proportional to** space (Figure S4A).

表示一事物与另一事物之间成正比关系，较正式的短语是 is directly proportional to，或直接用 is proportional to，而不是 is positively proportional to。

句 100）的参考修改方案：

✓ Doubling the area of the well doubled the cell number at which proliferation reached zero, suggesting that in this setting, carrying capacity **is** (**directly**) **proportional to** space (Figure S4A).

表示反比关系的短语是 is inversely proportional to。例如：

- The wavelength **is inversely proportional**

to the frequency when the propagation velocity is constant.

101）**Figure 2D is** the corresponding external quantum efficiency（EQE）of a typical light-soaked device accompanied by an electroluminescence（EL）spectrum before and after light soaking.

学术论文中表述或引用图表时，使用的较正式的谓语动词通常是 show(s)，list(s) 或 express(es)，而不是 are（is）。

句 101）的参考修改方案：

✓ Figure 2D **shows**（**lists**，**expresses**）the corresponding external quantum efficiency（EQE）of a typical light-soaked device **accompanied** by an electroluminescence(EL) spectrum before and after light soaking.

✓ As **shown**（**listed**，**expressed**）**in** Figure 2D，the corresponding external quantum efficiency（EQE）of a typical light-soaked device **is accompanied** by an electroluminescence（EL）spectrum before and after light soaking.

102）This equation shows that ***Reb*** has **close affinity with** the large blood vessel's diameter d，**blood velocity of flow** v，heat power E and **warm-up** period t.

affinity 的常规意思是亲和力、吸引力，也可表示密切关系，但在此句中用 close relationship 会更好。如果一定要用 affinity，则其后接 to，而不是 with。

句 102）的参考修改方案：

✓ This equation shows that **parameter *Reb*** has **a close relationship** with the large blood vessel's diameter d，**velocity of blood flow** v，heat power E and **heating** period t.

✓ This equation shows that **parameter *Reb*** has **affinity to** the large blood vessel's diameter d，**velocity of blood flow** v，heat power E and **heating** period t.

注意：表示两事物之间有某种关系时，通常可以用两个句型：there be ... relationship between...and...；has（have)... relationship with...。例如：

- Using publicly available sequencing data，we found that **there was** a fractional power law **relationship between** the number of amplicons required **and** the sensitivity of detection，with a plateau at ~60 amplicons.
- Using publicly available sequencing data，we found that the number of amplicons required **has** a fractional power law **relationship with** the sensitivity of detection，with a plateau at ~60 amplicons.

103）However，this often causes middle school students **some problems**，**as** stress，anger and loneliness，etc.

此句用“some problems，as”不太妥当。表示引起诸如……的问题或用来举例时，较正式的是使用固定短语 such...as...或 such as...。

句 103）的参考修改方案：

✓ However，this often causes middle school students **such problems as** stress，anger and loneliness，etc.

✓ However，this often causes middle school students **some problems**，**such as** stress，anger and loneliness，etc.

104）Only a select set of genes is differentially expressed between the three regions—a limited level of premitotic diversity that **is consistent to** the postmitotic model.

本句的短语 is consistent to 使用不当。与 consistent 搭配的介词是 with（be consistent with），不是 to（be consistent to），表示“与……一致”“与……是一致的”。

句 104）的参考修改方案：

✓ Only a select set of genes is differentially expressed between the three regions—a limited level of premitotic diversity that **is consistent with** the postmitotic model.

105）This article is organized **as following**. First, we introduce some common business designs and describe the related governance patterns.

用 follow 表示"如下"并充当副词修饰谓语动词时，正式用语是 as follows，而不是 as following，后者通常充当定语，修饰名词。

句 105）的参考修改方案：

✓ This article is organized **as follows**. First, we introduce some common business designs and describe the related governance patterns.

106）The cutoff frequency **depends on** the estimated heave frequency **proportionally**, and the heave frequency is obtained online from the *z*-axis acceleration **with** fast Fourier transform (FFT).

此句使用了 depends on ... proportionally，表面意思是甲事物（The cutoff frequency）成比例地决定于乙事物（the estimated heave frequency），表述上有点绕，不如换一句话，就是甲事物与乙事物成比例（甲事物随着乙事物的增加而增加，减少而减少），而表达这一意思有一个正式短语，这就是 is proportional to。同时，还可以考虑将介词 with 改为现在分词 using，并在 fast Fourier transform 前增加一个不定冠词。

句 106）的参考修改方案：

✓ The cutoff frequency **is proportional to** the estimated heave frequency, and the heave frequency is obtained online from the *z*-axis acceleration **using a** fast Fourier transform (FFT).

107）A new double-roller clamping spinning process **is suitable to form the** large thin-walled cylindrical **part** with complex curved **flange**.

此句错用了固定短语 is suitable to，正确的应该是 is suitable for，后面跟动名词。另外，large thin-walled cylindrical part 前面的定冠词多余，名词 part 和 flange 均应该用复数。

句 107）的参考修改方案：

✓ A new double-roller clamping spinning process **is suitable for forming** large thin-walled cylindrical **parts** with complex curved **flanges**.

108）Since the mechanism has unlimited rotational capability around axis Y, α is fixed as **0** without loss of generality **and** only the ranges of β and t are evaluated.

此句用了阿拉伯数字 0，不太正式，在学术体中用单词 zero 来表示数字 0 较为正式。另外，and 前后为并列句，其前面应该加逗号。

句 108）的参考修改方案：

✓ Since the mechanism has unlimited rotational capability around axis Y, α is fixed as **zero** without loss of generality, **and** only the ranges of β and t are evaluated.

6.4.3　词义搭配不当

109）Excessive efforts have been directed toward the use of CD derivative to **improve** the **low aqueous solubility** and **chemical instability** of PGEs.

此句中 low aqueous solubility 和 chemical instability 的意思是"低水容性""化学上的不稳定性"，做 improve（提高）的宾语，但词义不搭配，因为可以说"提高水容性"

"提高稳定性"，但不能说"提高低水容性""提高不稳定性"。表示提高、改善和使好转这类意思时，对象应该是"水容性""化学稳定性"，而不是"低水容性""化学上的不稳定性"。

句 109）的参考修改方案：

✓ Excessive efforts have been directed toward the use of CD derivative to **improve the aqueous solubility** and **chemical stability** of PGEs.

110）From the 35 rats, there are 5 rats whose **symptoms complete recovered**, 18 **partial recovered** and 12 **not recovered**.

此句语序是中文式的，用一般现在时也不妥。此外，说症状（symptoms）恢复（recover），词义不搭配，本意应是"病情""健康"恢复，或症状"消失""消退"，因此说症状消退（subside）、消失（disappear）更恰当。还有，句中 complete 和 partial 是形容词，用来修饰动词（recovered）是错误的，应改为副词；not recovered 不确切，未能表达出"无变化"这一本意。

句 110）的参考修改方案：

✓ **The symptoms subsided completely** in 5 of the 35 rats, **partially in** 18 and **unchanged in** 12.

111）This **work was performed under** National Natural Science Foundation of China.

perform 一词是执行、完成之意，表示通过做事而将事完成了，但 work（工作）一词太笼统，一般不与 perform 搭配。此句的表意是"本研究得到了国家自然科学基金资助"，将 work 用 study 替换，而 performed 用 supported 替换，才能准确地表出基金资助之意。

句 111）的参考修改方案：

✓ This **study was supported by** National Natural Science Foundation of China.

实际中 perform 一词使用较为广泛，在"事"明确的情况下，如检查、分析，或使用某种技术、进行某种操作等，可以用 perform 表示完成了这类事情。例如：

- **Sequencing was performed** on Illumina platforms (MiSeq or HiSeq 2500) using 76-cycle paired-end runs adapted to double-indexed libraries.
- No **statistical tests were performed** in the main paper.
- To better understand how the three core pathways relate to one another and to measure the relative strengths of their association with outcome, we **performed** a **statistical analysis** of competing models.
- Our **analyses were performed** on samples from participants of a clinical trial with predetermined in-and exclusion criteria.

112）Creative design and analysis of some deployable structures and mechanisms **were presented in Refs.** [**1-11**], but deployable membrane antenna structures **were not concerned in these literatures**.

此句由两个并列的被动语态分句组成。分句一大意是"在一些文献中提出了一些可展开结构的创新设计和分析方法"，谓语是 were presented，暗含的动作发出者是人（即作者，人提出、作者提出）。分句二的结构同分句一，大意是"在这些文献中没有考虑可展开薄膜天线结构"，谓语是 were not concerned，暗含的动作发出者同句一，也是人，但动词 concerned（涉及、关系到）在语义上与人不能搭配，不能说"人没有涉及""人没有关系到"。如果用 consider（考虑）一词，不仅语义合适，而且还与人能搭配，完全可以说"人没有考虑"。另外，分句二的 in these literatures 与分句一的 in Refs.［1-11］重复，可去掉。

句 112）的参考修改方案：

✓ Creative design and analysis of some deployable structures and mechanisms **were presented in Refs.［1-11］**, but deployable membrane antenna structures **were not considered**.

113） **The reason** why pyroxylin is unsuitable for coating material **is because** it increases the viscosity.

此例中 The reason 和 is because 在语义上有冲突，二者不搭配，应将 because 改为 that，直接表述“原因是什么”（that 引导的表语从句），而不是“原因是因为什么”。

句 113）的参考修改方案：

✓ **The reason** why pyroxylin is unsuitable forcoating material **is that** it increases the viscosity.

114）A new family of UP-equivalent PMs with high rotational ability and **weak requirements on** geometrical conditions is disclosed using displacement subgroup theory.

此例中形容词 weak 和名词 requirements 不搭配，不能说“弱需求”，应将 weak 改为 low，表意为“低需求”，而不是原来的“弱需求”。另外，介词 on 改为 in terms of 更合适。

句 114）的参考修改方案：

✓ A new family of UP-equivalent PMs with high rotational ability and **low requirements in terms of** geometrical conditions is disclosed using displacement subgroup theory.

6.4.4　词义不准、错用或笼统

115）**The Electronic Stability Program shorted as ESP** is a vehicle dynamics system that replies on the vehicle's braking system and supports the driver in critical driving situations.

本例子中误用 shorted 一词来表示“缩写”。表示缩短或缩写的词是词根为 short 的动词 shorten，而不是 shorted。short 是形容词，不能做动词。shorted 做形容词，表示“短路的”；做动词，表示短路或使短路。

句 115）的参考修改方案：

✓ **The Electronic Stability Program shortened as ESP** is a vehicle dynamics system that replies on the vehicle's braking system and supports the driver in critical driving situations.

✓ **The Electronic Stability Program**（**ESP**）is a vehicle dynamics system that replies on the vehicle's braking system and supports the driver in critical driving situations.

✓ **The Electronic Stability Program**（**ESP**）is a vehicle dynamics system that relies on a vehicle's braking system to support the driver in critical driving situations.

116）It **is founded** that the proposed improved EMD can be successfully applied to the fault diagnosis case of rotating machine simulation test rig of high-speed trains.

此句中 is founded 表“被发现”之意，但表示“发现”的动词原形是 find 一词，find 的过去分词是 found，不是 founded；founded 为动词 found 的过去分词，表示“创建”“建立”，而不是“发现”。

句 116）的参考修改方案：

✓ It **is found** that the proposed improved EMD can be successfully applied to the fault diagnosis case of rotating machine simulation test rig of high-speed trains.

117）Occurring of induced defense response of root **may due** to the increase of secondary metabolites in shoot after maize seedlings are subjected to mechanical damage.

此句中 may due 在形式上近似谓语（情态动词 + 动词原形），实际上不是，因为 due 不是动词，而是形容词、副词或名词。这里将形容词 due 误用为动词，因此应在 due 的前面补上系动词 be。

句 117）的参考修改方案：

✓ Occurring of induced defense response of root **may be due** to the increase of secondary metabolites in shoot after maize seedlings are subjected to mechanical damage.

118）**Mohebbi，et al** [8] studied the evolution of redundant strains in a single-roller flow forming process in one pass, and concluded that **the frictional work** can be neglected **in comparison to the redundant work**.

此句中的 in comparison to 使用不当。这个固定短语常用于表达"把……比作（甚至可以用来比喻）"，虽然有比较意味，但不用于比较两种事物；而与其相似的另一个固定短语 in comparison with 则用于对两种事物进行对比或比较，意思是"与……相比"。本例中，将 frictional work 与 redundant work 进行比较，而不是比作（比喻）关系，因此应改用 in comparison with。另外，本句中的人名 Mohebbi 与 et al 之间的逗号多余，而且 et al 是个缩写词，其后应该有一个表示缩写的小圆点；frictional work 和 redundant work 这两个词语前面的定冠词多余。

句 118）的参考修改方案：

✓ **Mohebbi et al.** [8] studied the evolution of redundant strains in a single-roller flow forming process in one pass, and concluded that **frictional work** can be neglected **in comparison with redundant work**.

119）In the simulation, the whole DRCS process is **finished** within 2 s **and** the stresses and strains at 1 s are chosen **to analysis**.

此句中的 finished 不如改用另一个词 completed。这两个词都表示结束、完成之意，但 finished 表明已达到所做事情的终点，着重圆满地结束或完成已着手的事，而 complete 含有从头到尾使其完全齐备，使其成为一个完美的整体，侧重完成预定的任务或使某事完善，补足缺少的部分等。本句所表示的语义是说完成一个过程，用 completed㊀更适合。另外，连词 and 前面可以加逗号；to analysis 本为动词不定式结构，因此用名词 analysis 错误，应改为动词 analyze，如果用名词 analysis，则可以将 to 改为介词 for。

句 119）的参考修改方案：

✓ In the simulation, the whole DRCS process is **completed** within 2 s, **and** the stresses and strains at 1 s are chosen **for analysis**.

㊀ 2012 年，在伦敦举行的语言大赛中，圭亚那选手 Samsunder Balgobin 在回答比赛中最后一道问题"你如何用一种容易让人理解的方式解释 complete 与 finished 的区别"时，给出了他的答案：When you marry the right woman, you are COMPLETE.（娶对老婆，你这一生就完整了。）When you marry the wrong woman, you are FINISHED.（娶错老婆，你这一生就完蛋了。）And when the right one catches you with the wrong one, you are COMPLETELY FINISHED.（让大老婆逮到你跟小三，你就彻底地完蛋了。）担任比赛的全世界一流的语言学家将桂冠戴在了这位选手的头上，奖品是环球游和一箱 25 年窖龄的朗姆酒，全场观众和评委为他起立鼓掌长达五分钟。（资料来源参见文献[97]）

120）Finally，**the tensile** tests **were done** and the relevant results were obtained.

此句中的谓语动词 done 的语义太宽泛、太笼统，任何行为、事件都可包含进去，如果换成稍具体一些的别的动词，如 performed，finished 等，则表达效果会提升。另外，tensile 前面的定冠词多余。

句 120）的参考修改方案：

✓ Finally，**tensile** tests **were performed** and the relevant results were obtained.

6.4.5　自定义简称首次直接使用

科技论文中应该用公知公用的标准的术语和符号，对于简称（缩略语、字母词）、代号等，除了相关专业的读者能清楚理解的以外，在首次出现时均应写出其全称和简称，再次出现时才可直接用简称。

121）**CPM** and **PERT** are the two widely used project network techniques for planning and scheduling construction activities as efficient tools for construction project management.

此例中直接使用简称 CPM 和 PERT，分别代替全称 Critical Path Method 和 Program Evaluation and Review Technique。如果在本句的前文没有给出过这两个简称的全称，那么句 121）就有简称使用不当的问题，因为首次出现直接用 CPM 和 PERT 时，恐怕没有多少人能理解其所指，除非再去论文中查找其全称或通过别的方式弄懂其意思。

句 121）的参考修改方案：

✓ **CPM**（**Critical Path Method**）and **PERT**（**Program Evaluation and Review Technique**）are the two widely used project network techniques for planning and scheduling construction activities as efficient tools for construction project management.

✓ **Critical Path Method**（**CPM**）and **Program Evaluation and Review Technique**（**PERT**）are the two widely used project network techniques for planning and scheduling construction activities as efficient tools for construction project management.

然而，对于软件、算法或程序名之类的相当于标准名称的简称或简写，如果在论文中用其全称，特别当此全称较长时就会显得啰唆、冗长，难记、难懂。对于这种名称，不论在论文标题还是正文中，首次出现时均可直接使用其简称，但除了在标题中以外，在缩略语之后通常应给出其全称。例如：

- Parallel **AFT** Tetrahedral Mesh Generation for **JAUMIN**（标题）
- Unstructured mesh application software programming framework **JAUMIN**（**J adaptive unstructured mesh applications infrastructure**）supports rapid development of multiple petascale parallel application softwares，and has been successfully applied to structural mechanics analysis and optimization design of major scientific device，fission energy and other areas.
- This paper introduces tetrahedral mesh generation method by **AFT**（**advancing front technique**）which seamlessly integrates into **JAUMIN** and provides parallel tetrahedral mesh generation for those application softwares based on **JAUMIN**.

需要注意的是，写作中应把握好一个“度”，合理区分出哪些简称可以首次直接使用，哪些不可以，对于后者，在行文中首次出现时，要么先简称后全称，要么先全称后简称。

122）This paper presents the method of **VMC** formation based on **SS** in **RMS**.

此例中 VMC、RMS 分别是 virtual manufacturing cell、reconfigurable manufacturing system 的简称，是制造领域较为常见的术语，首次出现时可以直接使用简称，但 SS 是 similarity science 的缩略语，纯属作者为了写作方便而自己定义的，别人不太可能知道其具体所指，因此不能直接使用，如果确是出于写作的方便而使用，那么首次出现时就必须先交代其全称。

句 122）的参考修改方案：

✓ This paper presents the method of **VMC** formation based on **similarity science** in **RMS.**

✓ This paper presents the method of **VMC** formation based on **SS**（**similarity science**）in **RMS.**

✓ This paper presents the method of **VMC** formation based on **similarity science**（**SS**）in **RMS.**

✓ This paper presents the method of **virtual manufacturing cell**（**VMC**）formation based on **similarity science** in **reconfigurable manufacturing system**（**RMS**）.

✓ This paper presents the method of **VMC**（**virtual manufacturing cell**）formation based on **similarity science** in **RMS**（**reconfigurable manufacturing system**）.

6.5　逻辑

科技论文是用来表述客观事物的，客观、真实是基本要求，除非语境或修辞需要，否则语句表达必须符合客观事理或人的正常思维规律，这就是逻辑。

6.5.1　语义不通或不准确

123）Summarizing the aforementioned analysis, the calculation flowchart of the drilling load is depicted in Fig. 18, and the parameters of the drill bit are **shown** in Table 2.

此句 and 后面的语句表达“参数见表 2”之意，使用动词 shown（显示）并无不妥，但如果改用动词 listed（列出），表意更加准确，更合逻辑。

句 123）的参考修改方案：

✓ Summarizing the aforementioned analysis, the calculation flowchart of the drilling load is depicted in Fig. 18, and the parameters of the drill bit are **listed** in Table 2.

124）Any possible harmful effects on nearby population should be **paid attention to** after a leakage of radioactive wastewater.

此句使用 pay attention to 的被动语态，语法上无问题，但逻辑和习惯上不妥，应将 pay attention to 中的 attention 提出来做主语。

句 124）的参考修改方案：

✓ **Attention should be paid to** any possible harmful effects on nearby population after a leakage of radioactive wastewater.

125）Space-borne membrane antenna structures are mainly classified as **parabolic and planar membrane** antenna structures.

此句的表意是将空间运载薄膜天线的结构分为抛物面（parabolic）、平面（planar）两类，含这种天线要么属于抛物面类、要么属于平面类之意，有二者之一的语义逻辑，而 parabolic 和 planar 之间用了连词 and，表达不出此意，如果用连接词 either…or，就能表示出这种细微的语义差别。

句 125）的参考修改方案：

✓ Space-borne membrane antenna structures are mainly classified as **either parabolic or planar membrane** antenna structures.

126）A new **method** of **using** NASTRAN **software** to **analyze and predict** the complex eigenvalue of friction noise is developed.

此句把使用（using）软件（NASTRAN）的主体（动作发出者）表述为一种新方法，即方法（method）使用（using）软件（software）来分析和预测（analyze and predict），而按表意，这一主体应该是“人”（即作者研究团队），这样就存在语义不通的逻辑问题。

句 126）的参考修改方案：

✓ A new **method** is developed to **analyze and predict** the complex eigenvalue of friction noise **by using** NASTRAN **software**.

127）Membrane materials used in membrane antennas are **commonly** polyester（PET）film and polyimide（PI）film［37-39］.

此句的表意是“薄膜天线中常用的薄膜材料是聚酯膜和聚酰亚胺膜”，“常用”就是“经常使用”，对应的英语是 commonly used 或 used commonly，其中副词 commonly 用来修饰动词 used。然而，此句错将 commonly 和 used 分开，将 commonly 放在系动词 are 之后，即 commonly 修饰句子的谓语（commonly 充当表语），造成语义不通。

句 127）的参考修改方案：

✓ Membrane materials **commonly** used in membrane antennas are polyester（PET）film and polyimide（PI）film［37-39］.

128）**From** Table 2，compared with polyester film，**polyimide film** has **larger modulus**，higher tensile strength，lower thermal expansion coefficient，smaller elongation at break **and** stronger anti-ultraviolet radiation ability.

此句开头的介词短语（From Table 2），暗含的主语（逻辑主语）应该是人（作者），与句子的主语 polyimide film 不一致，造成逻辑问题。另外，larger modulus 前遗漏了不定冠词，and 前遗漏了逗号。

句 128）的参考修改方案：

✓ **As shown in** Table 2，compared with polyester film，**polyimide film** has **a larger modulus**，higher tensile strength，lower thermal expansion coefficient，smaller elongation at break，**and** stronger anti-ultraviolet radiation ability.

✓ **From** Table 2，**we found** that compared with polyester film，**polyimide film** has **a larger modulus**，higher tensile strength，lower thermal expansion coefficient，smaller elongation at break，**and** stronger anti-ultraviolet radiation ability.

129）The work reported has been **partially completed**.

此例中 partially completed 表示“部分完成”，含“未完成”之意，而 completed 表示“完成”之意，前面再用 partially（部分）来修饰，这样“完成”与“部分”在语义上有冲突，逻辑有矛盾。

句 129）的参考修改方案：

✓ The work reported is still **uncompleted**.

130）**Dynamics** of **such** rigid-flexible coupling system **as spacecraft structure** with large membrane antenna **structure**，**needs** to be further studied，**so does dynamics** of antennas in deployment process.

用来列举、举例时，such as 通常是一个整体短语，但本句错将 such 前置而与 as 分离，形成 such…as 结构，即在句子主语 Dynamics 的后置修饰语 rigid-flexible coupling system 的前面加了 such 一词，破坏了语义的正常逻辑。此外，spacecraft structure 是列举的对象，即它是 rigid-flexible coupling system 的一种具体类别，暗含这样的系统有很多，因此 system 应该用复数；同时，

spacecraft structure 的后置介词短语 with large membrane antenna structure 中的 structure 也应该用复数。另外，句子最后用了 so does… 结构，使得全句成为由两个句子组成的并列句，这样，前一个句子的主体性下降，而后一个句子的辅体性上升，两句成为平级关系，语义逻辑关系发生变化，实际上后一个句子是前一个句子的补充、说明成分，宜将后一个句子表述为状语，其中 dynamics 之前还应补一个定冠词。

句 130）的参考修改方案：

✓ Dynamics of rigid-flexible coupling **systems**, **such as spacecraft structures** with large membrane antenna **structures**, **needs** to be further studied, **along with the dynamics** of antennas in the deployment process.

131）**Moreover** the authors **carried out** many **finite element simulations** and **experimental researches** and **obtain** the effects of main process parameters **such as** the roller radius, the spacing between two rollers **and** the feed rate of **roller** on the DRCS process.

此例中 finite element simulations 和 experimental researches 在语义上顺序颠倒，逻辑上不通，应该是先有实验研究，后有有限元模拟。另外，Moreover 后面应有停顿，需加一个逗号；第一个连词 and 前后的谓语动词的时态（carried out，obtain）应一致，统一改为过去式为好；第二个谓语 obtain 前的连词 and 的前面应加逗号；such as…用来列举，与前面的中心词（parameters）为非限定性的关系，其间应该加逗号停顿一下；such as 后面的部分为三个列举项，最后一项的 and 前面也应该加逗号；roller 应为复数形式。

句 131）的参考修改方案：

✓ **Moreover**, the authors **carried out many experimental research** and **finite element simulations**, and **obtained** the effects of main process parameters, **such as** the roller radius, the spacing between two rollers, **and** the feed rate of **rollers** on the DRCS process.

132）Packing **method** for membrane antennas should be studied, **in which** antenna storage volume and packing complexity can be reduced with no electronic device **damaging occurring**.

本句逗号后面的部分为非限定性定语从句（in which…），但所表达的意思（降低天线存储容量和封装复杂度而不发生电子设备损坏）是前面主句意思（应研究薄膜天线的封装方法）的目的，因此用定语从句导致语义逻辑关系发生变化，应改为目的状语从句（so that…）。另外，两个现在分词连用（damaging occurring）不合适，将 damaging 改为名词 damage 更合适。

句 132）的参考修改方案：

✓ Packing **methods** for membrane antennas should be studied, **so that** antenna storage volume and packing complexity can be reduced with no electronic device **damage** occurring.

6.5.2 本文、本研究未明确区分

133）**In the study**, Figures 5, 6, 8 and Tables 1, 2, 4 are listed to support the experiment.

此句中用 In the study 不如用 In the paper，因为从语义来看，在论文中展示图表比在研究中展示图表更加准确严密，逻辑更加通顺。

句 133）的参考修改方案：

✓ **In the paper**, Figures 5, 6, 8 and Tables 1, 2, 4 are listed to support the experiment.

✓ **The paper** list Figures 5, 6, 8 and Tables 1, 2, 4 to support the experiment.

✓ Figures 5, 6, 8 and Tables 1, 2, 4 are listed to support the experiment **in the paper**.

134) **In the paper**, an SZG4031 towing tractor is used as the sample vehicle.

此句中 In the paper 表意欠佳，因为在 paper 中使用 tractor 不如 study (research) 中使用更贴切、准确，逻辑更通畅，因为是“人”在“研究”中而不是在“论文”中使用 tractor。

句134）的参考修改方案：

✓ **In the study**, an SZG4031 towing tractor was used as the sample vehicle.

✓ **The study** used an SZG4031 towing tractor as the sample vehicle.

✓ An SZG4031 towing tractor was used as the sample vehicle **in the study**.

135) **This article aimed** to investigate the expression of CCR7 in osteosarcoma and discuss the underlying relationship between the expression and clinical significance.

136) **The aim of the present paper** is to summarize the history of the Yusho incidence that occurred in the western part of Japan in 1968.

这两句中 This article aimed 和 The aim of the present paper 表意欠佳，因为说 article、paper 的目的不如说 study (research) 的目的更符合逻辑（是研究的目的，而非文章的目的）。

句135）的参考修改方案：

✓ **This study aimed** to investigate the expression of CCR7 in osteosarcoma and discuss the underlying relationship between the expression and clinical significance.

句136）的参考修改方案：

✓ **The aim of this study** is to summarize the history of the Yusho incidence that occurred in the western part of Japan in 1968.

✓ **This study aims** (**aimed**) to summarize the history of the Yusho incidence that occurred in the western part of Japan in 1968.

注意：在能清楚表达出是由作者所做（提出）的工作（理论、方法等）时，特别是在摘要中，没有必要使用或重复使用 the paper (in the paper) 和 the study (in the study) 之类的词语。另外，这两类词语在表意上有差异：the study、the research 侧重作者所做的研究工作，而 the paper、the article 侧重提出这项研究工作的模式是论文，即读者正在拿到或阅读到的。如果不加区别随意混用这类词语，就容易混淆其间的语义差异。

6.5.3　国内、国外未明确区分

137) The contributing factors **at home and abroad** were analyzed by retrospective analysis so as to provide new thought and reasonable interventions for further studies.

138) At present, **domestic and foreign** scholars mainly use experimental method, analytical method and numerical method to carry out impact effect research of gas jet.

此两句中分别用了 at home and abroad 和 domestic and foreign 来表述“国内外”之意，这是一种惯用逻辑思维，以作者所在国为“国内”，其他国为“国外”的角度来写的，表面上似乎没有什么问题。但从英语论文主要面向国际这一角度来看，这类词通常不应该出现，特别是当论文的作者来自多个国家时（论文的作者有多位，其中有的来自同一国家，有的可能来自另一国家，有的还可能“身兼数国”），使用诸如 at home and abroad、domestic and foreign、our country 之类的词时，容易出现指代不明或逻辑混乱的错误。因此，当表意清楚时，这类词均可去掉。

句 137）、138）的参考修改方案：

✓ The contributing factors were analyzed by retrospective analysis so as to provide new thought and reasonable interventions for further studies.

✓ At present, scholars mainly use experimental method, analytical method and numerical method to carry out impact effect research of gas jet.

139）The research status of intelligent manufacturing **in our country** is analyzed and prospected.

140）**Domestic** scholars usually use two-dimensional numerical simulation model to carry out the research.

141）**Foreign** scholars usually use three-dimensional numerical simulation model to carry out the research.

这三句分别用 in our country、Domestic、Foreign 表示"在我国""国内""国外"，暗含从作者所在国家的角度来表达的意味，当读者所在国家与作者所在国家不同，或论文的作者来自多个国家时，就会造成指代不明或逻辑错误。写作行文必要时应明确交代国名。

句 139）~141）的参考修改方案（以中国作者为例）：

✓ The research status of intelligent manufacturing **in China** is analyzed and prospected.

✓ **Chinese** scholars usually use two-dimensional numerical simulation model to carry out the research.

✓ Scholars **outside China** usually use three-dimensional numerical simulation model to carry out the research.

142）**Overseas** scholars usually use three-dimensional numerical simulation model to carry out the research.

Overseas 是"海外的；外国的"之意。此句中的 Overseas 实际所指可能不太清楚，可以按照表意进行修改。

句 142）的参考修改方案：

✓ Scholars **outside China** usually use three-dimensional numerical simulation model to carry out the research.

✓ Scholars **outside the mainland of China** usually use three-dimensional numerical simulation model to carry out the research.

6.5.4　先旧后新逻辑颠倒

143）The use of land, water and minerals has increased more than tenfold during the past two centuries. Future increases in population and economic development will intensify this pressure. **Major environmental changes** varying from disruption of local ecosystems to disturbance of the biosphere are the likely **cumulative impacts** of human activities.

此例最后一句中，前面的 Major environmental changes 为新信息，后面的 cumulative impacts 为上文提过的旧信息，新旧信息倒置，不符合正常的"先旧后新"的逻辑顺序。

句 143）的参考修改方案：

✓ The use of land, water and minerals has increased more than tenfold during the past two centuries. Future increases in population and economic development will intensify this pressure. **The cumulative impacts** of human activities are likely to lead to **major**

environmental changes, varying from disruption of local ecosystems to disturbance of the biosphere.

修改后，将前面提过的“旧信息”置于句子的前面，后面提出的“新信息”置于句子的后面。用此原则写出的句子，不仅可以逐步深入地表达作者想要表达的观点，而且符合读者的阅读预期及按顺序展开的逻辑思维方式，思路不至于突然中断。

144）**PSO** is emerging evolutionary computation technology based on swarm intelligence, which has been applied successfully in many fields. To further improve the global search ability of **PSO**, a **particle swarm optimization based on pyramid model** (**PPSO**) is presented to solve **optimization problems such as the layout design** of an international commercial communication satellite (INTELSAT-Ⅲ) **cabin**.

此例中的核心缩略语 PSO 首次出现时没有写出全称，而后面的 PPSO（改进的 PSO）出现时给出了 PPSO 全称，逻辑上存在欠缺，不如 PSO 首次出现时就写出全称通顺，而且 PPSO 出现时直接引用 PSO 更能体现出 PPSO 和 PSO 的内在逻辑关系，更符合人们由不知到知、由知之不多到知之甚多的学习式、认识式的思维过程。同时，将 such as 后面所列举的对象（the layout design）直接表述为要解决的问题，这样就要去掉 such as，并将 the layout design 和 such as 前面的 optimization problems 合并重新表述为 the layout optimization problems against the background。

句 144）的参考修改方案：

✓ **Particle swarm optimization**（**PSO**）is emerging evolutionary computation technology based on swarm intelligence, which has been applied successfully in many fields. To improve the global search ability of **PSO**, a **multi-population PSO based on pyramid model** (**PPSO**) is presented. Then, the **PPSO** is applied to solve **the layout optimization problems against the background** of an international commercial communication satellite (INTELSAT-Ⅲ) **module**.

6.6　句法

6.6.1　结构不妥当

145）In most specimens **there is** more biotite than hornblende.

此句为 there be 句型，是英语中的常用句型，意思是“有”，表示“人或事物的存在”或“某地有某人或事物”（there 在此句型中是引导词，已经没有副词“那里”的含义）。In most specimens 为状语，与 there is 连用，形成“状语 + 谓语”结构，意思是“在大多数标本中有……”，如果改用主谓结构，即将 In most specimens 改为 most specimens 做主语，there is 改为动词 contain 做谓语，表意是“大多数标本中含有”，那么句子的结构会更紧密，表意更直截了当。

句 145）的参考修改方案：

✓ Most specimens **contain** more biotite than hornblende.

146）**But** further studies **are still necessary considering** the cable deformations during the adjustment.

按语义，此句分词短语 considering…是 studies 的修饰成分，二者应相邻，但其间插入了表语成分 are still necessary，使得句子结构不够紧密。但用分词短语 considering 做 studies 的后置修饰语时，含正在考虑的意思，进行时的意味浓，不如改为现在时意味的定语从句。况且，转折连词与后面句子之间有停顿更好，但 but 一词后面不宜直接加标点，这样改用 however 就可以了。

句 146）的参考修改方案：

✓ **However**，further studies **that consider** the cable deformations during the adjustment **are still necessary**.

语法规则：句中相关部分（如修饰语与被修饰语）应尽量相邻，关系明确，使句子结构紧密、清晰，表达准确、简短。

147）**The heave compensation** is widely applied to decouple the heave motion in offshore installations. **Heave compensation** is always divided into two categories：passive heave compensation and active heave compensation.

此例中两个句子的主语相同，语义上本是连贯的，但用两个单句来分别表述，在某种程度上就破坏了这种连贯性，应以两个分句组成一个复句的结构形式来表述，这样分句二的主语就可以从形式上省略而承接分句一的主语，并以 and 来关联两个分句。此外，还可以考虑将句子开头的定冠词去掉。

句 147）的参考修改方案：

✓ **Heave compensation** is widely applied to decouple the heave motion in offshore installations，and is always divided into two categories：passive and active heave compensation.

148）β_2 is the penetrating **helix angle of external cylinder of the drill bit**，β_3 is the penetrating **helix angle of inner cylinder of the drill bit**.

此句中有两个参数 β_2 和 β_3，其名称分别是钻头外、内柱面的贯穿螺旋角，包含三个主体名词，分别是钻头（drill bit）、柱面（cylinder）、螺旋角（helix angle），这三个词的语义关系是"钻头上有柱面，柱面上有螺旋角"，因此表达为 helix angle of external cylinder of the drill bit 从语义上讲并无不妥。然而，这一结构中出现了两个 of，有"绕"的语感，如果将第二个 of 去掉，将其前后成分交换位置，不仅语义没有变化，而且结构上会更紧密。

句 148）的参考修改方案：

✓ β_2 is the penetrating **helix angle of the drill bit external cylinder**，β_3 is the penetrating **helix angle of the drill bit inner cylinder**.

149）It can be seen **that Region *A* shows tensile tangential strain**，**compressive radial and thickness strain**，and **the absolute value of the tangential strain is the biggest**.

此句中 that 后面有两个并列的句子，因第二个句子（the absolute value…the biggest）前面缺少 that，容易让人误解为这第二个句子是第一个句子（Region *A* shows … thickness strain）中的谓语动词 shows 的三个对象（画线部分）中的第三个，即三个并列成分中的第三个。另外，tensile tangential strain 和 compressive radial and thickness strain 是两个并列的成分，其间可以直接用连词 and 连接；表示数值大小宜用 large 而不是 big，通常 big 多用来表示尺寸的大小。按以上思路修改后，句子结构会更紧凑，语义也会非常明晰。

句 149）的参考修改方案：

✓ It can be seen **that Region *A* shows tensile tangential strain and compressive radial and thickness strain**，and **that the absolute value of tangential strain is the largest**.

150）Finally，**the tensile** tests **were done** and the true stress-strain curve of the material formed by DRCS **was obtained and compared** with the true stress-strain curve of the initial material, as shown **as** Fig. 15.

此句中除了谓语动词 done 表意笼统（不如改为 performed 明确）外，还存在句子结构欠合理的问题。句子出现了三个谓语动词被动式（done、obtained、compared），前一个和后两个有不同的主语，因混在同一句中表述，显得结构臃肿，层次不清，可读性较差。可以考虑将前一个谓语和后两个谓语分别放在不同的句子中来表述，各自有其相应的主语，这样结构调整后，表达效果大为提升。

句 150）的参考修改方案：

✓ Finally, **tensile** tests **were performed**. The true stress-strain curve of the material formed by DRCS **was obtained and compared** with the true stress-strain curve of the initial material, as shown **in** Fig. 15.

151）Expressing the position vectors of points A_i （$i = 1, 2, 3$） in **global** coordinate system as a suitable form that is easy to understand.

此句不成句，明显缺谓语，可考虑修改为主动句或被动句，有时也可用祈使句。另外，global 前面缺少定冠词。

句 151）的参考修改方案：

✓ **We express** the position vectors of points A_i （$i = 1, 2, 3$） in **the global** coordinate system as a suitable form that is easy to understand.

✓ The position vectors of points A_i （$i = 1, 2, 3$） in **the global** coordinate system **are expressed** as a suitable form that is easy to understand.

✓ **Express** the position vectors of points A_i （$i = 1,2,3$） in **the global** coordinate system as a suitable form that is easy to understand.

6.6.2 语义不一致

（1）集体名词做主语

152）The **data show** that a chemical change takes place.

153）The **data** in Fig. 4 **has** five segments of turbulent information.

data 是集体名词，其单数形式是 datum，但很少用。

前一句中的主语 data 指数据、数字的整体，表示单数，其后谓语动词用复数形式 show 有误，应为单数形式 shows；后一句中的主语 data 指很多数据，侧重指数据、数字中的成员，表示复数，其后谓语动词用单数形式 has 有误，应为复数形式 have。

句 152）、153）的参考修改方案：

✓ The **data shows** that a chemical change takes place.

✓ The **data** in Fig. 4 **have** five segments of turbulent information.

语法规则：集体名词（如 class、data、family、group、public、team 等）做主语时，如果作为一个整体看待，谓语动词用单数；如果作为整体中的个体成员看待，谓语动词用复数。

（2）“a（the）number of + 复数名词”做主语

154）There **is a large number of workers** in the factory, who **are** from America.

155）**The number of the researchers** in the university **are** 680.

前一句中的 a large number of workers 指很多工人，其前系动词用单数形式 is 有误，应为复数形式 are；后一句中的 The number of the researchers 指研究者的数量，是一个整体数字，其后系动词用复数形式 are 有误，应为单数形式 is。

句 154）、155）的参考修改方案：

✓ There **are a large number of workers** in the factory, who **are** from America.

✓ **The number of the researchers** in the university **is** 680.

语法规则："a number of + 复数名词" 做主语时，谓语动词用复数形式；"the number of + 复数名词" 做主语时，谓语动词用单数形式。

（3）"every（each、much、little、many、few、a few 等）+ 复数名词" 做主语

156）**Every** eventuality **and every** possibility **were** taken into account.

此句主语为 every…and every…结构，谓语应该用单数，were 需改成 was。

句 156）的参考修改方案：

✓ **Every** eventuality **and every** possibility **was** taken into account.

语法规则：名词前由 every、each、much、little 修饰时谓语为单数，由 many、few、a few 修饰时谓语为复数；由 every…and every…，each…and each…，no…and no…结构做主语时谓语为单数。

（4）"relationship between…and…" 做主语

157）**The relationship between** financial sustainability **and** socio-economic development **were** analyzed.

此句的主语是 The relationship（单数），其后 between…and…部分是它的后置修饰语，句子的谓语动词的数应与主语一致，而不是与主语的修饰语的数一致。因此用 were 有误，应改为 was。

句 157）的参考修改方案：

✓ **The relationship between** financial sustainability **and** socio-economic development **was** analyzed.

表示二者之间的关系用介词结构 "relationship between…and…"，三者之间的关系用介词结构 "relationship among…，…，and…"，但需要注意的是，不论用这两种结构的哪一种，relationship 通常不用复数形式。

158）**Relationships among** force，velocity，and displacement **are** analyzed based on the field investigation and literature in this paper.

此句开头的 Relationships 应改为单数，相应地谓语中的 are 改为单数 is。

句 158）的参考修改方案：

✓ **Relationship among** force，velocity，and displacement **is** analyzed based on the field investigation and literature in this paper.

6.6.3　就近不一致

159）The novel 5-DOF leg is able to reach different centers of rotation，providing **either** the concave **or** convex arcs that **satisfies** the basic principle of displacement of walking machines.

160）**Not only** the material property and the fractal parameters **but also** the radius of the asperity **have** a close relationship with the critical contact areas.

此两句的谓语动词 satisfies 和 have 应该与最靠近它们的那个主语（convex arcs 和 the radius）的单、复数形式相一致，因此应分别改为 satisfy 和 has（convex arcs 为复数，the radius 为单数）。

句 159）、160）的参考修改方案：

✓ The novel 5-DOF leg is able to reach different centers of rotation，providing **either** the concave **or** convex arcs that **satisfy** the basic principle of displacement of walking machines.

✓ **Not only** the material property and the fractal parameters **but also** the radius of the asperity **has** a close relationship with the critical contact areas.

语法规则：谓语动词与其最靠近的那个主语的单、复数形式相一致。当由 or，either…or，neither…nor，not…but，not only…but also 连接两个并列主语时，谓语动词应与最靠近它的那个主语的单、复数形式一致。

以上是应该就近一致的情况，但也有就“远”一致的情况。例如：

161）The color metallograph，scanning electron microscope **methods as well as microhardness was** used to observe the curving interface zones of Al/Cu.

此句中 as well as 的前面成分的中心词为复数形式 methods（The color metallograph 和 scanning electron microscope 是两个名词性结构，做 methods 的修饰语），谓语动词的数应与这个 methods 保持一致，即应该为复数，因此用 was 有误，应改成 were。

句 161）的参考修改方案：

✓ The color metallograph，scanning electron microscope **methods as well as microhardness were** used to observe the curving interface zones of Al/Cu.

语法规则：用 as well as 连接几个成分（名词或名词结构）做主语时，谓语动词的数取决于其所连接的前一个成分的数，即与其前一个主语的数保持一致；但由 both…and…连接两个成分做主语时，谓语动词用复数。

6.6.4　独立结构成分错用

162）**When they are necessary**，the solutions were deaerated by bubbling nitrogen.

163）**Judged from the spectral changes**，exhaustive photolysis of compound 4 had occurred.

164）**Assumed that distance *d* is induced by the norm**，***M*** is a symmetrical and positively defined matrix.

这三句中逗号前面的部分为独立结构成分，与后面的主句没有语法关系，但第一句中的引导词 when 后面有多余的成分（they are），后两个句中的引导词形式错误，错将现在分词形式（Judging、Assuming）写成过去分词（Judged、Assumed）。

句 162）~164）的参考修改方案：

✓ **When necessary**，the solutions were deaerated by bubbling nitrogen.

✓ **Judging from the spectral changes**，exhaustive photolysis of compound 4 had occurred.

✓ **Assuming that distance *d* is induced by the norm**，***M*** is a symmetrical and positively defined matrix.

独立结构成分的引导词常见的有 assuming，concerning，considering，failing，given，judging，provided，providing，regarding，taking，when 等，注意这种结构与其后的主句无语法关系。

6.6.5　定语从句使用不当

定语从句使用不当主要包括两个方面：不该用定语从句而用了定语从句（可用简洁的分词短语代替）；定语从句中的关系代词不合适。

165）A **lowpass filter** is used to correct **the phase error which is caused by the highpass filter**.

此句的名词短语 the phase error 用定语从句 which is caused by…修饰，显然比不上直接用过去分词短语 caused by…修饰简洁，也就是说，短语充当修饰语的优先级要高于定语从句。另外，低通滤波器、高通滤波器的正式用语应为 low-pass filter 和 high-pass filter。

句 165）的参考修改方案：

✓ A **low-pass** filter is used to correct **the phase error caused by** the **high-pass filter**.

166) **Anything which** is hot radiates heat.

167) Copper is **the first** metal **which** man learned to use.

168) **The smallest** living things **which** can be seen under a microscope are bacteria.

这三个定语句错用了关系代词 which，应改为 that，因为先行词为不定代词、疑问代词，或被序数词、形容词的最高级等修饰时，关系代词应该用 that（但不定代词 something 后也可用 which）。

句 166）~168）的参考修改方案：

✓ **Anything that** is hot radiates heat.

✓ Copper is **the first** metal **that** man learned to use.

✓ **The smallest** living things **that** can be seen under a microscope are bacteria.

169) This means that the max stress of **CBFE which** arc is according with rotational direction is bigger than the max stress of **CBFE which** arc is disaccording with rotational direction.

此句中有两个定语从句，两处关系代词 which 错用，应改为 whose，做 arc 的定语。（CBFE 是 curved beam flexure element 的缩写，中文名是曲梁柔性单元。）

句 169）的参考修改方案：

✓ This means that the max stress of **CBFE whose** arc is according with rotational direction is bigger than the max stress of **CBFE whose** arc is disaccording with rotational direction.

170) The two operating situations, **which EMCVT was used in traditional vehicle and hybrid vehicle** are analyzed.

此句逗号后面的部分是由关系代词 which 引导的定语从句，但关系表达错误，which 前遗漏了介词 in。也可不用定语从句。

句 170）的参考修改方案：

✓ The two operating situations, **in which** EMCVT was used in traditional and hybrid vehicles respectively, are analyzed.

✓ We analyzed the two operating situations, **in which** EMCVT was used in traditional and hybrid vehicles respectively.

✓ The two operating situations are analyzed—EMCVT was used in traditional and hybrid vehicles respectively.

✓ We analyzed the two operating situations—EMCVT was used in traditional and hybrid vehicles respectively.

171) Wang **made** a systematic research on this topic, and this type of method **was** called the "simulation based optimization", **which can directly obtain the optimal plan corresponding to a limited sample size**.

此句中定语从句（which can...size）修饰先行词（simulation based）optimization，即 which 指代 optimization，这样语义上就是 optimization can directly obtain the optimal plan...（优化能获得最优方案），用名词 optimization 做谓语动词 obtain 的主语，逻辑上显然不通，正确的表意应该是“由优化而获得了最优方案”，因此需要在关系代词 which 前加介词 from，加介词后此定语从句变成无主语句，故而需要转变为被动语态，这样就有主语了。另外，还可考虑将动词 made 替换为 conducted。

句 171）的参考修改方案：

✓ Wang **conducted** a systematic research on this topic, and this type of method **is** called the "simulation based optimization", **from which the optimal plan corresponding to the**

limited sample size can be obtained.

6.7　主语一致

动词短语（如分词短语、不定式短语、动名词短语）及省略主语的从句做修饰语时，其逻辑主语与句子的主语不统一（不相同、不一致）时就形成悬垂修饰，即修饰语与句子主语没有相关性。但可以使用独立结构成分，这种成分是位于句首用来修饰整个句子而不是句中某个部分的修饰成分，完全独立，形式上不受句中任何部分所制约。

6.7.1　悬垂分词

172）**Using** our method to establish bi-level optimization model of robust design, **the preferences** of f and Δf are given in Table 1.

173）**Referring** to Eq.（12）and Eq.（14）, **the overall power output** from batteries P_b can be **shown** as follows.

以上两句中现在分词 Using 和 Referring 的主语指人（作者），与句子的主语 the preferences 和 the overall power output 不是同指，形成悬垂分词修饰。

句 172）、173）的参考修改方案：

✓ **Using** our method to establish bi-level optimization model of robust design, **we** give the preferences of f and Δf in Table 1.

✓ **Referring** to Eq. 12）and Eq.（14）, **we** can **express** the overall power output from batteries P_b as follows.

174）**Comparing** to ant colony optimization（ACO）, **max-min ant system**（MMAS）has three differences as follows.

此句中现在分词 Comparing 为悬垂分词，若改为过去分词 Compared，则分词的主语与句子的主语 max-min ant system 就一致了，因而可以避免悬垂分词修饰。

句 174）的参考修改方案：

✓ **Compared** to ant colony optimization（ACO）, **max-min ant system**（MMAS）has three differences as follows.

175）When **confronted** with these limitations, **the experiments** were discontinued.

此句的过去分词 confronted 的主语指人，与句子的主语 the experiments 不是同指，形成悬垂分词修饰。

句 175）的参考修改方案：

✓ When **confronted** with these limitations, **we** discontinued the experiments.

based on 是英语科技论文中常用的短语，使用不当也会出现悬垂分词修饰。

176）**Based on** fast corresponding, a general requirement modeling **method is proposed**.

此句不存在悬垂修饰，但 based on 使用不当。这个短语只限于修饰与其紧邻的先行词，故此句表示成“Based on…”修饰主语“method”，即“method based on…”，与本意“Based on…”修饰谓语“…is proposed”相违背。

句 176）的参考修改方案：

✓ **On the basis of** fast corresponding, a general requirement modeling method **is proposed**.

✓ A general requirement modeling method **is proposed based on** fast corresponding.

177）So **based on** constitutive equations, **the investigation** of lead-free soldered joints properties under thermal loading **become** a major concern.

此句存在的问题同句 176），错将“based on”表示成修饰“the investigation”，

与本意修饰 become 相违背。

句 177）的参考修改方案：

✓ So, **on the basis of** constitutive equations, the investigation of lead-free soldered joints properties under thermal loading **become** a major concern.

但不可改为以下句子，因为这样修改后，"based on" 就修饰 concern 了：

× So the investigation of lead-free soldered joints properties under thermal loading become a major **concern based on** constitutive equations.

178）**Based on** customer requirements, **we** make the modularization configuration, establish the semantic network description of configuration knowledge and gain the product structure tree that satisfies the customer requirements.

此句 Based on 的主语与句子的主语 we 不一致，形成悬垂分词修饰，可将 Based on 改为 Basing our decision on。

句 178）的参考修改方案：

✓ **Basing our decision on** customer requirements, **we** make the modularization configuration, establish the semantic network description of configuration knowledge, and gain the product structure tree that satisfies the customer requirements.

179）**Based on** the fuzziness of design process, the open-loop rigidity reasoning systems that are primarily logic reasoning are expanded to the recursive close-loop flexible reasoning systems to revise the reasoning results **based on** practical instance.

此句中第一个 Based on 使用不当，可改为 On the basis of。

句 179）的参考修改方案：

✓ **On the basis of** the fuzziness of design process, the open-loop rigidity reasoning systems that are primarily logic reasoning are expanded to the recursive close-loop flexible reasoning systems to revise the reasoning results **based on** practical instance.

6.7.2 悬垂不定式

180）**To demonstrate** the proposed method, **the design** of the two-bar structure acted by concentrated load is presented.

181）**To overcome** the deficiencies in the methods of gaining and describing the requirement information, **a requirement information cell** (**RIC**) **method** for describing general requirement information is investigated.

182）**In order to know** the population dynamics of Hefei City, **the quadrat investigation** was conducted on its age structure.

此三句不定式短语 "To demonstrate…" "To overcome…" "In order to know…" 分别修饰句子的主语 the design、a…method、the quadrat investigation，但动词 demonstrate、overcome、know 的逻辑主语是人，不是 design、method、investigation，因此均形成悬垂不定式修饰。

句 180）~182）的参考修改方案：

✓ **To demonstrate** the proposed method, **we** presented the design of the two-bar structure acted by concentrated load.

✓ **To overcome** the deficiencies in the methods of gaining and describing the requirement information, **we** investigated a requirement information cell (RIC) method for describing general requirement information.

✓ **In order to know** the population dynamics of Hefei City, **we** conducted the quadrat investigation based on its age structure.

6.7.3 悬垂动名词

183）After **determining** the mapping of A_i from L_1 to L_p, **the next array dimension** can be processed in the same way.

此句中介词 After 引导的是带有动名词 determining 的修饰性短语，此短语的主语应是句子的主语 the next array dimension，但 dimension 无法执行 determining 这个行为，因此形成悬垂动名词修饰。

句 183）的参考修改方案：

✓ After **determining** the mapping of A_i from L_1 to L_p, **we** can process the next array dimension in the same way.

6.7.4 悬垂省略主语的从句

184）When **developing** a new product, **right knowledge** may be got from repository or expert by designers.

此句逗号前面的部分为省略主语 designers 的从句，被省略的这个主语与逗号后面的主句的主语 right knowledge 不一致，形成悬垂省略主语的从句修饰。

句 184）的参考修改方案：

✓ When **developing** a new product, **designers** can get right knowledge from a repository or experts.

✓ When **designers** develop a new product, **they** can get right knowledge from a repository or experts.

✓ When **designers** develop a new product, **right knowledge** can be gotten from a repository or experts.

185）When **simple**, we can solve **this type of problem** quickly using the direct method.

此句的 When simple 是省略主语 this type of problem 的从句，被省略的这个主语与主句的主语 we 不一致，形成悬垂省略主语的从句修饰。

句 185）的参考修改方案：

✓ When **simple**, **this type of problem** can be solved quickly by using the direct method.

6.8 语态

不少人误认为英语科技论文宜优先用被动语态来表达意思和组织句子。据有关资料表明，英语科技论文中被动语态的使用曾在 1920—1970 年比较流行，但由于主动语态的表达更为准确，而且阅读容易，因而目前大多数科技期刊提倡用主动语态。*Science*、*Cell* 和 *Nature* 等名刊的论文中，主动语态和第一人称的使用十分普遍。

6.8.1 主动语态未优先使用

186）**The fact** that such processes are under strict stereoelectronic control **is demonstrated** by our work in this area.

此句为被动语态形式，共 18 个词。若改为主动语态形式，则只需 14 个词，简洁了不少。

句 186）的参考修改方案：

✓ **Our work** in this area **demonstrates** that such processes are under strict stereoelectronic control.

187）**A wavelet network** suiting to approach multi-input and multi-output system **is constructed**.

此句用被动语态，谓语太短而主语特长，头重脚轻。若改用主动语态，则显得自然，阅读容易。

句 187）的参考修改方案：

✓ **We constructed** a wavelet network

suiting to approach multi-input and multi-output system.

6.8.2 前后分句语态不一致

188) The elastic deformation of wheelset caused by wheel flat **will increase** the vibration acceleration of vehicle components, but the rigid body dynamics model **is mainly adopted** in the the related research at present. (车轮扁疤所诱发的轮对弹性变形会导致车辆系统部件振动加速度增大，但目前相关研究主要采取刚体动力学模型。)

此例由两个分句组成。前后分句的语态不一致，分句一为主动语态，分句二为被动语态，按同一复句前后分句的语态宜一致的原则，可将分句二改为主动语态。

句 188) 的参考修改方案：

✓ The elastic deformation of wheelset caused by wheel flat **will increase** the vibration acceleration of vehicle components, but at present the related research mainly **adopts** the rigid body dynamics model.

189) In practice, only after these problems **are rationally explained or solved**, **did** the engineers use the optimal plans.

此句由两个分句组成，后一分句为倒装句，其时态应承袭前一分句用现在时而不用过去时。

句 189) 的参考修改方案：

✓ In practice, only after these problems **are rationally explained or solved**, **do** the engineers use the optimal plans.

6.8.3 整段句子全部用被动语态

190) Five methods for deploying and forming large parabolic membrane antennas **are compared and analyzed** including inflation, inflation-rigidization, elastic ribs driving, SMP-inflation and electrostatic forming, and membrane material properties for parabolic membrane antennas **are presented**. Additionally, the development of structure configurations, tensioning system design and dynamic analysis of large planar membrane antenna structures **are discussed**. Finally, future directions of large space membrane antenna structures **are summarized** and technical difficulties **are proposed** based on design and analysis, membrane materials and processes, membrane packing and surface accuracy maintaining.

本例为一篇综述的摘要，每个句子都用被动语态（笔者将这种连续使用被动语态的现象称为“一‘被’到底”），主语较长，谓语很短，表达显得呆板、生硬，若改为主动和被动语态相结合的形式，把相关谓语动词归类放在同一句，并进行修辞锤炼，表达效果会大大提升。

句 190) 的参考修改方案：

✓ This review **presents** the development and detailed comparison of five methods for deploying and forming large parabolic membrane antennas, **compares** properties of membrane materials for parabolic membrane antennas, and **investigates** recent advances in configurations, tensioning system design, and dynamic analysis for planar membrane antenna structures. Finally, **future trends** for large space membrane antenna structures **are pointed out**, and **technical problems are proposed**.

修改后为两组句子。第一组句子包含三个单句，主语同为 This review，谓语为三个动词，分别为 presents、compares 和 investigates（笔者将这种几个谓语动词连用的现象称为“连谓”），均为主动语态；第二组句子由两个句子组成，分别为 future trends ... are pointed out 和 technical problems are proposed，均用被动语态。这样修改后，语义层次相对集中、清晰，便于理解，主动、被动语态相结合，语言和谐、生动，表达效果明显提升。

6.9 文体

文体分为口语和书面语，而书面语又分为普通语和学术语。有的词语适于口语而不适于书面语，有的适于普通语而不适于学术语，有的则只适于学术语。

6.9.1 普通语用于学术语

191）However，we **can't** rule out the possibility that Cys976 located in the C-terminal region of sKLB，which **could't** be modelled owing to weak electron density，may form a disulfide bond with the nearby Cys523.

192）However，this formation scenario **isn't** possible if Schiller and colleagues' proposal is correct.

193）It is known that these substances exist in certain foods，and that an adequate supply of them is necessary to health，but they **haven't** yet been isolated in a pure condition，although several workers claim to have done so successfully.

以上三句属于学术语，其中的否定词 can't、could't、isn't、haven't 为普通语，不够正式，应改为与其完全同义只是形式不同的 can not（cannot）、could not、is not、have not，这几个为学术语。

句 191）~193）的参考修改方案：

✓ However，we **can not** rule out the possibility that Cys976 located in the C-terminal region of sKLB，which **could not** be modelled owing to weak electron density，may form a disulfide bond with the nearby Cys523.

✓ However，this formation scenario **is not** possible if Schiller and colleagues' proposal is correct.

✓ It is known that these substances exist in certain foods，and that an adequate supply of them is necessary to health，but they **have not** yet been isolated in a pure condition，although several workers claim to have done so successfully.

194）Because **ref**[8] on an entire subject area has already been searched and summarized，focused topics can be investigated more nimbly.

此句中的 ref[8] 表示引用了参考文献[8]，其中 ref 为 reference 一词的简写，当引文序号以上标的形式出现时，从文体上来讲用 ref 不大合适。学术论文中引用参考文献的正式用词除了 reference 之外，还有 literature、report。

句 194）的参考修改方案（引文序号为上标）：

✓ Because **the reference**[8] on an entire subject area has already been searched and summarized，focused topics can be investigated more nimbly.

✓ Because **literature**[8] on an entire subject area has already been searched and summarized，focused topics can be investigated more nimbly.

✓ Because **the previous report**[8] on an entire subject area has already been searched and summarized，focused topics can be investigated more nimbly.

195）For manuscripts utilizing custom algo-rithms or software that are central to the paper but not yet described in the published **reference**[2-8]，software must be made available to editors and reviewers upon request.

引文序号平排时，reference 可以用缩写的形式 Ref 或 Ref.（引用多篇文献时用 Refs 或 Refs.），这种缩写形式含有专有名词的意味，与文后参考文献表（References）相对应。

句 195）的参考修改方案（引文序号为平排）：

✓ For manuscripts utilizing custom algorithms or software that are central to the paper but not yet described in **Refs.**［**2-8**］, software must be made available to editors and reviewers upon request.

196）Design parameters for inflatable antennas generally include inflation pressure, membrane thickness, elastic modulus of membrane material, boundary conditions, temperature **and so on**.

此句使用了 and so on 表示“等等、诸如此类”之意。与此短语同义的还有另一个缩写词 etc，与 and so on 没有区别，不过 etc 多用于书面语，而 and so on 则较为普通，况且 etc 在形式上更简洁。

句 196）参考修改方案：

✓ Design parameters for inflatable antennas generally include inflation pressure, membrane thickness, elastic modulus of membrane material, boundary conditions, temperature, **etc**.

6.9.2　口语用于学术语

197）It was **quite** important to accurately analyze the loss property of aircraft hydraulic systems.

此句中的副词 quite 使用不妥。该词在口语中常用，但在学术论文中不宜使用，应改为 very。

句 197）的参考修改方案：

✓ It was **very** important to accurately analyze the loss property of aircraft hydraulic systems.

198）If such **thing happens**, a second method must be considered.

此句中的名词 thing、动词 happens 使用不妥：thing 一词在口语中常用，在学术论文中较少用，可代替它的常用词有 event、condition、complication 等；happen 一词的本意是“发生”，但侧重“偶然发生”，用 occur 更正规。

句 198）的参考修改方案：

✓ If such **an event occurs**, a second method must be considered.

但要注意，happen 一词有时不能或不宜换为 occur。

199）To confirm this scenario, one must rule out the possibility of trivial diffusive edge modes that **occur** to exhibit the quantized conductance value for some particular length.

200）In the case of ordinary chondrites, this would have **occured** in the inner Solar System at a late stage in its formation, after the time by which Mars had accreted most of its mass.

句 199）中的 occur 和句 200）中的 occurred 使用不当，这两句均有“碰巧、偶然发生”的意味，因此应分别改为 happen、happened。况且，happen 与不定动词组合（happen to do sth），表示“碰巧、偶然发生某事”，而 occur 与不定动词组合（occur to do sth），表示“突然做某事”。

句 199）、200）的参考修改方案：

✓ To confirm this scenario, one must rule out the possibility of trivial diffusive edge modes that **happen to** exhibit the quantized conductance value for some particular length.

✓ In the case of ordinary chondrites, this

would have **happened** in the inner Solar System at a late stage in its formation, after the time by which Mars had accreted most of its mass.

201）Some progress has also been made in this domain over the past 10 years in China. However, **up to now**, no membrane antennas have been applied in space **yet** except **American** inflatable **antenna with a diameter of 14 meters having** space flight experience in 1996.

此句中的短语 up to now、名词 meters 使用不太妥当：up to now 是 up until now 的口语化，虽然二者的意思相同，均表示“直到现在、迄今为止”，但前者是口语词语，较随意，后者是书面词语，较正式，学术语中宜用后者；meters 是物理量 diameter（直径）的单位的名称，其标准单位符号是 m，在学术语中表示量值（数值 + 单位）时，宜用标准单位符号而不用量名称做单位。此句还有两处有语言提升的空间：连词 yet 多余；分词短语（having…）本用来修饰说明天线（antenna），但其间被介词短语（with a diameter of 14 meters）分隔，表达效果下降，改为定语从句好一些。

句 201）的参考修改方案：

✓ Some progress has also been made in this domain over the past 10 years in China. However, **up until now**, no membrane antennas have been applied in space, except **an American** inflatable **antenna with a diameter of 14 m that had** space flight experience in 1996.

6.10　积极修辞

6.10.1　语义重复

语义重复有语境和形式两类：语境类是指句中出现了不必出现的词语（冗词），尽管其同义词语未出现（同一语义在同一句中常用一个词语表达，如果该词语不出现不影响语义的表达，那么就不用出现）；形式类是指用几个同义词语来重复表达，多因无意或疏忽造成，并非出于某种修辞的需要。

202）The **vessel motion** can be divided into **horizontal motion** and **vertical motion**.

203）Section 4 discusses the current views and **methods** of possible **ways to address** these problems, and puts forward some opinions on **the development** trend **them**.

句 202）出现了三个 motion，从形式上看，第二个 motion 多余，又因为此句的上一句（以上未给出）已出现过 vessel motion，因此第一个 motion 前面的 vessel 也多余。句 203）虽然未出现同形词，却出现了同义词，分别为 methods 和 ways，应去掉 methods，考虑到 methods 一词的专业性，ways 还可用 methods 替换。另外，表示某某方法，应该用 of 介词结构（method of），而不是动词不定式（method to），这样 ways to address 就应改为 methods of addressing；句子结尾的 them 与上文无语法关系，必须去掉，但考虑到其语义和前面词语的语义相关性，可将 development 前面的定冠词改成 their，或在 them 前加 of。

句 202）、203）的参考修改方案：

✓ The **motion** can be divided into **horizontal and vertical motion**.

✓ Section 4 discusses the current views and possible **methods** of addressing these problems, and puts forward some opinions on **their development** trend.

204）**In the paper**, we propose a robust and fast image denoising method.

此句为某论文的摘要的第一个句子，其中 In the paper 属冗词，因为在不出现它的情况下，也能清楚表明是在 the paper 中提出了“…method”，这种暗含的语义是由语境明确确定的。也可以用 paper 做主语而将主语 we 去掉。

句 204）的参考修改方案：

✓ **We** proposed a robust and fast image denoising method.

✓ **This paper** proposes a robust and fast image denoising method.

205）This paper looks into the **future prospect** of the new artificial intelligence technology in the 21st century.

此句中 prospect 表示前景、前途，已含未来、今后之意，在其前面再用表示未来、今后的 future 一词来修饰，就造成语义重复，因此 future 属于冗词。这种重复实际上是由用另一个词对与其同义的某个词进行多余的说明、解释所造成的，即 future 与 prospect 组合造成语义重复。

句 205）的参考修改方案：

✓ This paper looks into the **prospect** of the new artificial intelligence technology in the 21st century.

206）The prepared **process** of alloy powder by titanium powder and titanium hydride powder **were** heated at **a temperature of 500 ℃** for 2 h and then **a temperature of 600 ℃** for 3 h, and at **a temperature of 750 ℃** for 3 h, respectively, in 4.0×10^{-3} Pa vacuum conditions with heating rate of 5 ℃ • min^{-1}.

此句中℃是摄氏温度的单位，已明显含温度之意，因此在其前面出现 temperature，就造成语义重复，即 a temperature of（三处）冗余。但 5 ℃ • min^{-1} 前面的 heating rate 不在冗余之列，因为在不出现 heating rate 的情况下，通常较难知道℃ • min^{-1} 在句中是做哪个量的单位。

句 206）的参考修改方案：

✓ The prepared **process** of alloy powder by titanium powder and titanium hydride powder **was** heated at **500 ℃** for 2 h and then **600 ℃** for 3 h, and at **750 ℃** for 3 h, respectively, in 4.0 μPa vacuum conditions with heating rate of 5 ℃ • min^{-1}.

207）**PRC reaction** was applied to detect the genomic DNA segment.

此句中 PRC 是 polymerase chain reaction（聚合酶链反应）的缩写，最后一个词是 reaction，再在其后放一个 reaction 就语义重复了，也可将后一个 reaction 写成其他词语。

句 207）的参考修改方案：

✓ **PRC** was applied to detect the genomic DNA segment.

✓ **Polymerase chain reaction** was applied to detect the genomic DNA segment.

✓ **PRC technique** was applied to detect the genomic DNA segment.

✓ **PRC method** was applied to detect the genomic DNA segment.

208）**PSO algorithm** is emerging evolutionary computation technology based on swarm intelligence, which has been applied sucessfully in many fields.

此句中 PSO 是 particle swarm optimization 的缩写，其中 optimization 表“优化”，实际上已含“算法”之意，即表示“优化算法”，因此再在其后放 algorithm 一词，就语义重复了。

句 208）的参考修改方案：

✓ **PSO** is emerging evolutionary computation technology based on swarm intelligence, which has been applied successfully in many fields.

✓ **Particle swarm optimization** is emerging evolutionary computation technology based on swarm intelligence, which has been applied successfully in many fields.

✓ **Particle swarm optimization (PSO)** is emerging evolutionary computation technology based on swarm intelligence, which has been applied successfully in many fields.

✓ **PSO (particle swarm optimization)** is emerging evolutionary computation technology based on swarm intelligence, which has been applied successfully in many fields.

209) In order to determine the shape of **membrane needed to be cut**, an inverse solution to **configuration** of a membrane structure in the unstressed state should be obtained based on its desired configuration under **stressed state**.

此句的 membrane needed to be cut 表示"需要切割的薄膜"之意，其中的 needed 属于冗词，去掉后不影响语义。另外，有两处遗漏定冠词。

句 209) 的参考修改方案：

✓ In order to determine the shape of **membrane to be cut**, an inverse solution to **the configuration** of a membrane structure in the unstressed state should be obtained based on its desired configuration under **the stressed state**.

210) Our study **discussed and studied** the value **of** and way to access the state's prescription drug monitoring program **carefully**.

211) Five U. S. Centers for Disease Control and Prevention (CDC) guideline-recommended safe prescribing strategies were **discussed and studied in detail**.

此两句均有语义重复的问题，因为 discussed、studied、carefully、in detail 都含"仔细、详尽"的语义。可从 discussed、studied 中选其一而去掉另一个（多数情况下用 discussed 更合适，studied 较为笼统），并将 carefully、in detail 统统删除。

句 210)、211) 的参考修改方案：

✓ Our study **discussed** the value and way to access the state's prescription drug monitoring program.

✓ Five U. S. Centers for Disease Control and Prevention (CDC) guideline-recommended safe prescribing strategies **were discussed**.

实际中，只要语义重叠或相同的两个词语连用，就会出现语义重复。例如：

① **completely** filled、**free** gifts、**final** destination、**first** initiated、**past** history、research **work**、as **already** stated、bright blue **in color**;

② limit condition、sketch map、layout scheme。

以上第一组词语中，每个词语的首词的意思已在其后第二个词语中暗含了，连用就会语义重复，应根据情况去掉冗余的修饰语，或选用其中一个合适的词；第二组词语中，每个词语的前后两个词的语义相同或相近，选择一个即可。这样，这两组词语可改为：

① filled、gifts、destination、initiated、history（或 past）、research（或 work）、as stated、bright blue;

② limit（或 condition）、sketch（或 map）、layout（或 scheme）。

6.10.2　语句不简洁

提倡使用词语的简化形式，即用简明的词语替代与其同义的较长的短语，有助于简化语句表达。

212) **In order to** improve the modeling accuracy, a hypothesis that divides the drill-soil interaction into four parts: cuttings screw conveyance, cuttings extruding, cuttings bulldozing, and in situ simulant cutting, is proposed to establish a novel model based on the passive earth pressure theory.

213) The proposed research provides the **instruction** to adopt a suitable drilling strategy to match the **rotary motion and penetrating motion so as to** increase **safety** and reliability of drilling control in lunar sampling **mission**.

in order to 和 so as to 均为固定短语，意思是"为了；以便"，表示目的，这两句中二者都可以用 to 代替，表意不变，但词数减少，语句变得简洁。另外，句 213）还存在一些问题，如 rotary motion and penetrating motion 中前一个 motion 多余，safety 前遗漏了定冠词，motion 和 mission 应为复数形式。

句 212）、213）的参考修改方案：

✓ **To** improve the modeling accuracy, a hypothesis that divides the drill-soil interaction into four parts: cuttings screw conveyance, cuttings extruding, cuttings bulldozing, and in situ simulant cutting, is proposed to establish a novel model based on the passive earth pressure theory.

✓ The proposed research provides the **instructions** to adopt a suitable drilling strategy to match the **rotary and penetrating motions**, **to** increase **the safety** and reliability of drilling control in lunar sampling **missions.**

214) **By simulating calculation**, the flow field in the molten pool **is not a steady state**, but **periodic changed**, **during oscillating twin-roll strip casting process.**

此句开头的 By simulating calculation 为状语，中间为两个单句，最后的 during…process 又为状语。按语义，这两个单句及其后的状语所述的意思是"模拟计算（simulating calculation）的结果"，这样还不如直接用 simulating calculation 做句子的主语，而将其后的语句改为宾语从句。另外，simulating calculation 语义重复，可将 calculation 去掉；第二个句子的时态应与第一个句子的时态一致，改为一般现在时（changes）；periodic 为形容词，不能修饰谓语动词 changes，须改为副词 periodically。此外，is not a steady state 有待提升，可以改为 is not in steady state。还有，oscillating 前遗漏了冠词。

句 214）的参考修改方案：

✓ **Simulations showed that** the flow field in the molten pool is not **in** steady state but **periodically changes**, during **the** oscillating twin-roll strip casting process.

215) Fig. 2 **is the schematic diagram** of the DRCS process.

此句中 schematic 和 diagram 两个词都有图的意思，表意重复，语句不简洁，可将 diagram 去掉。不过，diagram 去掉后，句意是"图 2 是 DRCS 过程的原理图"，语义仍有重复，一个插图本身就是图，而插图的功能是表述和展示，说"图是一种什么图"，不如说"图显示（描述、展示、说明）了什么"更加直接、明快。按以上思路修改，语言会变得简洁明快。

句 215）的参考修改方案：

✓ Fig. 2 **shows a schematic** of the DRCS process.

英语科技论文中常见短语的简明词语有很多，表 6-1 列出了一些供参考，写作中从语句简洁的角度，提倡用简明词语。

表 6-1　英语科技论文常见短语及其简明词语

较长短语	简明词语	较长短语	简明词语
a majority of	most	has been shown to be	is
a number of	many，several，some	has the capability of	can，is able to
a small number of	a few	if conditions are such that	if，when
accounted for the fact that	because	if it is assumed that	if
all of	all	in a satisfactory manner	satisfactorily，adequately
an innumerable number of	innumerable，countless，many	in all cases	always，invariably
are found to be	are	in case	if
are in agreement	agree	in close proximity	near
are known to be	are	in connection with	about，concerning
are of the same opinion	agree	in consequence of this fact	therefore，consequently
as a consequence of	because of	in length	long
as far as our own observations	we observe	in（my，our）opinion	（I，we）think
at a rapid rate	rapidly	in order to	to
at the present time（moment）	now，at present	in spite of the fact that	although
at this point in time	now，currently	in the case of	in，for
based on the fact that	because	in the course of	during，while
by means of	by，with	in the event that	if
carry out	perform	in the near future	soon
caused injuries to	injured	in the vicinity of	near
contemporaneous in ages	contemporaneous	in those areas where	where
covered over	covered	in view of the fact that	because
definitely proved	proved	is in a position to	can，may
despite the fact that	although	is known to be	is
due to the fact that	because，due to	it appears that	apparently
during that time	while	it has been reported by WANG	WANG reported
during the course of	during，while	it is clear that	clearly
during the time that	while，when	it is often the case that	often
exposed at the surface	exposed	it is possible that	possibly
fall off	decline	it is possible that the cause is	the cause may be
fewer in number	fewer	it is this that	this
for a distance of 10 km	10 km	it is worth pointing out that	note that
for the purpose of examining	to examine	it would appear that	apparently
for the reason that	because	it would thus appear that	apparently
future plans	plans	lacked the ability to	could not
give rise to	cause	lager amounts of	much

（续）

较长短语	简明词语	较长短语	简明词语
large in size	large	reported in the literature	reported
lager numbers of	many	results so far achieved	result so far，result to date
located in，located near	in，near	serves the function of being	is
look after	watch	small in size	small
masses are of large size	masses are large，large masses	subsequent to	after
of great importance	important	take into consideration	consider
of such hardness that	so hard that	the majority	most
on account of	because	the question as to whether	whether
on behalf of	for	the treatment having been	after treatment
on the basis of	from，by，because	there can be little doubt that	this probably is
on the grounds that	because	through the use of	by，with
original source	source	throughout the entire area	throughout the area
oval in shape，oval-shaped	oval	throughout the whole of the experiment	throughout the experiment
owing to the fact that	because，due to		
prior to（in time）	before	two equal halves	halves
prove up	test	was of the opinion that	believed
referred to as	called	with the result that	so that

6.10.3 语义不连贯

句子的主语在表意上要清楚、连贯，通常应避免在同一复句中前后分句主语不一致或介词短语的逻辑主语与句子的主语在逻辑上不一致，造成句子冗长、迂回，语义不完整、不连贯。

216） Then **keep** the **external load constant** and **lower temperature to below** the glass transition temperature，**and** the antenna remains in the same state after unloading.

由 and 连接并列句时，各分句的句式通常应该一致。此句前面的分句为祈使句（keep the external load…），后面的分句为陈述句（the antenna remains…），前后分句的句式不相同，造成语义不连贯。再仔细考察一下，在语义上，前面的分句所述内容为后面的分句所述内容的条件，因此将前面的分句表述为状语更合适，后面的分句自然就是一个单句了。另外，还有其他语病：第一个连词 and 后面的 lower 和 below 语义重复，按语义逻辑，是使温度低于（keep temperature to below），而不是使较低的温度低于（keep lower temperature to below），因此 lower 属于冗词；Then 后面有停顿，应加逗号；lower 去掉后，其后的 temperature 前面应加一个定冠词。

句 216）的参考修改方案：

✓ Then，**by keeping** the **external load constant** and **the temperature lower to** the glass transition temperature，the antenna remains in the same state after unloading.

✓ Then，**by keeping** the **external load constant** and **the temperature below** the glass transition temperature，the antenna remains in the same state after unloading.

217）**These vugs** carry no gold and **the tenor of the vein** has not been affected by **them**.

218）Compared with Gunawan's method, **a solution** obtained by applying our method is **better** and **our method** doesn't reject some robust solutions.

句 217）中，and 后面的句子的主语 the tenor of the vein 与前面的句子的主语 These vugs 不一致；句 218）中，and 后面的句子的主语 our method 与前面的句子的主语 a solution 不一致，改成一致时，还要连带修改或调整句子开头的过去分词状语。这种不一致语病，会造成语句冗长、迂回，语义自然难以连贯。

句 217）、218）的参考修改方案：

✓ **These vugs** carry no gold and do not affect the tenor of the vein。

✓ **Applying our method** may obtain **a better solution** than applying Gunawan's method, and doesn't reject some robust solutions.

219）**Through analysis of** software ANSYS, **the interface connection** accords with displacement and forces equation condition.

此句前面的介词短语 Through analysis of software ANSYS 做状语，表示"人通过对某软件分析"，逻辑主语是"人"，而后面句子的主语 the interface connection 是"物"，二者不一致，语义不连贯。考虑语义的连贯性，句子的后面部分（the interface connection…equation condition）是"分析"的结果，可将介词 Through 去掉而直接用 analysis 做主语，用 showed（显示、表明）做谓语，"the interface connection…equation condition"做宾语。

句 219）的参考修改方案：

✓ **Analysis** of software ANSYS **showed that** the interface connection accords with displacement and forces equation condition.

✓ **Analyzing** with software ANSYS **showed that** the interface connection accords with displacement and forces equation condition.

220）**The high voltage safety management system is based on PIC Microchip** which makes it easily to share information among controllers by CAN, and **we can operate it easily using the flexible program technology**.

此句中两个粗体部分为同一句中的两个并列句，前面的句子用了被动语态，主语为 The high voltage safety management system，因后面的句子用了主动语态，将本应承接下来的该主语变成了宾语（it），使得前后句语义不连贯。同一复句中前后分句的语态通常应该一致，若将后一句子变为被动语态，问题就解决了。

句 220）的参考修改方案：

✓ **The high voltage safety management system is based on PIC Microchip** which makes it easily to share information among controllers by CAN, and **can be easily operated by the flexible program technology**.

221）**Finally**, drilling experiments **are conducted to validate the proposed model**.

此句中的 Finally，表明其后所述是最后一项工作，但仔细品味一下（通过钻井实验验证了该模型的有效性），会感觉意思不完整，按正常逻辑，作者验证完模型的有效性后，还应总结或提出点什么，例如总结或提出相关的结果、结论。

句 221）的参考修改方案：

✓ **Finally**, drilling experiments **are conducted to validate the proposed model** and **the relevant results are presented**.

222）In addition，one may think that the parallel module of the Exechon robot，which is a 2-UPR-SPR PM，belongs to the UP-equivalent PMs family. We **will** demonstrate that mechanism is not a UP-equivalent PM **in the following section**.

此句群有两个句子。后一个句子承接前一个句子，意思是在下一节我们将做什么，显然状语 in the following section 放在句尾不如放在句首，更能使这种承接变得直接、顺畅，语义更连贯。另外，in the following section 指明是本论文的某一或某几节，行文导向是论文，因此宜用现在时，而不是将来时。

句 222）的参考修改方案：

✓ In addition，one may think that the parallel module of the Exechon robot，which is a 2-UPR-SPR PM，belongs to the UP-equivalent PMs family. **In the following section**，we demonstrate that the mechanism is not a UP-equivalent PM.

223）The organization of this paper is as follows. **Section 2** presents a brief introduction of notations and properties of the UP-equivalent motion. **Section 3** discusses the limb bonds of the UP-equivalent PMs. **Section 4** constructs the UP-equivalent PMs in a special category. The rotational capability of the proposed mechanisms is illustrated **via an example** in **section 5**. **Section 6** discusses the application of the proposed PMs. The conclusions are summarized in **section 7**.

此句群有七个句子，介绍文章的章节安排，各章介绍语句都以章节号做主语的主动语态表述时，前后句式就一致，语义就会连贯。以上句五和句七用了被动态，某种程度上破坏了这种一致性和连贯性。当然，从错综修辞的角度看，如果有意不要这种一致性和连贯性，也是可以的。

句 223）的参考修改方案：

✓ The organization of this paper is as follows. **Section 2** presents a brief introduction of notations and properties of the UP-equivalent motion. **Section 3** discusses the limb bonds of the UP-equivalent PMs. **Section 4** constructs the UP-equivalent PMs in a special category. **Section 5 illustrates** the rotational capability of the proposed mechanisms. **Section 6** discusses the application of the proposed PMs. **Section 7 summarizes** the conclusions.

6.10.4 修饰密集

用长系列修饰语，即连续用多个形容词，或多个名词，或多个形容词、名词来修饰名词，会造成前面一边倒的密集型修饰语，不够地道。

224）The chlorine containing **high melt index propylene based** polymer.

此句中多个词，如形容词 high、名词 index 和 propylene（丙烯）等一起做 polymer（聚合物）的修饰语，修饰语较为密集且一边倒，语言表达不够规范。修改思路是，用连字符连接名词短语中的名词，或使用介词结构，形成修饰单元，将修饰单元置于中心语之后。

句 224）的参考修改方案：

✓ The chlorine containing **propylene-based** polymer **of high melt index**（高熔融指数含氯丙烯基聚合物）。

修改后，介词短语 of high melt index 置于中心语 polymer 之后，另一部分 propylene-based（连字符结构）仍然位于该中心语之前，语言效能得到提升。

225）Another disadvantage is that it is difficult for inflatable antennas to maintain the required surface accuracy under **space alternating temperature conditions.**

当中心名词前有几个名词、形容词做修饰语时，可以考虑将其中一个名词变为介词结构而置于中心名词的后面。本句中名词 conditions 前有三个修饰语，分别是名词 space、形容词 alternating 和名词 temperature，应将名词 space 变为形容词 spatial，那么 conditions 前面的 3 个修饰语就是 spatial、alternating 和 temperature，显得臃肿、啰唆，不够地道。这时可以考虑将 space 变为介词结构而后置（in space），修改思路同句 224）。

句 225）的参考修改方案：

✓ Another disadvantage is that it is difficult for inflatable antennas to maintain the required surface accuracy under the alternating temperature conditions **in space.**

226）**Chinese characteristics new types** of **machines** were presented by the research group of Professor Tan in 2019.

当一个中心词前有两个名词或其短语做修饰语时，当名词（或短语）1 是名词（或短语）2 的一部分或所具有的性质、特点时，应使用“with + 名词（或短语）1”短语，放在中心词后。本句中，中心词 machines 前有两个名词短语做修饰语，分别是 Chinese characteristics 和 new types，因为 Chinese characteristics 是 new types 的一部分或所具有的性质、特点，所以可形成 with 结构（with Chinese characteristics），放在中心词（machines）的后面。

句 226）的参考修改方案：

✓ **New types** of machines with **Chinese characteristics** were presented by the research group of Professor Tan in 2019.

✓ **A new type** of machine with **Chinese characteristics was** presented by the research group of Professor Tan in 2019.

227）However, **the surface accuracy maintaining ability** depends on further improvement of rigidizable material thermal stability, which is not good enough at present.

此句主语部分 their surface accuracy maintaining ability 的中心词是 ability，此中心词前面的词语均为其修饰语，修饰成分较多，语义关系不很准确。本意是 maintain surface accuracy（保持表面精度，动宾结构，突出动作），却表述为 surface accuracy maintaining（表面精度保持，主谓结构，动作不明显），后者的动作性效果不突出。可以将前置分词结构（surface accuracy maintaining）改为后置不定式结构（to maintain surface accuracy），动作性效果会显著提升。

句 227）的参考修改方案：

✓ However, **the ability to maintain surface accuracy** depends on further improvement of rigidizable material thermal stability, which is not good enough at present.

6.10.5　主体弱化

句子的主要意思应该用句子的主要成分如主语或主句来表达，而且连词（引导词）应选用恰当，如果用次要成分如状语或从句来表达，或连词选用不恰当，语义轻重就会发生转化，主要意思的表达效果就会弱化或下降。

228）**The process** of inflation-rigidization **is that the membrane reflector is rigidized** after being inflated to the desired parabolic **shape**, and **inflation pressure is released**.

本句的意思是，在膨胀强化（inflation-rigidization）过程中发生了两件事（即两个现象），一是将薄膜反射器膨胀到所需的抛物线形状后进行硬化（the membrane reflector is rigidized），二是膨胀压力得到释放（inflation pressure is released），其主体意思在于强化过程中发生的现象，过程（The process）应是状语，而现象应是主语，但以上句子将过程表达为主语，现象表述为表语，主要意思弱化。

句 228）的参考修改方案：

✓ **In the process** of inflation-rigidization, **the membrane reflector is rigidized** after being inflated to the desired parabolic shape and **inflation pressure is released**.

229）**Since** the penetrating speed and the rotary speed are low, the drilling process is regarded as a quasi-static process.

本句前一分句中连词 Since 后面的部分意思是“渗透速度和转速较低”，后一分句的意思“钻井过程被认为是一个准静态的过程”，前者为表原因的从句，后者为表结果的从句，因果关系非常明确和肯定，但用 since 来引导这一原因从句时，表达效果会有所下降，明显比不上用 because 来引导的效果强烈。

句 229）的参考修改方案：

✓ **Because** the penetrating speed and the rotary speed are low, the drilling process is regarded as a quasi-static process.

230）**Comparing with** the traditional vertical twin-roll strip casting process, **this paper** focuses the flow distribution in the pool.

本句中将表示辅体意思的状语放在前面，而将表示主体意思的主句放在后面，即状语在主句前，主体有所弱化，通常比不上主句在主句后的句式，能更加突出句子的主体意思。在论文的摘要写作中，通常宜用“主句在前状语在后”的句式。

句 230）的参考修改方案：

✓ **This paper** focuses on the flow distribution in the pool, **in comparison with** the traditional vertical twin-roll strip casting process.

6.10.6 主谓分家

通常，句子的谓语应靠近其主语，但当句子组织欠严密时，容易使谓语远离主语，即主谓分家。

231）**The decolorization** in solutions of the pigment in dioxane, which were exposed to 10 h of irradiation, **was no longer irreversible**.

此句的主语 The decolorization 与谓语 was no longer irreversible 之间被定语从句 which were…irradiation 分隔，使语义紧密的二者在形式上相距较远，表意松散。

句 231）的参考修改方案：

✓ When the pigment was dissolved in dioxane, **decolorization was no longer irreversible** after 10 h of irradiation.

✓ **Decolorization was no longer irreversible** after 10 h of irradiation when the pigment was dissolved in dioxane.

这两个方案的区别在于用主句（重要事实）还是从句（辅助说明）开头。方案一以从句开头，主句的主语与谓语在形式上相邻，表达得到提升，但仍有提升空间；方案二以主句开头，先陈述重要的事实，辅助从句紧跟其后，表达效果进一步提升。

6.10.7 状语前置

状语（如表时间、目的、地点、方式、条件、原因等）置于句首或句中会冲淡主句的重要性，造成句式欠佳，这就是状语位置安排不当。通常，主要观点及每个支撑论点用不同的语句分开写，表示主要观点的语句宜放在前面，表示状语的语句应放在后面。

232）**From data obtained experimentally**, power consumption of telephone switching system was determined.

233）**After CIK transfusion**, 6 cases' liver function（ALT and/or BIL）got much better, and the other 6 cases continued normal.

234）**To ensure sheet metal quality as well as assembly quality**, CMMs are widely used in automotive industry production.

此三句的本意均是侧重表达逗号后面的主句的意思，逗号前面的部分仅为陪衬性的状语成分（前面两句为介词短语做状语，分别为方式状语和时间状语，第三句为不定式短语做目的状语），不如放在主句后面的表达效果好。

句 232）~234）的参考修改方案：

✓ Power consumption of telephone switching system was determined **from data obtained experimentally**.

✓ Six cases' liver function（ALT and/or BIL）got much better and the other six cases continued normal **after CIK transfusion**.

✓ CMMs are widely used in automotive industry production **to ensure sheet metal quality as well as assembly quality**.

6.10.8　修饰语移位

这里的修饰语移位是指修饰语句与所修饰的中心词分离而处于不合适的位置，属于句子结构错位。修饰语主要有定语和状语。副词通常做状语，也是一种重要的修饰语，在句子中有句首、句中、句尾三种位置。多数副词位于谓语动词后，有宾语时通常在宾语前，但宾语较短时可放在宾语后。如果句子有系动词、情态动词、助动词，副词通常位于这些动词后、行为动词前。修饰语也包括定语从句。

235）The heave compensation **is applied widely** to decouple the heave motion in offshore installations.

236）The analysis of the soil rupture is based on the passive earth pressure theory which **has been used** to predict bulldozing force **successfully**.

此两句中的副词 widely、successfully 应分别放在系动词（is、has been）后面，行为动词 applied、used 前面。另外，句 236）中由 which 引导的定语从句，在语义上与先行词是非限定性关系，因此 which 前应加逗号。

句 235）、236）的参考修改方案：

✓ **Heave** compensation **is widely applied** to decouple the heave motion in offshore installations.

✓ The analysis of the soil rupture is based on the passive earth pressure theory, which has been **successfully** used to predict the bulldozing force.

237）Drilling experiments indicate that the drilling loads calculated by the proposed model **match** the experimental results **well**.

此句中的副词 well 位于句尾，宜放在谓语动词 match 的后面。

句 237）的参考修改方案：

✓ Drilling experiments indicate that the drilling loads calculated by the proposed model **match well** the experimental results.

238）The gap between the electrodes and membrane **can be adjusted** by the electrostatic force so as to **control** the reflector shape accuracy **actively**.

239）**Before** the DRCS process, thin-walled cylindrical workpiece is fixed on the internal expanding clamp, which **expands** radially under an axial compressive load **to**

clamp the cylindrical workpiece **tightly**.

240） The optimal compromise plan proposed by Meeker was widely used in engineering and **became** the "benchmark" for most of the improved plans **subsequently**.

副词用来修饰动词或谓语，句 238）结尾的副词 actively，是修饰谓语 can be adjusted，还是动词 control，需要考究一番。按语义，actively 是用来修饰 control 的，位于 control 之前更合适。句 239）结尾的副词 tightly，是用来修饰 clamp 的（不是用来修饰 expands），位于 clamp 之前更合适。句 240）结尾的副词 subsequently 从形式上看应该是修饰谓语动词 became 的，但从语义考究，它是用来修饰名词短语 improved plans，但修饰名词或名词短语应该用形容词而不是副词，因此将 subsequently 改为 subsequent 而移到 improved 一词的前面。

句 238）~240）的参考修改方案：

✓ The gap between the electrodes and membrane **can be adjusted** by the electrostatic force so as to **actively control** the reflector shape accuracy.

✓ **Prior to** the DRCS process, **a** thin-walled cylindrical workpiece is fixed on the internal expanding clamp, which expands radially under an axial compressive load **to tightly clamp** the cylindrical workpiece.

✓ The optimal compromise plan proposed by Meeker was widely used in engineering and became the "benchmark" for most of the **subsequent** improved plans.

241） A coring **device** is installed in the hollow auger **which has been introduced in the literature**.

按表义，此句中的定语从句（which...）所修饰的中心词是 device，本应放在该中心词的后面（而且是非限定性的），但误放在 auger 一词的后面，自然让人理解为该定语从句所修饰的中心词是 auger，因而造成语病。

句 241）的参考修改方案：

✓ A coring device, **which has been introduced in the literature**, is installed in the hollow auger.

242） A more comprehensive mathematical model **to study the influence of different nozzles to the temperature field and the distribution of solidified shell was established**.

按表义，此句中的不定式短语（to study...shell）所修饰的中心词是谓语动词（was established），本应放在该谓语动词的后面充当目的状语，但放在了主语的中心词（model）的后面，成为名词短语（A more...model）的后置修饰语，句子结构变化导致语义变化。

句 242）的参考修改方案：

✓ **In addition**, a more comprehensive mathematical model **was established** (**in order**) **to study the influence of different nozzles to the temperature field and the distribution of solidified shell**.

243） In order to study the vibration characteristic of high speed vehicles due to wheel flat more accurately, based on the mature and widely known vehicle-track coupling model and the rigid-flexible coupling model of vehicle system, **a corresponding analytical model is improved**, where the influences of elastic deformation of main vehicle parts and track are comprehensively considered. (为更准确地研究车轮扁疤对高速车辆振动特性的影响，基于目前成熟且广泛已知的车辆-轨道耦合模型和车辆系统刚柔耦合模型，综合考虑车辆主要部件的弹性振动和轨道弹

性振动的影响，建立改进的车辆-轨道动力学模型。）

此句为复合句，主句为第二、三个逗号之间的部分（a corresponding…improved），前面有两个较长的短语，其中短语一（In order to…more accurately）做状语，短语二（based on…vehicle system）本用来修饰主句的谓语（is improved），应位于此谓语的后面，目前放在主句的前面不合适；后面有一个 where 引导的定语从句，修饰中心词 model，表示“在……模型中”（in…model）。整体看，主句很短，不够突出，淹没在前面较长的状语和前移来的修饰语以及后面的定语从句中。如果将短语二放在 is improved 后面，则它分隔中心词和定语从句，使原来紧密的中心词和定语从句间的关系变得松散，即修饰语后移时，容易使人误以为那个定语从句修饰的中心词是在 based on…vehicle system 语句中结尾的词语（system）。因此，须将此定语从句分离出来直接作为一个独立的单句。

句 243）的参考修改方案：

✓ To study the vibration characteristic of high speed vehicles due to wheel flat more accurately, **a corresponding analytical model is improved** based on the mature and widely known vehicle-track coupling model and the rigid-flexible coupling model of vehicle system. In the model the influences of elastic deformation of main vehicle parts and track are comprehensively considered.

其中最后一句也可改为：

This model comprehensively considers the influences of elastic deformation of main vehicle parts and track.

6.10.9　长句泛滥

所谓长句泛滥，是指未按表意仔细简化、合理组织语句，用了较多的长句，甚至用了特长句，有时还将语义不太紧密或上下文不大关联的句子写在一个句子中，或把语义逻辑关系较为明显、结构层次较为清晰的几个自然不同的句子写在一个句子中，即把本应写成几个较短的句子写成了一个长句甚至特长句，进而造成语义不清。SCI 论文写作的趋势是，句子越来越短，据有关资料，17 世纪初一个英语句子平均有 40 ~ 60 个单词，20 世纪初平均有 21 个单词，20 世纪 70 年代平均有 17 个单词，目前平均有 12 ~ 17 个单词。

244）New descriptors of local environment and atomic state, the *X* and *Y* indexes, can accurately reflect electron distribution around atoms in different chemical microenvironments, **therefore** when these are applied to characterize local chemical environment and atomic self-state, **a satisfactory result was obtained to simulate and predict ^{13}C chemical shift of 22 natural amino acids and 4 non-natural amino acids.**

此句 therefore 前面的部分为一较长的单句（前面部分），后面的部分为一个更长的时间状语从句（后面部分），这两部分语义逻辑、结构层次较为清楚，写在一个长句中意义不大，写成两句更合适。另外，对于后面部分，因从句（when these are applied to…self-state）在前，主句（a satisfactory result was obtained…amino acids）在后，就造成后面部分与前面部分的衔接不够紧密。这样，就可考虑将后面部分的从句分离出来，直接组织为一个句子（以下修改后的句二，简称句二），并以主语 these 开头，与前面部分自然形成主承前主的语法关系，上下衔接自然紧密而顺畅；分离后剩下的句子（后面部分的主句）自然成为另一句子（以下修改后的句三），并与句二紧密关联，直接表达句二所述动作（are applied to）的结果（a satisfactory result was obtained）。这样此例就可重组为三个并列的较短的单句。

句 244）的参考修改方案：

✓ **New descriptors** of local environment and atomic state, the *X* and *Y* indexes, can accurately reflect electron distribution around atoms in different chemical microenvironments. **These** are applied to characterize local chemical environment and atomic self-state. **A satisfactory result was obtained** to simulate and predict ^{13}C chemical shift of 22 natural amino acids and 4 non-natural amino acids.

245) The gear transmission is grade seven, the gear gap is 0.000 12 radians, the gear gap has different output values corresponding to any given input value, nonlinearity of the gear gap model can be described by using the phase function method, the existing backlash block in the non-linear library of the Matlab/zdimulink toolbox can be used, the initial value of gear gap in the backlash block is set to zero.

此句为由 6 个单句组成的并列复句，整体上形成了特长句。可按其中各单句之间的关联程度进行重组，单句之间为一般关系时用句点分隔（取代原来的逗号），紧密关系时重组为一句，或用分号分隔。这样就有以下修改方案，将长句化为了短句。

句 245）的参考修改方案：

✓ The gear transmission is grade seven. The gear gap, which is 0.00012 radians, has different output values corresponding to any given input value. The nonlinearity of the gear gap model can be described by using the phase function method. The existing backlash block in the non-linear library of the Matlab/zdimulink toolbox can be used; the initial value of gear gap in the backlash block is set to zero.

修改后为 4 组句子。原来并列的第二句和第三句合并为一个非限定性定语从句，与前后的句子用句点分隔；原来并列的最后两句（第五句和第六句）之间的逗号用分号替代；原来的第四句和第五句之间的逗号用句点替代。

246) **K9 optical glass is used as** confinement medium, one of whose sides connected with the sample was coated with black paint 86-1 (The thickness is about 0.025 mm), **meanwhile, for preventing martensite** induced by laser quench from accruing temper phenomena under the heating produced by laser plasma explosion, **the large acoustic impedance organic file is inserted** between the metal and the coating, whose thickness is about 0.5 mm.

此句为特长句，需要重组。连接副词 meanwhile 用来连接句子，被连接的前后句子的语态应该一致，此例也做到了语态一致（前面句子的主体为被动结构 K9 optical glass is used as…；后面句子的主体也为被动结构 the large acoustic impedance organic file is inserted…），但后面句中将很长的目的状语（for preventing martensite … explosion）放在了句首，不论在形式还是语义上均破坏了这种一致性。因此，应将该目的状语移到句子的后面，使前后两个句子以一致的被动结构开始，接着再出现相应的状语部分。而且这两个句子不必写到一起形成一个复句，这样，meanwhile 前面的逗号就可改为句号，meanwhile 改为 Meanwhile。基于以上思路，此例就可重组为由几个短句组成的句群。

句 246）的参考修改方案：

✓ **K9 optical glass is used as** confinement medium, one of whose sides connected with the sample was coated with black paint 86-1

(The thickness is about 0.025 mm). Meanwhile, **the large acoustic impedance organic file is inserted** between the metal and the coating, whose thickness is about 0.5 mm, for preventing martensite induced by laser quench from accruing temper phenomena under the heating produced by laser plasma explosion.

247) Where m is the mass of the heavy disk mounted at the mid-span of a massless elastic shaft, e is the eccentricity of the mass center from the geometric center of the disk, φ is the angle between the orientation of the eccentricity and the ξ axis, ξk and ηk are the stiffness coefficients in two principal directions of shaft respectively, c is the viscous damping coefficient of the shaft and the disk, ω is the rotating speed.

此句对多个物理量（符号）给予解释，每个符号的解释部分均相当于一个单句，但因为各部分之间用逗号分隔，使得整段语句形成一个特长句，不太妥当，应重组为正常的短句形式，各符号的解释部分改用分号分隔。

句 247）的参考修改方案：

方案一（量符号与注释语之间用系动词 is 或 are）：

✓ Where m is the mass of the heavy disk mounted at the mid-span of a massless elastic shaft; e is the eccentricity of the mass center from the geometric center of the disk; φ is the angle between the orientation of the eccentricity and axis ξ; ξk, ηk are the stiffness coefficients in two principal directions of shaft respectively; c is the viscous damping coefficient of the shaft and the disk; ω is the rotating speed.

方案二（量符号与注释语之间用长破折号）：

✓ Where m—Mass of heavy disk mounted at the mid-span of a massless elastic shaft; e—Eccentricity of the mass center from the geometric center of the disk; φ—Angle between the orientation of the eccentricity and axis ξ; ξk, ηk—Stiffness coefficients in two principal directions of shaft respectively; c—Viscous damping coefficient of the shaft and the disk; ω—Rotating speed.

248) It is shown from effect of different SVM models that **these models with inputs in which sound signal is included have a high percentage of accuracy in detection of defects**, but **when these models with inputs which include voltage signal, their detection accuracy will be reduced**.

此句为 It 做形式主语、that 引导的主语从句。从句由以下两个复合句组成，以连词 but 关联：

ⅰ. these models <u>with inputs in which sound signal is included</u> have a high percentage of accuracy in detection of defects

ⅱ. when these models <u>with inputs which include voltage signal</u>, their detection accuracy will be reduced

句ⅰ画线部分为修饰语，所修饰的中心语是 these models（即句子主语）。此修饰语由介词短语 with inputs（处于主要位置）及对该介词短语进行修饰的定语从句 in which sound signal is included（处于次要位置）组成。句ⅰ本应写为长度较小的句子，但因为嵌套两部分修饰语，而且处于主要位置的修饰语是短语，远比处于次要位置的修饰语（句子）短，语感较差，阅读较为困难。可以考虑将此定语从句改为分词短语来修饰：with inputs <u>including sound signal</u>。

句ⅱ形式上为时间状语从句，逗号前面部分为从句部分，但缺少谓语。其中画线部分的结构基本同句 i 画线部分，也存在语感较差、阅读困难的问题。可考虑顺接句 i，不再提及 these models，而直接改为以 the inputs 做主语、include 做谓语、voltage signal 做宾语的完整的时间状语从句。

另外，show 的结果直接说会更加简洁明快，用 It is shown…that…句型有冗余啰唆之感，况且 that 后面用两个复合句也不大合适。鉴于此，可将原来的 It is shown…that…句型改为简洁的主动语态句型（主语 Effect of different SVM models，谓语 shows，宾语 the following result），后加冒号，引出两个结果语句。

句 248）的参考修改方案：

✓ Effect of different SVM models shows the following result: **the models with inputs including sound signal have a high percentage of accuracy in detection of defects**, but **when the inputs include voltage signal, the detection accuracy will be reduced**.

实际写作中，有时根据需要可以组织长度较短的句子（如不多于 60 个单词），但长度较短的句子若包含复合陈述或从句中嵌套从句时，也可能会很长，这些复合陈述或嵌套从句容易引起阅读困难，甚至造成与主要观点混淆，因此应该设法通过积极修辞来避免这种情况。

参考文献

REFERENCES

[1] 梁福军. SCI论文写作与投稿 [M]. 北京：机械工业出版社，2019.

[2] 梁福军. 英文科技论文规范写作与编辑 [M]. 北京：清华大学出版社，2014.

[3] 梁福军. 科技论文规范写作与编辑 [M]. 4版. 北京：清华大学出版社，2021.

[4] 梁福军. 科技论文规范写作与编辑 [M]. 3版. 北京：清华大学出版社，2017.

[5] 梁福军. 科技论文规范写作与编辑 [M]. 2版. 北京：清华大学出版社，2014.

[6] 梁福军. 科技论文规范写作与编辑 [M]. 北京：清华大学出版社，2010.

[7] 梁福军. 科技语体语法与修辞 [M]. 北京：清华大学出版社，2018.

[8] 梁福军. 科技语体标准与规范 [M]. 北京：清华大学出版社，2018.

[9] 梁福军. 科技语体语法、规范与修辞 [M]. 北京：清华大学出版社，2016.

[10] 李达，李玉成，李春艳. SCI论文写作解析 [M]. 北京：清华大学出版社，2012.

[11] 李海燕. 英语语法、词汇专项训练 [M]. 北京：机械工业出版社，2018.

[12] 苏前辉. 实用英语语法与修辞 [M]. 北京：北京师范大学出版社，2015.

[13] 李龙，洪玲. 英语语法 [M]. 北京：北京理工大学出版社，2018.

[14] 张俊东，杨亲正. SCI论文写作和发表：You Can Do It [M]. 北京：化学工业出版社，2015.

[15] 张向阳. 实用大学英语语法教程 [M]. 2版. 南京：东南大学出版社，2012.

[16] 左边草. 英语语法学习指南 [M]. 广州：中山大学出版社，2007.

[17]《英汉大词典》编委会. 英汉大词典：全新版 [M]. 北京：商务印书馆国际有限公司，2005.

[18] WAITE M. 牛津袖珍英汉双解词典（第11版）[M]. 米晓瑞，等译. 北京：外语教学与研究出版社，2018.

[19] 百度百科. 英语语法 [EB/OL]. [2019-07-08]. https://baike.baidu.com/item/%E8%8B%B1%E8%AF%AD%E8%AF%AD%E6%B3%95/5949894?fr=aladdin.

[20] 百度文库. 英语语法 [EB/OL]. [2019-07-08]. https://wenku.baidu.com/search?word=%D3%A2%D3%EF%D3%EF%B7%A8&lm=0&od=0&fr=top_home&ie=gbk.

[21] 百度文库. 英语词汇 [EB/OL]. [2019-07-08]. https://wenku.baidu.com/search?word=%D3%A2%D3%EF%B4%CA%BB%E3&ie=gbk.

[22] 有道词典 [CP/OL]. 北京：网易有道公司，2019.

[23] GAO W, EMAMINEJAD S, NYEIN H Y Y, et al. Fully integrated wearable sensor arrays for multiplexed *in situ* perspiration analysis [J/OL]. Nature, 2016, 529 (7587): 1586027. https://doi.org/10.1038/nature16521.

[24] KLEINSTIVER B P, PATTANAYAK V, PREW M S, et al. High-fidelity CRISPR-Cas9 nucleases with no detectable genome-wide off-target effects [J/OL]. Nature, 2016. doi: 10.1038/nature16526.

[25] SEKAR A, BIALAS A R, RIVERA H D, et al. Schizophrenia risk from complex variation of complement component 4 [J/OL]. Nature, 2016. doi: 10.1038/nature16549.

[26] BAKER D J, CHILDS B G, DURIK, et al. Naturally occurring $p16^{Ink4a}$-positive cells shorten healthy lifespan [J/OL]. Nature, 2016, 530 (February). doi: 10.1038/nature16932.

[27] SILVER D, HUANG A, MADDISON C J, et al. Mastering the game of Go with deep neural networks and tree search [J/OL]. Nature, 2016, 529 (January). doi: 10.1038/nature16961.

[28] SUTHERLAND W J, WORDLEY C F R. A fresh approach to evidence synthesis [J]. Nature, 2018, 558 (21 June): 364-365.

[29] WEISS D J, NELSON A, GIBSON H S, et al. A global map of travel time to cities to assessine qualities in accessibility in 2015 [J/OL]. Nature, 2018. doi: 10.1038/nature25181.

[30] TELLEY L, JABAUDON D. A mixed model of neuronal diversity [J]. Nature, 2018, 555 (22 March): 452-457.

[31] PERRY I B, BREWER T F, SARVER P J, et al. Direct arylation of strong aliphatic C-H bonds [J/OL]. Nature, 2018, 560 (2 August): 70-75. https://doi.org/10.1038/s41586-018-0366-x.

[32] MARDIS E R. Many mutations in one clinical-trial basket [J]. Nature, 2018, 554 (8 February): 173-174.

[33] VAITES Laura Pontano, HARPER J Wade. Protein aggregates caught stalling [J]. Nature, 2018, 555 (22 March): 449-451.

[34] LEE S, CHOI J, MOHANTY J, et al. Structures of β-klotho reveal a 'zip code'-likemechanism for endocrine FGF signalling [J]. Nature, 2018. doi: 10.1038/nature25010.

[35] HOPKINS B D, PAULI C, XING D, et al. Suppression of insulin feedback enhances theefficacy of PI3K inhibitors [J/OL]. Nature, 2018. https://doi.org/10.1038/s41586-018-0343-4.

[36] MARIATHASAN S, TURLEY S J, NICKLES D, et al. TGFβattenuates tumour response to PD-L1blockade by contributing to exclusion of T cells [J]. Nature, 2018. doi: 10.1038/nature25501.

[37] SLON Viviane, MAFESSONI Fabrizio, VERNOT Benjamin, et al. The genome of the offspring of a Neanderthalmother and a Denisovan father [J/OL]. Nature, 2018, 561 (6 September 6): 113-116. https://doi.org/10.1038/s41586-018-0366-x.

[38] SLAYMAKER I M, GAO L, ZETSCHE B, et al. Rationally engineered Cas9 nucleases with improved specificity [J/OL]. Science, December 2015 [2015-12-01]. https://sciencemag.org/content/early/recent/1 December 2015/Page 1/10.1126/science.aad5227.

[39] FARIA Nuno R, AZEVEDO R do S da Silva, KraemerMoritz U G, et al. Zika virus in the Americas: Early epidemiological and genetic findings [J/OL]. Science, 2016, 352 (6283): 345-349 [2016-12-14]. http://science.sciencemag.org/. doi: 10.1126/science.aaf503.

[40] GUO D, SHIBUYA R, AKIBA C, et al. Active sites of nitrogen-doped carbon materials for oxygen reduction reaction clarified using model catalysts [J/OL]. Science, 2016, 351 (6271): 361-365 [2016-12-15]. http://science.sciencemag.org/. doi: 10.1126/science.aad0832.

[41] WATERS C N, ZALASIEWICZ J, SUMMERHAYES C, et al. The Anthropocene is functionally and stratigraphically distinct from the Holocene [J/OL]. Science, 2016, 351 (6269): 137, aad2622-1-aad2622-10 [2016-10-18]. http://science.sciencemag.org/. doi: 10.1126/science.aad2622.

[42] GARCEZ P P., LOIOLA E C, COSTA R M D, et al. Zika virus impairs growth in human neurospheres and brain organoids [J/OL]. Science, 2016, 352 (6287): 816-818 [2016-12-14]. http://science.sciencemag.org/. doi: 10.1126/science.aaf6116.

[43] MCMEEKIN D P, SADOUGHI G, REHMAN W, et al. A mixed-cation lead mixed-halide perovskite absorber for tandem solar cells [J/OL]. Science, 2016, 351 (6269): 151-155 [2016-11-19]. http://science.sciencemag.org/. doi: 10.1126/science.aad5845.

[44] TABEBORDBAR M, ZHU K, CHENG J K W, et al. In vivo gene editing in dystrophic mouse muscle and muscle stem cells [J/OL]. Science, 2016, 351 (6271): 407-411 [2016-12-19]. http://science.

sciencemag. org/. doi: 10. 1126/science. aad5177.

[45] DÍAZ S, PASCUAL U, STENSEKE M, et al. Assessing nature's contributions to people: Recognizing culture, and diverse sources of knowledge, can improve assessments [J/OL]. Science, 2018, 359 (6373): 270–272 [2018-1-18]. http://science. sciencemag. org/. doi: 10. 1126/science. aap8826.

[46] HUGHES M P, SAWAYA M R, BOYER D R, et al. Atomic structures of low-complexity protein segments reveal kinked β sheets that assemble networks [J/OL]. Science, 2018, 359 (6376): 698–701 [2018-2-8]. http://science. sciencemag. org/. doi: 10. 1126/science. aan6398.

[47] RIBAS A, WOLCHOK J D. Cancer immunotherapy using checkpoint blockade [J/OL]. Science, 2018, 359 (6382): 1350–1355 [2018-7-10]. http://science. sciencemag. org/. doi: 0. 1126/science. aar4060.

[48] COHEN J D, LI L, WANG Y, et al. Detection and localization of surgically resectable cancers with a multi-analyte blood test [J/OL]. Science, 2018, 359 (6378): 926–930 [2018-6-28]. http://science. sciencemag. org/. doi: 10. 1126/science. aar3247.

[49] TSAI H, ASADPOUR R, BLANCON J, et al. Light-induced lattice expansion leads to high-efficiency perovskite solar cells [J/OL]. Science, 2018, 360 (6384): 67–70 [2018-4-5]. http://science. sciencemag. org/. doi: 10. 1126/science. aap8671.

[50] WU S, FATEMI V, GIBSON Q D, et al. Observation of the quantum spin Hall effect up to 100 kelvin in a monolayer crystal [J/OL]. Science, 2018, 359 (6371): 76–79 [2018-1-4]. http://science. sciencemag. org/. doi: 10. 1126/science. aan6003.

[51] DOCTOR J N, NGUYEN A, LEV R, et al. Opioid prescribing decreases after learning of a patient's fatal overdose [J/OL]. Science, 2018, 361 (6402): 588–590 [2018-4-9]. http://science. sciencemag. org/. doi: 10. 1126/science. aat4595.

[52] HUANG Y, MAO K, CHEN X, et al. SIP-dependent interorgan trafficking of group 2 innate lymphoid cells supports host defense [J/OL]. Science, 2018, 359 (6371): 114–119 [2018-1-4]. http://science. sciencemag. org/. doi: 10. 1126/science. aam5809.

[53] HUGHES T P, ANDERSON K D, CONNOLLY S R, et al. Spatial and temporal patterns of mass bleaching of corals in the Anthropocene [J/OL]. Science, 2018, 359 (6371): 80–83 [2018-1-4]. http://science. sciencemag. org/. doi: 10. 1126/science. aam5809.

[54] SAMKHARADZE N, ZHENG G, KALHOR N, et al. Strong spin-photon coupling in silicon [J/OL]. Science, 2018, 359 (6380): 1123–1127 [2018-3-24]. http://science. sciencemag. org/. doi: 10. 1126/science. aar4054.

[55] TIAN F, SONG B, CHEN X, et al. Unusual high thermal conductivity in boron arsenide bulk crystals [J/OL]. Science, July 2018 [2018-7-5]. http://science. sciencemag. org/. doi: 10. 1126/science. aat7932.

[56] JAIN S, WHEELER J R, WALTERS R W, et al. ATPase-Modulated stress granules contain a diverse proteome and substructure [J/OL]. Cell, 2016, 164 (January 28): 1–12. http://dx. doi. org/10. 1016/j. cell. 2015. 12. 038.

[57] HIRANO T. Condensin-Based chromosome organization from bacteria to vertebrates [J/OL]. Cell, 2016, 164 (February 25): 847–857. http://dx. doi. org/10. 1016/j. cell. 2016. 01. 033.

[58] THOMSEN Alex RB, PLOUFFE Bianca, CAHILL III T J, et al. GPCR-G protein-β-arrestin super-complex mediates sustained G protein signaling [J/OL]. Cell, 2016, 166 (August 11): 1–13. http://dx. doi. org/10. 1016/j. cell. 2016. 07. 004.

[59] KUSEBAUCH U, CAMPBELL D S, DEUTSCH E W, et al. Human SRMAtlas: A resource of targeted assays to quantify the complete human proteome [J/OL]. Cell, 2016, 166 (July 28): 1–13. http://dx. doi. org/10. 1016/j. cell. 2016. 06. 041.

[60] CHARBONNEAU M R, O'DONNELL D, BLANTON L V, et al. Sialylated milk oligosaccharides promote microbiota-dependent growth in models of infant undernutrition [J/OL]. Cell, 2016, 164 (February 25): 1-13. http://dx. doi. org/10. 1016/j. cell. 2016. 01. 024.

[61] PEDMALE U V, HUANG S C, ZANDER M, et al. Cryptochromes interact directly with PIFs to control plant growth in limiting blue light [J/OL]. Cell, 2016, 164 (January 14): 1-13. http://dx. doi. org/10. 1016/j. cell. 2015. 12. 018.

[62] STROOPER B D, KARRA E. The cellular phase of Alzheimer's disease [J/OL]. Cell, 2016, 164 (February 11): 603-615. http://dx. doi. org/10. 1016/j. cell. 2015. 12. 056.

[63] LANDER E S. The heroes of CRISPR [J/OL]. Cell, 2016, 164 (January 14): 18-28. http://dx. doi. org/10. 1016/j. cell. 2015. 12. 041.

[64] WRIGHT A V, NUÑEZ J K, DOUDNA J A. Biology and applications of CRISPR systems: harnessing nature's toolbox for genome engineering [J/OL]. Cell, 2016, 164 (January 14): 29-44. http://dx. doi. org/10. 1016/j. cell. 2015. 12. 035.

[65] CUSANOVICH D A, HILL A J, AGHAMIRZAIE D, et al. A single-cell atlas of *In Vivo* mammalian chromatin accessibility [J/OL]. Cell, 2018, 174 (August 23): 1309-1324. https://doi. org/10. 1016/j. cell. 2018. 06. 052.

[66] GEE M H, HAN A, LOFGREN S M, et al. Antigen identification for orphan T cell receptors expressed on tumor-infiltrating lymphocytes [J/OL]. Cell, 2018, 172 (January 25): 1-15. https://doi. org/10. 1016/j. cell. 2017. 11. 043.

[67] ZHOU X, FRANKLIN R A, ADLER M, et al. Circuit design features of a stable two-cell system [J/OL]. Cell, 2018, 172 (February 8): 744-757. https://doi. org/10. 1016/j. cell. 2018. 01. 015.

[68] CHOPRA A, KUTYS M L, ZHANG K, et al. Force generation via β-cardiac myosin, titin, and α-actinin drives cardiac sarcomere assembly from cell-matrix adhesions [J/OL]. Cell, 2018, 44 (January 8): 87-96. https://doi. org/10. 1016/j. devcel. 2017. 12. 012.

[69] CASASENT A K, SCHALCK A, GAO R, et al. Multiclonal invasion in breast tumors identified by topographic single cell sequencing [J/OL]. Cell, 2018, 172(January 11): 1-13. https://doi. org/10. 1016/j. cell. 2017. 12. 007.

[70] MOHAMED T M A, ANG Y, RADZINSKY E, et al. Regulation of cell cycle to stimulate adult cardiomyocyte proliferation and cardiac regeneration [J/OL]. Cell, 2018, 173 (March 22): 1-13. https://doi. org/10. 1016/j. cell. 2018. 02. 014.

[71] LIU X S, WU H, KRZISCH M, et al. Rescue of fragile X syndrome neurons by DNA methylation editing of the *FMR*1 gene [J/OL]. Cell, 2018, 172 (February 22): 979-992. https://doi. org/10. 1016/j. cell. 2018. 01. 012.

[72] JANES M R, ZHANG J, LI L, et al. Targeting KRAS mutant cancers with a covalent G12C-specific inhibitor [J/OL]. Cell, 2018, 172(January 25): 578-589. https://doi. org/10. 1016/j. cell. 2018. 01. 006.

[73] KONERMANN S, LOTFY P, BRIDEAU N J, et al. Transcriptome engineering with RNA-targeting type VI-D CRISPR effectors [J/OL]. Cell, 2018, 173 (April 19): 1-12. https://doi. org/10. 1016/j. cell. 2018. 02. 033.

[74] SHI J, DENG G, KONG H, et al. H7N9 virulent mutants detected in chickens in China pose an increased threat to humans [J/OL]. Cell Research, 2017, 27 (12): 1409-1421. doi: 10. 1038/cr. 2017. 129.

[75] BRUHN Benjamin, BRENNY Benjamin JM, Sidoeri Dekker, et al. Multi-chromatic silicon nanocrystals [J/OL]. Light Science & Application, 2017, 6, e17007. doi: 10. 1038/lsa. 2017. 7.

[76] WU X, GUO Z, WANG H, et al. Mechanical properties of WC-Co coatings with different decarburization

levels [J/OL]. Rare Met. , 2014, 33 (3): 313-317. https://doi. org/10. 1007/s12598-014-0257-8.

[77] SHAO H, WANG Z, LIN T, et al. Preparation of TiAl alloy powder by high-energy ball milling and diffusion reaction at low temperature [J/OL]. Rare Met. , 2018, 37(1): 21-25. https://doi. org/10. 1007/s12598-015-0466-9.

[78] ZHOU C, TANG C, LIU F, et al. Regional moment-independent sensitivity analysis with its applications in engineering [J/OL]. Chinese Journal of Aeronautics, 2017, 30(3): 1031-1042. http://dx. doi. org/10. 1016/j. caj. 2017. 04. 006.

[79] Institute of Biochemistry and Cell Biology, Shanghai Institute for Biological Sciences, Chinese Academy of Sciences. Cell Research [J]. 2018, 28(4). Institute of Biochemistry and Cell Biology, Shanghai Institute for Biological Sciences, Chinese Academy of Sciences (CAS) and Springer Nature, 2018.

[80] Institute of Biochemistry and Cell Biology, Shanghai Institute for Biological Sciences, Chinese Academy of Sciences. Cell Research [J]. 2017, 27(12). Institute of Biochemistry and Cell Biology, Shanghai Institute for Biological Sciences, Chinese Academy of Sciences (CAS) and Springer Nature, 2017.

[81] Institute of Biochemistry and Cell Biology, Shanghai Institute for Biological Sciences, Chinese Academy of Sciences. Cell Research [J]. 2016, 26(4, 10). Institute of Biochemistry and Cell Biology, Shanghai Institute for Biological Sciences, Chinese Academy of Sciences (CAS) and Springer Nature, 2016.

[82] Changchun Institute of Optics, Fine Mechanics and Physics, Chinese Academy of Sciences (CAS). Light Science & Application [J]. 2017, 6(3, 4). Changchun Institute of Optics, Fine Mechanics and Physics, Chinese Academy of Sciences and Springer Nature, 2017.

[83] Changchun Institute of Optics, Fine Mechanics and Physics, Chinese Academy of Sciences (CAS). Light Science & Application [J]. 2016, 5(1). Changchun Institute of Optics, Fine Mechanics and Physics, Chinese Academy of Sciences and Springer Nature, 2016.

[84] The Nonferrous Metals Society of China, General Research Institute for Nonferrous Metals (GRINM). Rare Metals [J]. 2018, 37(1). GRINM Bohan (Beijing) Publishing Co. , Ltd. and Springer, 2018.

[85] The Nonferrous Metals Society of China, General Research Institute for Nonferrous Metals (GRINM). Rare Metals [J]. 2014, 33(3). Editorial Office of Rare Metals, 2014.

[86] The Chinese Anti-Cancer Association. Cancer Biology & Medicine [J]. 2016, 13(2). The Chinese Anti-Cancer Association and ELSEVIER, 2016.

[87] Cold and Arid Regions Environmental and Engineering Research Institute, Chinese Academy of Sciences; Science Press; The Geographical Society of China. Sciences in Cold and Arid Regions [J]. 2016, 8(5). Science Press, 2016.

[88] The Institution of Engineers Australia. Australian Journal of Water Resources [J]. 2013, 17(1). Engineering Media for Engineers Australia, 2013.

[89] Chinese Mechanical Engineering Society. Chinese Journal of echanical Engineering [J]. 2012, 25(1)-. Beijing: Editorial Office of Chinese Journal of Mechanical Engineering and Springer-Verlag Berlin Heidelberg, 2012-.

[90] Oxford Journals. Annals of Botany [J]. 2007, 99(5). Bristol, UK: Editorial Office of Annals of Botany, 2007.

[91] The European Federation of Neurological Societies. European Journal of Neurology [J]. 2007, 14(9). Edinburgh, UK: Blackwell Publishing, 2007.

[92] Chinese Academy of Sciences (CAS), National Natural Science Foundation of China (NSFC). SCIENCE CHINA Technological Sciences [J]. 2011, 54(7). Beijing: Science China Press; Heidelberg: Springer, 2011.

[93] Science Press, Springer Science, Business Media. Journal of Computer Science and Technology [J]. 2008, 23(2). Beijing: Editorial Office of Journal of Computer Science and Technology, 2008.

[94] Chinese Academy of Medical Sciences, ELSEVIER. Chinese Medical Sciences Journal [J]. 2011, 26(2). Beijing: Editorial Office of Chinese Medical Sciences Journal, 2011.

[95] The Geological society of China. Acta Geological Sinica [J]. 2008, 82(2). Beijing: Editorial Office of Acta Geological Sinica, 2008.

[96] Polish Academy of Sciences committee of Machine Design. The Archive of echanical Engineering [J]. Warszawa: Warszawska Drukarnia Naukowa PAN, 2009, LVI (4).

[97] 圣捷出国. complete 与 finished 的区别 [EB/OL]. [2019-09-30]. https://wenku.baidu.com/view/79783e9480eb6294dd886cb4.html.

[98] 中国科学技术期刊编辑学会. 科技期刊英文编辑研修班讲义 [G]. 北京：中国科学技术期刊编辑学会，2007.

[99] 中国科学技术期刊编辑学会. 科技期刊英文编辑研修班讲义 [G]. 北京：中国科学技术期刊编辑学会，2018.

[100] 中华人民共和国国家质量监督检验检疫总局，中国国家标准化管理委员会. GB/T 7714—2015 信息与文献 参考文献著录规则 [S]. 北京：中国标准出版社，2015.

后　记

POSTSCRIPT

日积跬步，终至千里，岁累溪流，乃成江海。截至本书成稿之日，我已在编辑岗位上奋战了26个春秋，“熬”成了资深编辑，工作之余还撰写了多部著作。我既是期刊、论文的编辑，又是他人著作的策划、加工编辑，同时还是自己著作的作者、编辑，我与编辑和写作结下了缘。

我不禁想起美国儿童图书作家、编辑多萝茜•哈斯（Dorothy Haas）曾经这样说：“编辑和写作是一个铜板的两面，编辑工作让我学会质疑，而写作则让我知道初稿的定稿是件非常困难的事情。而编辑的训练也让我不至于灵感一来就下笔不能自休。”“如果把写作和编辑比喻成一根管子的两端，彼此是对立的，但两者都建立在以文字表达情绪和感觉的基础上；虽然二者是同一个过程中的两个部分，不过却需要截然不同的技巧。”哈斯正因同时在编辑和写作这双重领域孜孜不倦、刻苦耕耘，才会有这样精辟的理解。我自踏入编辑部，就与稿子、论文打上了交道、结下了友情。我的体悟是，稿子只有内容好、结构好、语言美，才有发表的价值和意义。质量是稿子的灵魂，是论文的价值所在。我深愿为提升论文的质量进而为繁荣祖国的科技文化做点小小的贡献，这也是我持续撰写著作的动力源泉。

非常荣幸，受机械工业出版社的邀请，我撰写了《SCI论文写作与投稿》《英语科技论文语法、词汇与修辞　SCI论文实例解析和语病润色248例》高等学校英语科技论文写作教材。与高校教师讲写作不同，他们或许更侧重于实验设计、方案制定、结果分析、成果提炼等方面，而我的著作的特色是，从编辑的视角编写，更加注重如何将科学研究各个环节打磨为内容恰当、结构合理、逻辑通畅和语句优美的高品质论文。

在新著作付梓之际，欣慰之情再次不溢言表，感激之情亦涌入脑海，我应感谢很多人。首先，感谢美国密西西比州立大学刘宇澄（Yucheng Liu）教授、英国布鲁内尔大学程凯（Kai Cheng）教授和科技导报社副社长、副主编史永超编审为本书作序；同时，感谢名刊主编宋天虎研究员和编辑部主任白雨虹研究员、程磊编审，大学校长、博导刘清友教授，高校硕博研究生导师杜雪、范文慧、丁希仑、张树有、高亮、严如强、刘荣强、刘检华教授，行业协会领导李琛高级工程师为本书撰写推荐语；其次，感谢机械工业出版社高等教育分社韩效杰副社长、营销销售中心李双雷副主任为书稿选题出谋划策；再者，感谢机械工业出版社加工中心佟凤编辑为加工本书所提出的宝贵意见及所付出的辛勤劳动；最后，感谢家人孟晓丽教授级高工在本书撰写过程中给予的默默支持！

期待我的著作能带给读者不一样的感受和收获，对读者的写作有所帮助，引领读者高效地撰写出高品质的论文。同时也期待广大读者批评指正，助我写出更好、更多的著作和教材！

梁福军

2020-12-31